A Systems Approach to Agroecology

Pablo Tittonell

A Systems Approach to Agroecology

Springer

Pablo Tittonell
Agroecology Chair, Groningen Institute of Evolutionary
Life Sciences
Groningen University
Groningen, The Netherlands

Agroécologie et Intensification Durable (AïDA)
Centre de coopération Internationale en Recherche
Agronomique pour le Développement (CIRAD)
Université de Montpellier
Montpellier, France

ISBN 978-3-031-42937-8 ISBN 978-3-031-42939-2 (eBook)
https://doi.org/10.1007/978-3-031-42939-2

This Springer imprint is published by the registered company Springer Nature Switzerland AG
The registered company address is: Gewerbestrasse 11, 6330 Cham, Switzerland

Paper in this product is recyclable.

A mi familia

Foreword by Miguel Altieri

In his 1979 book *Agroecosistemas*, Robert Hart (QEPD) at CATIE in Costa Rica introduced the concept of the agroecosystem composed of interacting subsystems, with inputs and outputs. He conceived agriculture as a hierarchy of systems, from individual crops or livestock herds to regional and national agricultural systems. This concept of nested subsystems represented an approximation to understanding complex socio-ecological interactions across scales, such as the top-down effects of regional/national policies on farming systems, or conversely the upward effects occurring at the cropping system level which could influence regional food prices. Arguably, Hart's text was one of the first books in Spanish using a systems thinking approach.

As the field of agroecology evolved in Latin America, it was evident that it was no longer sufficient for students and professionals of agriculture to be experts in soils, entomology, plant pathology, sociology or other disciplines to understand the complexity of the agroecosystem. Although disciplinary training was necessary, transdisciplinary skills became essential, as problems in agriculture are not only scientific or technical but also social and political. Agroecology embraces such systemic approach; although firmly rooted in the sciences of ecology and agronomy, it integrates in a transdisciplinary way social, political and economic issues that affect agroecosystems.

Complex thinking is essential today as agriculture faces a series of global problems: energy shortages, water scarcity, environmental degradation, biodiversity loss, climate change, economic inequality, food insecurity and others. These problems cannot be addressed in isolation, as they are systemic in nature; that is, they are interconnected and interdependent. When one of these problems is aggravated, the effects spread throughout the system, exacerbating other problems. This is why Fritjof Capra introduced the concept of "systems view" involving a new kind of thinking – thinking in terms of relationships, patterns and context. Agriculturalists adopting such systems view do not focus only on the technical complexity of farming problems, but integrate the social and political dimensions of such problems.

In this book *A Systems Approach to Agroecology*, Pablo Tittonell calls for a holistic approach to agroecology, in order to deal with the social and ecological complexity of agroecosystem, their dynamics, uncertainties and sustainability. Divided into 10 chapters, Tittonell describes novel methods and tools to support codesign and evaluation of agroecological systems and their transitions. It provides guidelines on how to analyze temporal and spatial diversity in agroecosystems, but also the landscape structure and heterogeneity in which they are embedded. Such broader perspective is key given that landscape management forces alter ecological interactions and ecosystem services in agroecosystems. Various chapters in this book remind us that crop fields contain ecological networks of interacting species (soil micro-organisms, insects, plants, etc.) all linked in a spatial network connected to surrounding natural and semi-natural habitats via species movement, which in turn are influenced by management practices used across the landscape.

Capitalizing on his broad practical experience in Europe, Africa and Latin America, Tittonell draws on a combination of methodologies, ranging from participatory tools and field observations to mathematical simulation modelling to explain variability in agroecosystems and landscapes, and reconfiguration of agroecosystems undergoing agroecological transitions. Such methods are useful to gather strategic information for designing correct manipulation strategies of biotic interactions to optimize the functioning of agroecosystems with the aim of enhancing agricultural production, sustainability and resilience.

No doubt this book provides a methodological framework that can effectively help in directing the path of agroecological transitions towards environmental, economic and social sustainability. The book calls for an agroecological *science* that goes beyond researching principles and practices at the farm level, but that includes the assessment of livelihoods, ecosystem services and other socio-ecological outcomes influenced by ecological or political events and processes at scales much larger than a single farm.

University of California
Berkeley, CA, USA

Miguel Altieri

Foreword by Irene Cardoso

Agroecology is a subject of everybody because it is the science of food systems and everybody is somehow connected to food systems, at least as a consumer. We need agroecology to transform the hegemonic food system because it is not sustainable. It is based on fossil fuel and is one of the main threats to climate change. However, this unsustainable food system is supported by political and economic empires, which largely profit from it. In other words, it is a rich empire of a few that makes the planet sick!

Pablo Tittonell's book is a "bio-oil" or a "catalyser", as stated in Chap. 10, to make the "gears" of the agroecological transition run smooth and fast. The concept of transition is the subject of debate in the book, but it can give important contributions to the needed transformation of the food systems.

The book provides important knowledge for better understanding of the elements of agroecology and its dimensions as a movement, science and practice. The book also supplies important tools for the process of agroecological transition, from the definition of agroecosystems to dynamic models that allow quantifying and analysing trade-offs to inform decisions.

Nice practical examples from different scales can be found in the book. They are important to inspire us to join in a worldwide movement that will make sustainable food systems possible everywhere and for everybody. Not one being deserves to die of pesticides or other products that are used in agriculture with the excuse of the need to feed the world.

Universidade Federal de Viçosa
Viçosa, Brazil

Irene Maria Cardoso

Foreword by Jan Douwe van der Ploeg

Agroecology is progressively unfolding in many parts of the world and is doing so in a myriad of ways. This is occurring through the active engagement of strong social movements, the patient practices of peasants looking for pride and autonomy, partial incorporation in state policies, enthusiastic propagation by NGOs, the building of agroecological schools, the growing involvement of research institutions and countless other avenues. All this is having undeniable and clearly visible agronomic, economic, social, political and institutional impacts. And, as the general crisis in the main agricultural systems of this world deepens, it is becoming clearer that agroecology is not only an attractive promise – it is also the much needed alternative.

Within this scenario, this new textbook from Pablo Tittonell stands out as a timely and important contribution. While there is already a rich literature on agroecology, this textbook addresses a major *lacunae*. It retells the development of agroecology in *academic* terms and as an *academic* enterprise. It does so by clearly and convincingly showing that agroecology-as-science not only exists due to, and because of, its specific *object* (farming built on ecological principles and processes), but that it simultaneously has its own, distinctive *method*. I understand 'method' here as an extended and consistent methodology that both allows for the development of heterogeneous and dynamic practices and the elaboration of strong and elegant theories. The method proposed by Pablo Tittonell is powerful especially as it includes and builds on *panarchy* (which critically takes into account the *two-sided* interaction between different levels of aggregation), *nested subsystems* and *synecology* (the study of groups of organisms in relation to their environment). The importance of such methodological principles resides, in the first place, in that they clearly distinguish agroecology from the mainstream agronomy applied, developed and defended in the temples of industrialized agriculture. It is, in short, *different*. Secondly, these same methodological principles make it clear that agroecology, as science, is *superior* to mainstream (or 'traditional' as Tittonell calls it) agronomy. It is superior thanks to a better-equipped and more developed methodology. Thirdly, it means that agroecology cannot simply be incorporated in business-as-usual scenarios, as the main faculties of agriculture try to do. This is because it is *distinctively divergent.*

This textbook is scholarly rigorous in its exposition using, where needed, a mathematical language. This is required to bring forward the scientific strength and superiority of agroecology. It is also needed to bring together the rich work of many well-known pioneers in agroecology, the list of whom is too long to include here, into one well-consolidated whole. Many practitioners and young students may well find the language used difficult to understand on the first reading. Nonetheless, it offers them powerful instruments such as the notion of the 'land equivalent ratio' (that brings in the synergies resulting from the entwinement of different activities), the redefinition of efficiency as the 'emerging property of evolving networks' (instead of being located in a simple input-output ratio) and by solidly tying together the often sloppily used concepts of stability, reliability, resilience and adaptability. The same applies for systematically seeing diversity as an opportunity (instead of dealing with it as nuisance, as is done in mainstream agronomy). All these are, indeed, great tools for *practice*.

Tittonell's approach to agroecology, which goes from system *analysis* to *design*, is an important contribution that represents a milestone in the progressive unfolding of agroecology and will surely help drive the reconfiguration that our world badly needs. I very much recommend studying and using this book.

Wageningen University
Wageningen, The Netherlands

China Agricultural University
Beijing, China

Jan Douwe van der Ploeg

Preface and Acknowledgements

Writing this book took me 10 years, although I wasn't continuously writing during that period. The idea for the book came after I delivered a keynote speech at the 12th European Congress of Agronomy in Helsinki in 2012. An editor from Springer present in the audience approached me and proposed me to write about systems approaches in agriculture. Since then, I've experienced numerous life changes including raising my lovely son, changing jobs four times, living in three different countries and six different homes, starting multiple vegetable gardens, supervising 41 PhD candidates, building a house, traveling extensively, meeting countless inspiring farmers and colleagues, reviewing numerous papers, writing just as many myself, delivering many lectures and talks, and teaching thousands of hours in the classroom. In the remaining hours, I wrote this book.

When I began writing this book, agroecology was not as widely known globally as it is today. It was already a strong movement in Latin America, but its presence was largely confined to specific academic circles in the Americas and Europe. The first international symposium on Agroecology at the FAO had yet to take place. Back then, I faced criticism from colleagues in various organizations whenever I mentioned agroecology. Now many of them portray themselves as agroecology experts. In 2012, my students viewed me wearily or defiantly when I taught agroecology in the classroom. Today, they approach me eagerly to engage in research on agroecology. Times have certainly changed. We are currently witnessing the global expansion of agroecology, which seems unstoppable.

These changes meant that whatever I had written between 2012 and 2015, before taking a nearly five-year hiatus from writing, needed updating or substantial rewriting. It was during the pandemic lockdown that I found the time and focus to pick up this project again. Around the same time, Melanie van Overbeek from Springer contacted me after years of silence to inquire about my writing. Having to revisit my earlier work and update it in light of new methodological and socio-political developments was both challenging and highly motivating. Not least because, through reviewing the recent literature, I discovered a large number of new research groups contributing to agroecology from different parts of the world.

Over the past years, I have shared several chapters of this book with my students through courses and lectures. However, the versions presented here have been revised and updated with new material. Throughout the writing process, I had to make difficult decisions and leave out certain aspects of agroecology, particularly the practical aspects of agroecological design, that didn't strictly align with the focus on systems analysis. This book truly centres around systems analysis. Yet, it's challenging for me to separate systems thinking from agroecology as I have built my career with one foot in each realm. Nevertheless, those solely interested in systems analysis, and not specifically in agroecology, may also find the methods presented in this book valuable.

I would like to express my gratitude to all the farmers I have encountered in the past 25 years across four continents. I have learned and continue to learn agroecology from them. I am thankful to the PhD candidates I have supervised and the participants of my courses who constantly challenge my assumptions. I appreciate the journalists who critically and open-mindedly question the soundness of my statements, making them accessible to the wider public. I am grateful to the World Wildlife Fund for their financial support. On a personal note, I want to thank Simone and Dante for their unwavering support and for understanding the stolen hours dedicated to writing. I extend my thanks to Herman for his interest in this book and the writing process. I appreciate Stefanie for fuelling the writing of these pages with the best coffee in town. I am grateful to my colleagues at CIRAD, Wageningen, INTA, Groningen, and elsewhere for making work a pleasure. Santiago Chiquito and Santiago Grande, thank you for the countless hours and shared ideas. To Vero, thank you for joining me on this journey. Our collaboration has propelled us forward, allowing us to sail at a faster pace. To Rose, Japheth, Saidi, Roberto, and Caterina, thank you for allowing me to be part of your remarkable epics. Miguel, Irene, and Jan Douwe, thank you for being a source of inspiration. Lu, thanks for your encouragement.

Bariloche, Patagonia Pablo Tittonell
18 July 2023

Contents

Part I System Definitions

Part II Analysis-Oriented Approaches

Part I
System Definitions

Chapter 1
Why Agroecology, Why Systems, Why Now?

Abstract The world is not able to feed itself today, and will certainly not in the future, unless there is a radical transformation of the current food system. It is not just a matter of increasing agricultural production. Today we produce enough food to feed 10 billion people, yet 2 billon face food insecurity or hunger, while 1.2 billon are overweight and 4.7 million die every year of diet related diseases. This is the result of our current agriculture and food system, a model that emerged in the second half of the twentieth century and made agriculture dependent on fossil fuels and the chemical industry, favouring land concentration and the production of commodities for an industrial oligopoly. The currently dominant agri-food system is responsible for the continuous destruction of ecosystems, is a major driver of biodiversity loss due to direct and indirect effects, contributes to global warming and to the disruption of global biogeochemical cycles of nitrogen and phosphorus. It keeps large numbers of rural dwellers in poverty, and contributes to the depopulation of rural areas. Lack of a systems perspective has led to agricultural technologies being developed by reasoning at the scale of a single field or individual plant population. Such technologies – monocultures, fossil fuels, agrochemicals, genetically modified plants – form the basis of the current narrow model of agricultural production and are responsible for its vulnerability and social inequity. Agroecology proposes a new model for agriculture and food production that propends to distributive justice, biodiversity and ecosystem restoration, and food sovereignty. Agroecology relies on social and ecological principles to guide the design of sustainable agri-food systems. It proposes practical knowledge to make agriculture more sustainable in the North and the South, and to restore soils and make food production more resilient in the face of global changes. Agroecology's ability to simultaneously address the food, biodiversity, and environmental crises relies on embracing whole landscape and food systems approaches. This involves landscape-level restoration, collective action, governance and community involvement, as well as considering the broader context of food systems. In other words, system's thinking and systems analysis methods, or a systems approach to agroecology.

P. Tittonell, *A Systems Approach to Agroecology*,
https://doi.org/10.1007/978-3-031-42939-2_1

1.1 Feeding the World

A quick glance at human history reveals that linear extrapolation of the status quo is rarely the best way to anticipate events in the mid to long-term future. However, current trends in population growth, economic development, food and energy consumption and waste, and natural resource degradation indicate that we are heading towards difficult times. Nowadays, the year 2050 is used as a reference point in the future to imagine possible global scenarios regarding demography, climate, biodiversity, agricultural production, and food and energy demands. One of the most alarming scenarios suggests that by 2050, there will be 9 billion people on this planet, and the food demand will increase more than proportionally in relation to the population. This is due to the emergence of new economies and the consequent increase in the average purchasing capacity of future consumers. Under this scenario, it is estimated that world food production will need to increase by 70% to meet the demands of an increasingly wealthier and urban population. This population is expected to consume, on average, more animal products such as meat, eggs, and dairy products (FAO 2018). More recent estimates, based on meta-analysis of existing literature, indicate that food demand will increase between 30% and 60% by 2050, encompassing different scenarios, including climate change (van Dijk et al. 2021). Changes in consumption patterns are projected to have a potentially greater impact on food demands than changes in population growth.

Based on these global estimates and the limited scope for expanding the current agricultural area, it is implied that the necessary increase in food production will have to be achieved through further agricultural intensification. Moreover, the projected increase in food demand is often used to justify any form of agricultural intensification. Common narratives used nowadays to support the need for agrochemicals, patented seeds, or genetically modified crops include "humanitarian priorities," "necessary trade-offs between nature and humans," or "land sparing by increasing yields to create space for nature," among others. Additionally, some argue that promoting agroecology is unethical in a world facing hunger.[1] However, are the assumptions underlying these estimates and opinions sound, logical, and evidence-based? What kind of agricultural intensification would be necessary to feed and shelter the world population in the future? Let's take a look at some global trends in search of answers to these questions.

[1] I do not cite these sources in order to avoid giving them a greater centrality than they actually deserve.

1.2 Global Trends: Food, Biodiversity and the Environment

Although it may seem obvious that food production will need to increase by 2050, this is only one aspect of the solution to global food security. It is true that this increase must come from intensifying current agricultural production. The potential for expanding the current agricultural footprint is limited – around 15% – and further expansion would lead to undesirable consequences for biodiversity and our global climate. In this context, the term 'agriculture' encompasses all forms of terrestrial food production, including livestock, aquaculture, greenhouse horticulture, agroforestry, and more. The critical question that remains unanswered is what type of agricultural intensification is necessary to meet future global food demands. Simply extrapolating trends and speculating about the future will not provide a satisfactory answer to this question; we must begin by examining our present situation.

Nowadays, enough food is being produced at a global scale to feed every person on Earth. Estimates by the Committee on World Food Security (https://www.fao.org/cfs/en/) of the FAO indicate that farmers produce approximately 2800 kilocalories per person per day, while the global average requirement is 2100 kilocalories per person per day (taking into account the wide diversity in food habits and physiological requirements across different world regions). Currently, we produce enough food to feed 10 billion people, even though the current global population is 7 billion. However, in 2021, 828 million people suffered from chronic undernourishment, particularly in countries and regions expected to experience the fastest population growth until 2050 (Fig. 1.1a). The breakdown of undernourished individuals in 2021 shows that 425 million are in Asia, 278 million in Africa, and 56.5 million in Latin America and the Caribbean, representing 9.1%, 20.2%, and 8.6% of the population in these regions, respectively (FAO 2022). The countries and regions within these continents that are unable to meet their food requirements also tend to have high levels of extreme poverty, wide inequalities, and armed conflicts, as well as poor infrastructure, political instability, and natural disasters, among other important factors. Therefore, food insecurity has multiple causes. Furthermore, the number of undernourished people has increased by nearly 150 million since the outbreak of the COVID-19 pandemic. In 2020, the United Nations estimated that 2.3 billion people were considered food insecure, while almost 3.1 billion could not afford a healthy diet.

Agricultural production, on the other hand, has increased significantly in many parts of the world over the past 50 years. However, in regions where farmers cannot afford agricultural inputs and technologies, especially where governments are unable to subsidize farmers, agricultural production has remained stagnant (see Fig. 1.2a). It is often claimed that intensive agriculture in regions like western Europe or the US is highly efficient. However, a system cannot be deemed efficient if it requires subsidies to remain profitable. High output levels are often seen as an indicator of efficiency, but achieving yields such as 13 tons of wheat per hectare, 12,000 kg of milk per cow per year, or 100 kg of tomatoes per square meter of greenhouse is only possible by relying on external inputs of energy, nutrients, and financial resources.

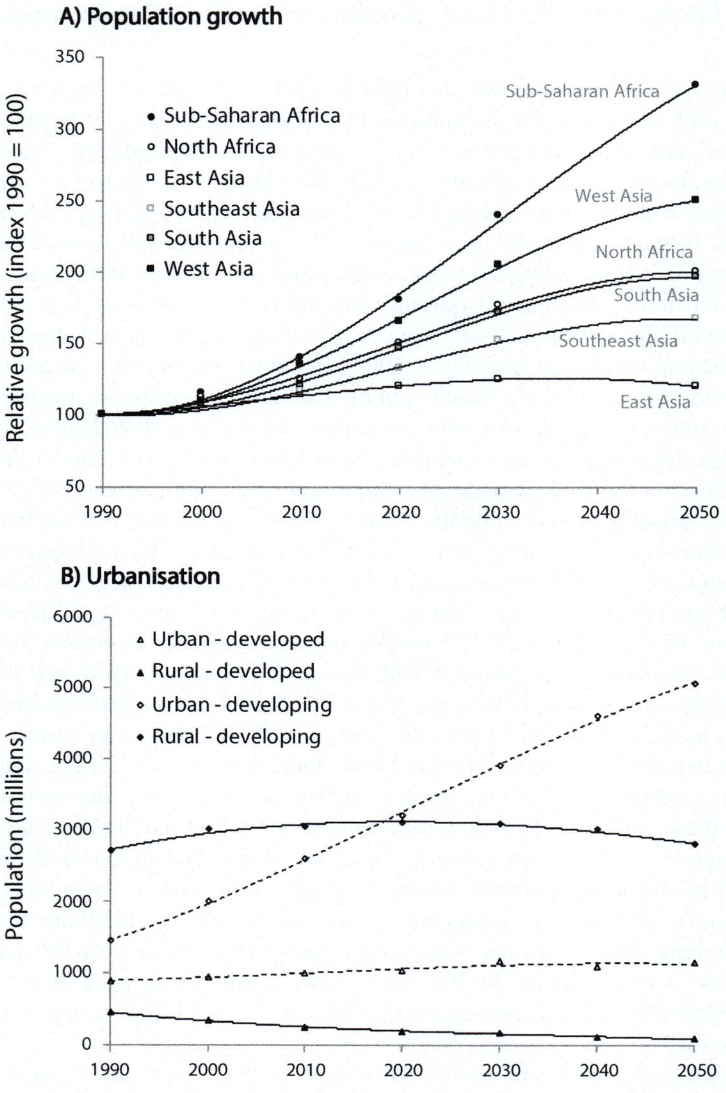

Fig. 1.1 World trends. Projected population growth in Africa and Asia (**a**) and number of people living in urban and rural areas in developed and developing regions (**b**) to the year 2050 (trend lines are 3rd order polynomial functions only valid within the range depicted in the figure). (Source: FAOStat)

Labour, which is often provided by underpaid or even illegal immigrants in these systems, is also a form of energy subsidy. Moreover, the environmental and health consequences of these systems, which are typically not considered when assessing efficiency, cannot be overlooked. For instance, in Europe, the costs associated with

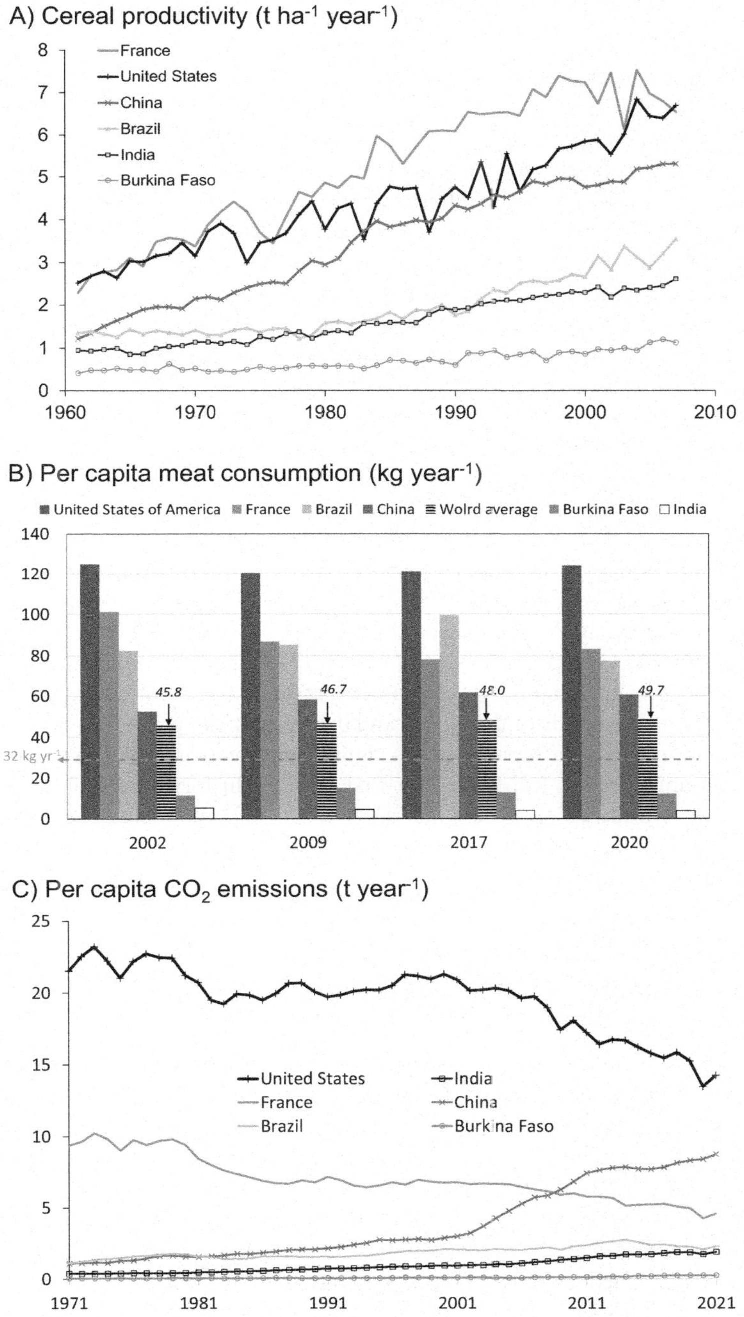

Fig. 1.2 Production and consumption trends in selected countries from Europe, America, Africa and Asia. (**a**) Changes in cereal productivity during the second half of the twentieth century. (Source: FAOStat); (**b**) Per capita meat consumption over the last 20 years, with the horizontal dashed line indicating the annual intake recommended by the Word Health Organisation. (Source: FAOStat); (**c**) Fossil fuel CO_2 emissions per capita over the last 50 years (EDGAR: Emissions Database for Global Atmospheric Research, European Commission, 2022)

pesticide use in terms of environmental impact and human health were estimated to be around 2.3 billion Euros, while the profits generated by the pesticide industry amounted to approximately 0.9 billion Euros (Basic 2020). This means that every Euro of profit the pesticide industry makes imposes a societal cost of 2.55 Euros in terms of environmental and health impacts. Similarly, when high-protein diets (e.g., containing 4.5% protein) are fed to beef cattle, they excrete 92% of the ingested nitrogen, resulting in nitrogen-rich manure or slurry that pollutes the air through ammonia release, as well as the soil and water table when applied to agricultural fields. Thus, while farmers profit from selling the meat that contains 8% of the N applied, society has to deal with the externalities produced by the other 92%. The same analysis can be applied to excessive phosphorus use, inefficient water use, biodiversity loss, and the subsequent decline in ecosystem services. It is evident that intensive agriculture is far from efficient.

Since 2012, the majority of the world's population has been residing in urban areas, and this trend is expected to continue and intensify in the developing world in the coming years (Fig. 1.1b). Urbanization reduces the autonomy of poor families in terms of food security. If this trend persists, the urban poor will increasingly rely on the market to access food. Conversely, the share of smallholder families that is nowadays largely, but not exclusively, focused on self-subsistence, will need to produce surpluses for sale in order to feed the growing urban population. Urbanization also brings about changes in consumption patterns. Present-day human diets are imbalanced in most parts of the world, and the overconsumption of animal products, particularly in wealthier regions, poses significant threats to both human health and the environment (Fig. 1.2b). For the first time in history, the number of obese individuals worldwide exceeds the number of undernourished. Approximately 65% of the global population lives in countries where obesity-related deaths surpass those caused by hunger.

1.2.1 Rebalancing Animal Protein Consumption

Rebalancing the production and consumption of animal protein is also an imperative to secure food for everyone in the future, not just because of the inefficiency of livestock production itself, but also because the impact the industry has on the environment, including global warming, on animal welfare and on human health. Barnosky (2008) calculated that the current global biomass of the megafauna (anything between a sheep and an elephant, thus including humans and their livestock) estimated at 1.9×10^{12} kg, exceeds by 10 times the carrying capacity of the Earth estimated at 0.18×10^{12} kg, based on global plant photosynthesis. Excluding livestock, current human plus wild megafauna biomass represent 0.46×10^{12} kg, about 2.5 times higher than Earth's carrying capacity. If we are able to maintain such large number of domestic megafaunal biomass today, it is because we have been relying on the photosynthetic energy accumulated over millions of years in fossil fuels since the onset of the industrial revolution. It is obvious that we will not be able to sustain these livestock numbers for much longer, relying on a non-renewable resource such as oil. Thus global livestock, of which 3.6 billion are ruminant, affect climate change through their own fermentative emissions and also through the use of fossil fuels to sustain their production.

However, the solution to achieving sustainability in the food system is not as straightforward as commonly presented, such as solely advocating for reduced meat consumption or adopting vegetarian or vegan diets. It is not as simple as that. Firstly, the animal industry encompasses more than just meat production; it includes dairy, eggs, fibre, fish, protein, and fats. Secondly, sustainability is a multidimensional concept, subjective, and context dependent. Becoming 'sustainable' entails much more than just reducing meat consumption. The large disparities in terms of CO_2 emissions registered between world regions cannot be solely ascribed to differences in meat consumption (Fig. 1.2c). Additionally, animals are a crucial component of the agroecosystem, that when properly managed can deliver a series of ecological services besides contributing to farming households' income and nutrition. Agroecosystems that raise animals on native or long-term pastures may contribute to biodiversity conservation, carbon sequestration, water regulation and animal welfare (cf. Tittonell 2021). But, most importantly, in a world that counts nearly 2 billion people who are food insecure, of which 0.8 billion are hungry, speaking of 'reducing' meat consumption seems almost unethical.

This is why I refer to speak of 'rebalancing' our consumption, not only of meat, but of all other animal products. In terms of meat, the World Health Organisation recommends 90 g per person per day to stay healthy, which translates into about 32 kg per year. Current world average meat consumption is 40 kg per person per year, already above the recommended intake (Fig. 1.2b). This average is the result of large disparities across world regions.

1.2.2 Biodiversity and Poverty

The publication of the 2019 report by the Intergovernmental Panel on Biodiversity and Ecosystem Services (IPBES 2019), plus the WWF Living Plant report published bi-annually (the latest in 2022), revealed the alarming rates at which many species are running towards extinction, next to the many that are already extinct. These reports speak of nearly one million species threatened with extinction, a figure that has been contested due to the many assumptions made in its calculation. Some 1.7 million species of plants and animals are described, but the total number of species is much larger. In the IPBES assessment they estimated the number of plant and animal species to be 8.1 million, of which 75% are insects. They assumed extinction rates of 10% for insects and 25% for non-insects, based on best current knowledge. Summed up, these yield 1,175,000 species threatened by extinction. It is beyond the point to discuss whether these figures are accurate of not. Even one tenth of that number of threatened species would already be alarming.

Biodiversity loss is the result of five major global drivers, namely climate change, land use change including deforestation, overexploitation (specially through fishery), invasive alien species, and the pollution of habitats and resources. Agriculture, and food production more generally, have a great accelerating impact on these drivers. Next to considering the intrinsic value of all living species, loss of biodiversity entails a loss of ecosystem services, with direct consequences for humanity. Similar to these recent biodiversity reports, the planetary boundaries assessment (Steffen et al. 2015) also pointed to three major environmental drivers that have reached critical values at Earth systems' level: loss of genetic diversity, and disruption of the biogeochemical cycles of nitrogen and phosphorus. Here again, agriculture and food production have a great responsibility, next to being accountable for one quarter of all greenhouse gas emissions. This figure includes also all the deforestation that goes on annually to create new agricultural land. There are, however, plenty of myths about deforestation in the global South.

About 70% of the current deforestation is done by large commercial farming operations to produce soy, oil palm, livestock, etc., and the remaining 30% by smallholder family farmers. Deforestation is often presented as a consequence of poor yields, implying that intensive agriculture can 'save the forest' by sparing land through achieving high yields on already deforested land. It is actually the opposite: it is the high yields and high profits that incentivise farmers to expand their agricultural land through investing in costly deforestation (e.g. Grau et al. 2013). Sometimes deforestation is presented as an unavoidable trade-off with regards to food security and human well-being. This is also a myth! Let us examine the case of Brazil, the only country in the word that in the 2000–10s achieved two of the Millennium Goals, reducing extreme poverty and eliminating hunger. As more than two thirds of the food consumed by Brazilians comes from family farming, the government of that time led by Lula da Silva created a Ministry of Agrarian Development, which focused on family farming and territorial development holistically. Several policies were launched to address food insecurity such as the famous *Fome Zero* (Null Hunger) program, relocation of urban people into rural areas, school catering using locally produced food, and a National Law on Agroecology and Organic production. The results were impressive. Figure 1.3 shows that while the number of hungry people went down in Brazil from 2003 to 2013, the rate of deforestation went also down during the same period. Indeed, when Lula da Silva came to power in Brazil in 2003, he set a target of reducing deforestation rates by 80% with respect to the average level in the previous 10 years (19,625 Km^2 $year^{-1}$), to be reached by the year 2020. The target was almost reached already in 2012. However, the congress impeachment of Lula's successor, Dilma Roussef, in 2016 interrupted the forest protection program. The later arrival of Jair Bolsonaro to power, supported by the agribusiness sector, incentivised deforestation again, reaching 16,557 Km^2 $year^{-1}$ in 2021. In 10 years (2003–2013), Brazil was able to

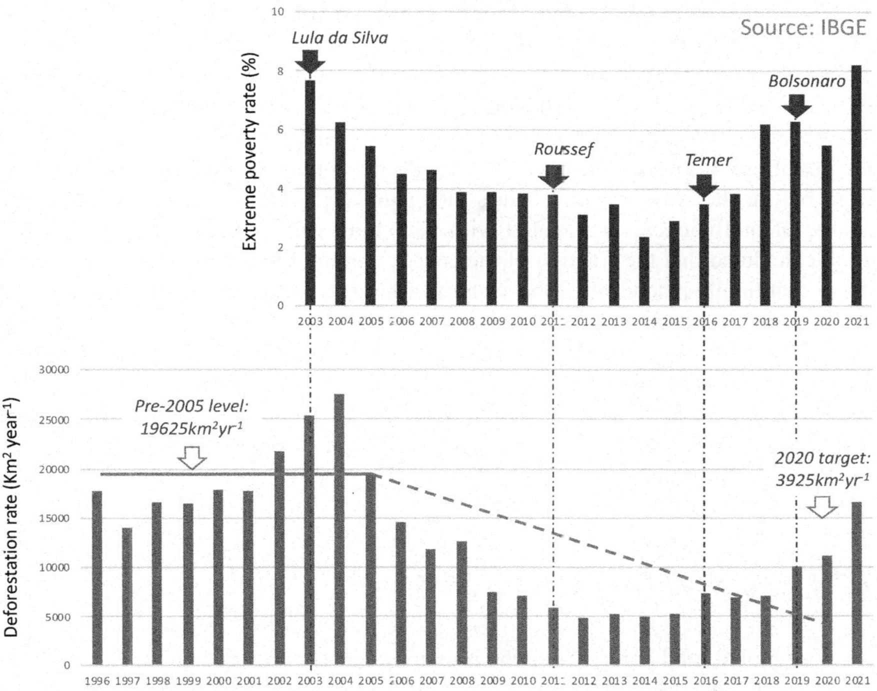

Fig. 1.3 Rates of extreme poverty (above) and of annual deforestation (below) in Brazil between 2003 and 2021. Black arrows at the top of the graph indicate the start of the four presidential administrations that governed Brazil during this period. The blue line indicates the reference deforestation rate, average from 1996 to 2005, and the dashed blue line the target set by Lula da Silva's administration. (Sources: IBGE (2022) and Silva Junior et al. (2021), *Nature Ecol Evol* 5, 144–145)

end hunger, reduce extreme poverty by 60% and deforestation by virtually 80%. Where are then the claimed trade-offs between food security and forest conservation? Rather, what happened after that period in Brazil illustrates how exposed 'sustainable development' can be to the uncertainties of politics, which may cause far more damage than climate change in a very short period of time.

1.2.3 Food System Inefficiencies

Finally, it is calculated that at least one third of the food produced worldwide never reaches a human stomach due to losses postharvest, during storage, processing, trading and consumption (FAO 2011). About 14% of the world food is lost between harvest and the retail market (FAO 2022), and 17% is wasted at the retail and consumer levels (UNEP 2021). Food loss and waste, which together amount to

1.3 billion tons per year, are the result of multiple processes, but an important one is the distance that food has to cover from production to consumption. Urban consumers build their diets on a large diversity of food items, often originating from more than one region, country or continent. International trade not only increases the rate of food waste through long distance transport and storage, but also by limiting the possibility to recycle food waste or by-products back to the agroecosystem. But even before the waste of food starts along the value chain, there is also another source of inefficiencies associated with very long value chains. In 2013, Cassidy et al. calculated that the fraction of the energy captured by crops through photosynthesis that finally reaches the food system varies also widely across regions and crop types. In the most 'productive' areas of the world, such as the US, western Europe, East China and parts of South America, barely 20–30% of the energy captured from solar radiation enters the food system in the form of food calories. Then food waste starts. Which means that the 30–50% food waste should be calculated from that fraction, not from total food production based on plant photosynthesis. Our current food system is intrinsically inefficient.

1.2.4 No Need to Wait Until 2050

Meeting global food demand in 2050 remains a great challenge, particularly because we are not yet able to meet such demand today. Current industrial agricultural production depends largely on external energy inputs and is responsible for a large share of deforestation, poverty, pollution, biodiversity loss and greenhouse gas emissions. About 70% of the energy contained in a grain of cereal produced under intensive agriculture comes from fossil fuels, which are by definition non-renewable. Production costs associated with this model of intensive agriculture are on the rise every year. Although food prices were increasingly volatile in the last decade and seem to be on an upward trend nowadays (the World Bank food price index reached a new high of 159.7 in 2022), farmers that practice intensive agriculture are heavily indebted and their production must be subsidized by other sectors of society. The international input and output market for agriculture is an oligopoly. A few multinational companies control the genetic diversity of our major crops, produce the agrochemical inputs used on them, and trade their harvest worldwide. In places where agriculture intensification did take place, and yields are currently high, the costs for biodiversity and the environment have been just too high. Yet, none of these companies is held responsible for the damage caused. Rather, the blame is thrown on farmer's shoulders. The trends shown earlier in Fig. 1.2a–c reflect current inequalities across world regions that industrial agriculture does not come to solve, but only to exacerbate. If we continue to subsidize and protect industrial farming, the number of hungry people will increase in the future, especially as resources such as water, nutrients, energy or land become scarcer, climate change accelerates, and soils continue to degrade.

1.3 Who Is Producing Our Food?

Most of the food produced and consumed in the world comes from smallholdings that farm less than a quarter of the cultivated land worldwide. The most conservative estimates point to 50% of the food consumed being produced on 20% of the land available (these estimates vary broadly,[2] see e.g. Lowder et al. 2016; Samberg et al. 2016; Herrero et al. 2017; IFAD 2016). Another way of looking at where food is produced is the use of proxy indicators, as done by international entities such as the World Food Program or the World Health Organisation. Cereal production is used as a very good proxy to gauge food production at global level, since 80% of the calories we ingest as humans comes from cereals (wheat, rice, maize, sorghum, barley, etc.), either as grain or flours, or transformed into milk, eggs, meat, fructose, malt, starch and a range of edible molecules. Cereal yields show great disparities across world regions (cf. Figure 1.2b). Figure 1.4 plots public FAO data on average cereal yield per country against the contribution of each country to global cereal production (cf. Tittonell et al. 2016). Yield and production figures are reported by every country to the FAO annually, and the methods used to estimate them are not the same in every country. Often yields are calculated as total production/ total area, and this way of estimating them may be subject to high variability specially when cultivated areas are small. In spite of their limited accuracy, these global trends provide interesting insights into where is food being produced at global level.

In Fig. 1.4a we see that 50% of the global production takes place in countries were average cereal yields are lower than 3.1 t ha^{-1}. Moreover, about 15% of the cereals are produced in countries where yields are very low, less than 1.9 ha^{-1} (and 1.3 t ha^{-1} on average – Fig. 1.4b). At the other side of the distribution, we can see that the contribution of all the countries where yields are very high, more than 5.9 t ha^{-1} on average, represents 12.5% of global cereal production. This includes the intensive, industrialised agroecosystems of the USA and western Europe, which are often portraited as being responsible for 'feeding the world'. Individual countries such as China, India and the USA contribute large amounts of food to the global production, but the largest share is contributed by the medium-high yield group that includes countries such as Brazil, Argentina, Mexico, The Russian Federation, Ukraine, Turkey and Vietnam. In terms of areas, most of the area under cereal production yields low to medium yields (Fig. 1.4b). The low yielding group, with an average yield of 1.3 t ha^{-1}, comprises large areas of farmland where soils are degraded, where most of the 828 million undernourished people live (more than half of the undernourished people on Earth are rural dwellers), and where the fastest population growth is expected in the future (cf. Fig. 1.2a).

[2] In part, these estimates vary when they consider either calories, or nutrients, or food volumes; or when they consider food produced versus food consumed; or depending on the farm size below which a farm is considered to be a 'smallholding' (often the value of 2 ha is used, and there are about 475 million farms below this size).

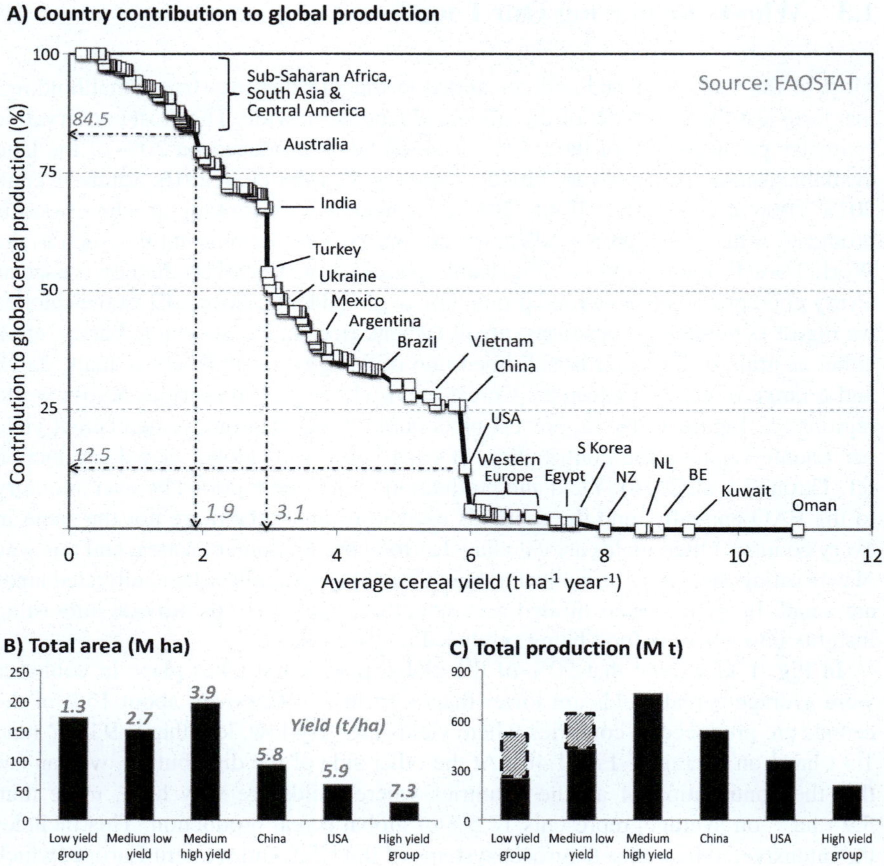

Fig. 1.4 (**a**) Cumulative contribution of each country to global cereal production against the average yield of cereals in each country. Results grouped into low yield (<1.9 t ha^{-1}), medium low (1.9–3.1 t ha^{-1}), medium high (3.1–5.8 t ha^{-1}), China, United States of America, and high yield (> 6 t ha^{-1}), and the respective areas (**b**) and total production (**c**) per category. Dashed bars in (**c**) indicate exploitable production gaps that could be filled if yields were doubled in the low yield group or increased by 50% in the medium low yield group. (Data available at: faostat.org. Modified from Tittonell et al. (2016))

1.3.1 Targeting Productivity Increases

When speculating about how to feed the world in the future, it is often claimed that food production will have to increase by 70% to meet global demands. Due to limited area available to expand agriculture, it is often assumed that yields will have to increase worldwide to achieve this extra 70% (which, by the way, is a contestable figure in the light of the food system inefficiencies discussed in previous pages). However, increasing yields in countries that fall within the high yielding group is challenging. On the one hand, because yields are already high, but also because these agroecosystems rely on subsidies of energy, nutrients and money that the world is no longer able to afford from an economic and, specially, from an environmental perspective. Increasing yields in countries where yields are already high will only worsen their impact on biodiversity and the environment, will exacerbate regional inequalities and lead to further resource concentration, and will certainly not contribute to feeding the world (they contribute barely 12,5% of our cereals today – Fig. 1.4a).

Increasing yields in the low yielding group, on the other hand, is agronomically, environmentally and economically sound and feasible. It will result in the production of extra food there where the demand is not yet covered, where population is expected to grow fast. It will generate income opportunities for poor farming families and contribute to national economies in which primary production represents up to 70% of their grow domestic product. Nowadays, a large share of the wheat consumed in the poorest countries of the world originate from Ukraine and the Russian Federation (e.g. in the year 2021: Eritrea: 55 and 45% of their imports, respectively; Madagascar: 51 and 32%, Tanzania: 48 and 22%; etc. see FAO State or World Food Security and Nutrition). It is obvious that increasing cereal production in the low yield group of countries will also render them more resilient, independent, and less exposed to the fluctuations of international markets. For example, doubling average yields in the low yield group, from 1.3 to 2.6 t ha^{-1}, which research shows to be a modest agronomic goal, will already represent large amounts of extra food being produced in places where people need it. If we now assume that yields could be also raised by 50% in countries of the medium low group (from 2.7 to 4 t ha^{-1}), another modest goal, the extra production by these two groups will approximately equal the current production of the USA and the high yielding group of countries combined (dashed bars in Fig. 1.4c). However, increasing yields in the low yielding group of countries remains socially and ecologically challenging, as discussed earlier. The technologies of the 'Green Revolution', which generally only work when subsidized, have proven to be useless at increasing productivity in these regions in spite of 50 years of trying to impose them through international 'development' cooperation.

1.3.2 Regenerating Degraded Soils

We need technologies and practices that fit the realities of smallholder farms in such challenging, often marginal environments. Technologies that are co-designed with farmers through a dialogue of wisdom among them and other actors of the food system, including scientists. Smallholder farming yields are not inherently low. Indeed, well-managed small farms tend to be more productive than large ones, as will be shown later in this book. One issue to be addressed urgently is the fact that the majority of the rural poor farm small pieces of land on degraded soils that have lost their fertility. Over half of the land used for agriculture worldwide is moderately or severely affected by degradation processes (UNCCD 2017). Between 2000 and 2015, Africa lost 22,4% of its agricultural soils, Asia 28%, South America 27%,

Oceania 35,5%, and this keeps on going as we speak. This impacts directly the lives of 1 billion people (and affects indirectly the ability of urban people to afford food). In the cases of Africa and Asia, the economic losses associated with soil degradation are estimated at US$65 billion and $84 billion per year respectively. While low agricultural productivity, food insecurity and soil degradation are multi-causal, a major driver of soil degradation is the pressure exerted by an increasing rural population in many parts of the world. Proportionally, there are more urban than rural people every year, yet in absolute terms the number of people in rural areas continues to increase. Rural population growth leads to land subdivision, impeding the practice of fallow rotations or shifting cultivation, traditionally used as strategies to maintain long-term soil fertility. Population growth leads also to achieving animal stocking rates that are often above the carrying capacity of the landscape, resulting in underfed animals, diseases, and low productivity, but also in the degradation of soils, water and vegetation resources. Many of these practices are not a conscious choice rural families make but rather a 'default' adaptation caused by poverty (Tittonell and Giller 2013).

1.4 What to Produce, How and by Who?

So far, we have discussed food production as a proxy of food security, and focused exclusively on tons of grain or total calories, without paying attention to the origin of such calories or more broadly to the composition of diets necessary for an adequate nutrition. On the other hand, we have only considered one of the pillars of food security, food availability, and we have not mentioned another related, much more comprehensive concept: food sovereignty. We have not paid any attention so far to what needs to be produced in terms of nutritional elements. We have discussed which countries produce most of food and how much more needs to be produced and where, but have not discussed how to increase food production sustainably. And finally, we have considered food production as if it was produced by 'countries', disregarding the most important factor in our food system: the farmer. Non-exhaustively, the following paragraphs provide an overview of concepts that can help us to start answering these questions: what needs to be produced, how and by who?

1.4.1 Food Security

Briefly, food security is defined by the UN organisations as consisting of at least four pillars: food availability, stability, access and utilisation. Food production is obviously related with food availability, although availability entails more than just production. Then, not only food needs to be available but also stable over time, minimising the negative effects of climatic variability on food production, but also those of market volatility or political instability. Food can be produced in plenty, stably over the years, and yet people not be able to access it. Access to food is a key pillar of food security. Even in countries and regions where large amounts of food are produced (and exported) there can be people in situations of hunger. Also at local scale, inequalities may lead to co-existing food secure and food insecure households due to differential opportunities or capabilities to access to food. The last pillar of food security is food utilisation, or the ability of human bodies to assimilate and utilise the nutrients in the food they ingest. Food utilisation may be hampered by the inability to cook food. Many food items need to be cooked for humans to be able to assimilate their nutrients. There are plenty of people on Earth who are unable to eat one hot meal a day due to lack of energy to cook their food (energy insecurity). Food utilisation also points to the need for a balanced diet, a nutritionally diverse intake of nutrient allows a good utilisation of the food calories, fibre and nutrients. Humans can barely survive on a calorie-only diet.

1.4.2 Human Nutrition and the Environment

A nutritionally unbalanced diet is one of the cornerstones of what is termed the 'double burden' of nutrition. Although hunger is an excruciating problem which is morally unacceptable, statistically speaking and only in quantitative terms obesity is an even more widespread problem, affecting some 1.2 billion people on Earth (IFPRI 2016). Globally, 13% of the adults are obese and 39% are overweight. The proportion of overweight children increased from 4% in 1975 to 18% in 2018 (World Health Organisation). Obesity is the result of the balance between calorie intake and expenditure, and a leading risk factor of premature deaths worldwide (8% of all deaths, or 4.7 million, were attributed to it in 2017). The increase in obesity rates is attributed to calories being more readily available in the diet, and to physical inactivity associated with more sedentary lifestyles. But obesity is not the only diet-related cause of death; heart diseases, cancer and type-2 diabetes are also major diet-related causes of death. Considered all together, poor diets were responsible for 22% deaths in 2017 (10.9 million deaths), which is more than all the deaths attributed to COVID-19 so far (6.9 million by June 5, 2023) or to car accidents (1.4 million deaths per year) worldwide. Because diet-related disorders are the number one cause of human deaths, and because diet composition is also a major driver of environmental impacts, nutritionists, ecologists and food specialists began to think together over the last decade about the global impact of diets on humans and the planet.

The first well-known report that emerged from such global exercises (Murray 2014) compared the needs of humanity in terms of major food items to conform a balanced diet, against the current production of these items at global scale (Fig. 1.5). In Fig. 1.5a, the bars indicate the current (2014) levels of production of the various food items necessary for a nutritionally balanced diet, one that minimises the risk of diet-related deaths. The dashed line across the graph shows the amount needed of each of these items to feed the entire world population. To homogenise units,

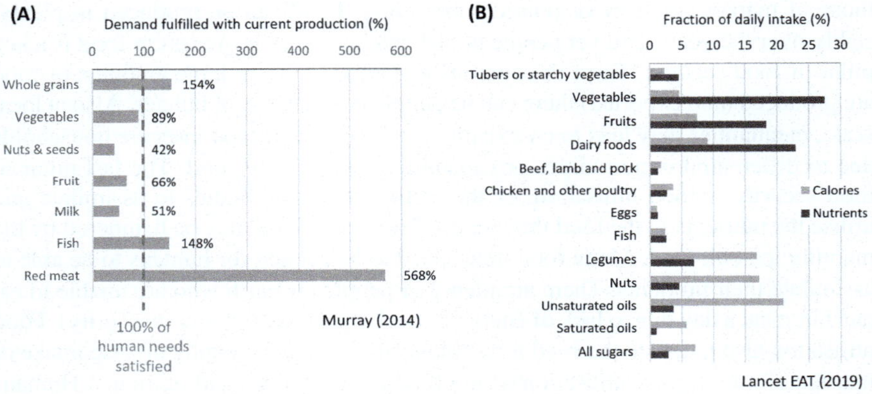

Fig. 1.5 (**a**) Proportion produced of the major food items of a balanced diet (grey bars) relative to their requirements at global level (dashed vertical line), based on the report of Murray (2014). (**b**) Recommended daily intake of different food items in a diet that balances nutrition and minimises environmental impact (assessed through the nine planetary boundaries – Steffen et al. 2015), expressed as a fraction of the total daily intake of calories (grey) and nutrients (black), as published in the Lancet-EAT report (2019)

everything is expressed as percentages. At the time of the study, the world was producing 54% more grain than globally needed for a balanced diet, 11% less vegetables, 50% less milk and about 35 to 55% less fruits, nuts and seeds. In contrast, the world was fishing or farming 48% more fish than needed for a balanced diet, and as much as 568% more red meat than necessary. This challenges the popular narrative sustaining that meat production will have to increase in the future to meet the demands of a growing and more affluent population. Today, we are still producing much more meat than needed for a balanced diet (yet, as mentioned earlier, there is a large number of people worldwide who cannot afford to eat meat) and this certainly contributes to diet-related disorders. Another message that can be drawn from this study is that, if the figures are reliable, we should increase (virtually double) the production of fruits, nuts and seeds to meet our global needs. Many of these species are perennial shrubs and trees. This offers opportunities to agroforestry or silvo-pastoralism, and to generally increase the proportion of trees in our production landscapes, contributing to both creating habitats for biodiversity and cooling the planet.

In 2019, a new report was published integrating the ideas of the previous one, calculating a balanced global diet to minimise death risks, but adding also a calculation of a diet that minimises global environmental impacts using the indicators of the planetary boundaries assessment (EAT-Lancet: Willett et al. 2019). The results indicate that a balanced global diet that minimises both human health risks and environmental impacts should consist of approximately 50% in volume of fresh fruits and vegetables, and about 20% of whole grains and 15% of plant-sourced protein (legumes and nuts), while minimising the intake of starchy crops such as potato or cassava, dairy products and meat (Fig. 1.5b). When expressing these values on the basis of the daily intakes of either calories or nutrients, then their relative proportions change. Vegetable, fruits, dairy and legumes provide most of the nutrients in such a balanced diet, while most of the calories come from legumes, nuts, unsaturated oils and dairy products. These calculations were based on an assumed *per capita* calorie intake of 2500 Kcal day^{-1}. It is important to notice that these are global estimates that consider also the diversity of human requirements across regions, cultures and lifestyles. They are not, however, guidelines to develop individual diet recommendations.

Globally speaking, current diets surpass the 'health' boundaries recommended for red meat by 288%, for tubers and starchy crops by 293%, and for eggs by 153%, whereas we are falling short of vegetables, fruits, nuts, edible seeds and legumes, still according to the Lancet-EAT report. Thus, at global level, we are not producing what we need to feed ourselves as humanity, not in terms of quantity, as we saw earlier, but of quality. The result of the dominant model of agriculture and food production is hunger and obesity, the double burden of malnutrition. Increasing future food production through the simplified agroecosystems of industrial agriculture, dominated by monocultures, will not contribute to addressing a key pillar or food security: nutritional diversity to sustain balanced diets.

1.4.3 Biodiverse Production Landscapes

Ensuring balanced diets depends as much on the intrinsic quality of the crop cultivars and animal landraces used to produce food, as on the diversity and integration of crops and animals grown and raised in a diverse agroecosystem. In other terms, a nutritionally diverse diet requires a nutritionally diverse agricultural landscape. Thus, next to the balance between what is produced and traded at global scale in terms of major food items versus global human needs, as illustrated in Fig. 1.5, assessing the nutritional quality of diets also requires examining (i) the nutritional quality of different agricultural products, and (ii) the nutritional diversity of the agroecosystem. The nutritional content, plus the content of water, fibre and other important metabolites in plant, animal and fungi tissues used for human nutrition are greatly influenced by the production system from where they originate. These traits are influenced by genetic characteristics, by the environment in which they grow or live and by management factors associated with the production, processing and transformation (and trading) systems. New varieties of wheat, for example, have greater contents of gluten and lower contents of antioxidants when compared with traditional cultivars but also, a single variety, will exhibit large variations in these elements when cultivated with different levels of nitrogen fertiliser, with or without irrigation, pure or in association with a legume crop, etc., etc. It is well known however that breeding of crop cultivars and animal landraces with the main objective of attaining high individual productivity levels has generally led to a decline in their nutritional quality.

Climate change, on the other hand, has also an impact on the accumulation of proteins and minerals in plants (cf. Soares et al. 2019). Traditional cultivars, even commercial ones, have also a greater genetic diversity than modern ones, which provides for better adaptation and resilience facing climate change. This is the case, for example, of 'California Wonder', a well-known and widely grown species of sweet pepper when compared against modern pepper cultivars (cf. Votava and Bosland 2002). Thus the characteristics of the agroecosystem, in terms of soil quality and overall ecological structural and functional integrity, interact with the characteristics of the genetic resources cultivated and raised in it to determine the nutritional quality and diversity of the food being produced. Additionally, biodiverse landscapes sustain a diversity of ecosystem functions, some of which are essential for its own functioning, while others represent ecosystem services of local and global importance. Biological pest regulation, water capture and storage, soil formation/ protection, and nutrient cycling, release and availability are key biodiversity-mediated ecosystem services necessary to sustain food production in agricultural landscapes. Maintaining landscape biodiversity is also essential to deliver cultural ecosystem services, such as a sense of belonging, traditional foods, space for spiritual and recreational functions, and traditional knowledge systems, practices and institutions.

1.4.4 Who Is Farming Nowadays, Who Will Farm in the Future?

It has been popularised in different sources that 70% of the workforce in agriculture is female, which would imply that farming is largely a feminine activity.[3] This estimation can be traced back to a 1972 UN report, but it has never been confirmed again. The most recent data produced by the UN (cf. UN Women report) points to 43% of the workforce in agriculture being provided by women, while less than 20% of the world land owners are women. About 60% of the employed women in sub-Saharan Africa work in agriculture, and up to 70% of them in South Asia. But of course these figures are difficult to estimate, especially in informal economies, and several assumptions are made in their calculation. For example, fetching water, which is an activity done mainly by women and children, is not computed as an agricultural activity yet it is crucial for the rural livelihood. Women in sub-Saharan Africa spend 40 billion hours a year fetching water; in countries such as Malawi, women spend on this activity 9.1 h per week while men only 1.1 h per week.

It is calculated that, since the land managed by women yields 20–50% more than land managed by men, providing equal rights of access to productive resources for women and men would result in overall greater land productivity. The UN Women

[3] I often implied this in my public lectures (e.g. TEDxEde 2014), as I felt it did justice to women's role in agriculture; unfortunately I was relying on evidence that was not solid or up to date.

report estimates that such an increase in productivity may lead to 10–15% increase in global food production. It is also true that in many farming systems and contexts there are activities that are more associated either with women or men, although this distinction is increasingly getting blurred. In 'traditional' rural societies, where they persist, men activities are often associated with livestock rearing, land preparation, and use of machinery or technology. In sub-Saharan Africa, the emergence of HIV-Aids led to a large proportion of female headed households, even grandmother-headed households in which women and young children performed all activities on the farm. I was confronted with this situation often enough when doing field work in East and Southern Africa 20 years ago (e.g. Tittonell et al. 2005, 2007). In places where men migrate seasonally to work in urban areas, virtually all activities on the farm are performed by women and children (e.g., Andersson 2001).

Women are also the custodians of an important aspect of agrobiodiversity: the knowledge associated with the traditional species and cultivars/landraces they keep, with their cultivation and management, with their multiple uses, with their maintenance and reproduction, with their legacy and adaptive capacity. The literature documents plenty of examples of this. For instance, women are able to recognise and manage a broader diversity of so-called local vegetables than men in Kenya (Figueroa et al. 2009), or to rear, sheer and process the wool of the almost forgotten, heirloom *Linca* sheep landrace in Patagonia (Reising et al. 2022), or to better recognise soil quality attributes through visual soil indicators in Nepal (Alomia-Hinojosa et al. 2018). In Brazil, the citizen-science exercise known as '*cadernetas agroecológicas*' (agroecological notebooks – Silva et al. 2020), in which women recorded all plant species and cultivars they grow, how, where, their yield, their uses, seed origin, income generated, among others, revealed plenty of previously unknown facts. For example, in Minas Gerais (Atlantic forest zone), one hectare of home garden, a highly diverse type of polyculture grown by women, produces the equivalent income as two hectares of coffee, which is the cash crop grown by men in that region; it produces between 6 and 15 minimum wage salaries per year. The analysis of the data also revealed nearly 1400 different accessions of maize being grown by women in home gardens.

Women's knowledge is essential for the design and management of biodiverse agroecosystems. However, women are often not officially 'in charge' of their farms, they often have less rights or less decision power over the land, not only when it comes to buying or selling but also on day to day management decisions. In certain contexts women are not allowed to own land legally, or livestock, or access education, or credit. Even in situations where these rights are granted by law, local traditional institutions and cultural constraints may prevent women for exercising such rights. This has been termed the 'gender normative climate' (Rietveld et al. 2023) or the prevailing set of customs and by-laws that shape gender equity in any given context. In most of the world, the dominant gender normative climate is, as one may expect, the result of patriarchy. And since men make most decisions on land use and farming (Michalscheck et al. 2020), and have been doing so at least since the onset of the green revolution (new cultivars, fertilisers, pesticides, irrigation), it is almost impossible not to associate modern agriculture with patriarchal normative climates and practices. Reverting the dominant gender normative climate appears as a pre-requisite to revert the deleterious practices of the green revolution. This prompted the movements in Latin America to state that there is not agroecology possible without feminism.

All over the world farmers are getting old. The average age of the farmers in parts of the world such as the European Union is reaching 60 years old as rural surveys indicate. But this is also the case in almost all farming regions of the world. This is the result of different demographic processes. For example, Novotny et al. (2021) show how farmer outmigration created a significant depopulation of former agricultural landscapes in Oaxaca, Mexico. As it is especially the young and middle-aged farmers who migrate, the old ones that remain are responsible for keeping the farms alive. But in most cases around the world, however, the increase in the average age of the farmer reflects the lack of interest in farming among the youth. In western Kenya, Tittonell et al. (2010) showed that although the rural youth did not get many opportunities to access urban jobs, they hang out in rural areas and preferred to dedicate their time to activities other than farming, such as operating a bicycle taxi, sand mining, or brick baking. Lack of succession is particularly evident in Europe, where farming has become increasingly unprofitable, dependent on subsidies, regulated, and seen as responsible of environmental damage, poor human nutrition, animal mistreatment, climate change and biodiversity loss. The rate of suicides among farmers is one of the highest when compared with other sectors of society in western Europe. As one farmer in Spain once told me: we spend more time filling in paperwork than driving our tractors. Who would want to be a farmer in such social, regulatory, economic (and cultural) context? Yet the farmers of tomorrow, those who will have to feed the world by 2050, *are* the children of today. A transformation of our agri-food system to rely on sustainable, biodiverse agroecosystems requires a new type of farmer. An agroecological transition needs to render farming and rural areas attractive to the youth. Why not considering the potential of new technologies, including robots, artificial intelligence, block-chain or nanotechnologies (or whatever new technology emerges in the future) to attract young people's innovation potential to rural areas to foster agroecology? A dialogue of wisdoms between traditional knowledge and frontier technologies, between the old and the new.

1.4.5 Food Sovereignty

Food security is a useful concept, especially when it is defined based on its four pillars (availability, stability, access and utilisation). Because it allows us to look beyond the common assumption that feeding the world depends exclusively on agricultural production, and that hence any form of agricultural intensification can be justified in the name of 'ending hunger'. Multinational agricultural input companies and their lobbyists – plus a large number of misinformed scientists – tend to use this seemingly altruist argument to justify the use of pesticides, genetically modified crops, or fossil fuel-derived fertilisers. Yet the concept of food security, comprehensive as it is, misses to capture an important question: who decides what needs to be produced, traded and consumed? Who decides how production should take place, where, when and under which type of practices? Understood differently in different contexts, the concept of food sovereignty emerged as a way of trying to provide an answer to all these questions. The peasants organisation *La Via Campesina* supposedly coined the term Food sovereignty at the 1996 World Food Summit, to highlight the role of smallholder farmers, their knowledge, autonomy and diversity as key pillars of any food policy. In February 2007, more than "500 representatives from more than 80 countries, of organizations of peasants, family farmers, fisher-folk, indigenous peoples, landless peoples, rural workers, migrants, pastoralists, forest communities, women, youth, consumers, environmental and urban movements" gathered together in the village of Nyéléni in Sélingué, Mali. They issued together a declaration (known as the Nyéléni declaration) that defines food sovereignty as "the right of peoples to healthy and culturally appropriate food produced through ecologically sound and sustainable methods, and their right to define their own food and agriculture systems. It puts the aspirations and needs of those who produce, distribute and consume food at the heart of food systems and policies rather than the demands of markets and corporations."[4]

The origin of the term is inaccurate, and although perhaps the Nyéléni declaration conveys the most broadly accepted definition nowadays, there are also other origins, uses and interpretations. In 1993, the French organisation *Union Paysans* marched to Geneva together with 8000 farmer from across Europe with a banner that read '*Souveraineté alimentaire*' (Heller 2013). Costa Rica agrarian movements used the term Food sovereignty in a 1991 document issued from a roundtable discussion on 'dumping' practices and export sovereignty (Alforja 1991). Back in 1983, the government of Mexico's National Food Program (PRONAL) first objective was 'to achieve food sovereignty'. There is also evidence of Central American governments using similar terms as back as in the 1960s. Since then, however, the term food sovereignty has been adopted and sometimes also adapted or reinterpreted by different actors worldwide, from governments to political movements, that ascribe new dimensions to it. The government of Morocco, for example, issued a Food Sovereignty plan to deal with the droughts that affected the country in 2021. The US Department of Agriculture uses the term to refer to an initiative they issued on revaluing Indian food. Nationalist movements, on the other hand, place emphasis on the second part of the term, as if food sovereignty was centrally a matter of national sovereignty (which, one may argue, in part also is).

Thus, to the discussion presented in previous paragraphs of this chapter around food production, sufficiency, access, availability, nutritional diversity, etc., we need to add a new dimension, the right of people to choose for food that is healthy, ecological, culturally appropriate. Producing and delivering this type of food for

---

[4]I am proud to have been invited to give a speech and a practical course on agroecology on the occasion of the 10th anniversary of the declaration in 2017, in the very village of Nyéléni.

everyone requires ecologically sound and culturally-sensitive methods and agroecosystems. Food cannot be commodified, which means that achieving food sovereignty requires redesigning the entire food system. One issue that remains underdiscussed in my view – although we have discussed it at length with different colleagues from the agroecology movement – is the matter on how to achieve food sovereignty of urban consumers. Or, as Timmermann et al. (2018) put it, food sovereignty versus consumer sovereignty. What is the room for manoeuvring that urban dwellers have, specially low income urban households, to exercise their sovereignty when it comes to purchasing their food where they can afford it, often not on agroecological markets of through the internet.[5] Is urban farming a promising option in this regard? Of course it is, but still a partial one. The megalopolis of today do not have enough urban land available to produce food for every inhabitant, and often the land is polluted with heavy metals and microplastics, or irrigated with unsafe water, or plainly unproductive. Urban agriculture in balconies, rooftops, gardens, squares or other public places has certainly a great untapped potential. We could be producing much more food in cities, no doubt. We could be recycling plenty of nutrients, water and organic matter from the urban waste back into food production systems. Urban agriculture offers ample opportunities for a circular food economy. Yet this is not enough to secure food for the millions of inhabitants that live in a modern city today, let alone to grant them food sovereignty.[6]

Supporters of food sovereignty tend to have a negative opinion on the concept of food security (cf. Edelman 2014). This is the result of large multinational lobbyist and governments using "food security" as an excuse to sell more inputs or seeds or machinery, or to impose certain models of agricultural production. But the fact that the concept is misused, often as a green-washing strategy, is not a good reason to discard or oppose it (beware that governments have also misused the term food sovereignty). Food security is one of the goals of food sovereignty. There is no food sovereignty without food security, and *vice versa*, food security is a goal while food sovereignty is a pathway to achieve it. Let us not create an unnecessary divide there, let us not fall in this trap.

1.5 What Is Agroecology?

Agroecology refers to the use of ecological principles for the design and management of sustainable agricultural systems (Altieri 1989; Gliessman 1990), a definition that was later enlarged to comprise not only the agroecosystem but the entire food

[5] A notable exception to this, which will be discussed in Chap. 10 of this book, is the new food paradigm created by the *Unión de Trabajadores de la Tierra* (Land Worker's Union) in Argentina.

[6] I do however believe that *peri*-urban agriculture, as practiced in concentric belts around cities, could have a major role to play at feeding cities to almost self-sufficiency, provided that the soils and climate of the surrounding landscapes allows acceptable levels of agricultural production.

system (Francis et al. 2003; Dalgaard et al. 2003). The ecological principles are those that can be observed in natural ecosystems, such as diversity, efficiency, recycling, natural regulation and synergies. These five principles were defined or implicit in the various publications by Miguel Altieri and Steve Gliessman, but their enumeration and labelling as single terms was proposed by me around 2013 (e.g. Tittonell 2014), used in a series of public lectures around the world, and then incorporated as such in the framework of the 10-elements that define agroecology (Barrios et al. 2020). Other schools of thought have proposed definitions that place a stronger emphasis on the social principles of agroecology, such as financial independence, democratic governance, social equity, diversity of knowledge, shared organisation, rural development and preservation of the social fabric (e.g. Dumont et al. 2016). Authors such as Eduardo Sevilla Guzmán, Victor Toledo or Peter Rosset took a sociological approach to agroecology, emphasising the role of social movements, traditions, local knowledge and collective action.

The influential work of rural sociologist Jan-Douwe van der Ploeg (1994), on farming styles among Dutch farmers, does not refer to agroecology explicitly but describes motivations, practices and knowledge systems that can be considered agroecological. Similarly, the work of authors such as René Dumont, Pierre Rabhi, Georges Toutain, Marc Dufumier, Dominique Soltner appeared since the 1970s in France is nowadays seen as part of the history of agroecology in this country, a precursor of the 'Agro-écologie' concept that emerged and evolved since the 2000s. To differentiate peasant and solidarity agroecology from government-led agroecology initiatives, the French movement *Agroécologie Paysanne* defines itself on the basis of four founding principles:

- A new peasant identity, with the vocation to produce food as the core of any agricultural activity
- Circular systems of carbon, water and nutrients at farm and food system scale
- A cooperative economy, with short value chains, with reduced dependence on the global market
- The democratization of governance both in agriculture and food systems

The early work of Altieri and Gliessman is often used to describe an approach to agroecology that is purely agronomic, circumscribed to the farm or the agroecosystem, supposedly ignoring collective action or larger transformative drivers. This is however not true, and may be the result of people being misled simply by the title of their original publications. Through their engagement with communities in Latin America, these authors have been always aware of the social principles and goals of agroecology. Already in 1989, Altieri published an article titled "Agroecology: A new research and development paradigm for world agriculture" (Agr. Ecosyst. Environ. 27, 37–46), and in 1995 he described agroecology as the 'science of natural resource management for poor farmers in marginal environments' in his book "Agroecology: The Science of Sustainable Agriculture" (CRC Press, Boca Raton). In the first chapter of this book Susanna Hecht writes about knowledge, social principles and the evolution of agroecological thought. Yet, this does not prevent people from continuing to highlight the differences between co-existing definitions of agroecology, and portrait them as if they were 'opposing views' within the agroecological movement.[7]

1.5.1 The Elements of Agroecology

Nowadays, agroecology is defined using 10 elements that include diversity, synergies, efficiency, resilience and recycling (reminiscent of the five ecological principles enunciated above); co-creation and sharing of knowledge (describing innovation approaches); human and social values, and culture and food traditions (context features); responsible governance, and circular and solidarity economy (relating to the enabling environment). Today we speak of elements, not of principles, as these may be interpreted as being normative (unlike organic agriculture, there are no strict norms in agroecology). The 10 elements of agroecology were used to develop a scoring system to assess the transition of farms or communities to agroecology, known as the Step 1 of the Tool for Agroecological Performance Evaluation (TAPE – Mottet et al. 2021). This tool was developed in response to a request by FAO member states, interested in knowing how important agroecology is at global level and how much it contributes to achieving the UN Sustainable Development Goals (SDG). To assess this, it is first necessary to know how many agroecological farmers are there in the world. Not an easy question to answer. Agroecology is non-certified, and there are farmers that may apply certain principles of agroecology but not all of them. The result is actually a gradient, degrees of agroecology, or degrees of progress along a trajectory of transition towards agroecology. The Step 1 of TAPE aims to estimate, through a series of scores, the degree of transition of a given farm, community or region towards agroecology.

During the development of TAPE, we performed the first field test of the tool by applying it to four then 25 family farms in northern Patagonia (Fig. 1.6). The four initial farms tested show very similar average scores for certain elements and widely different for others (Fig. 1.6a). The average score across all elements was 71% for the collective farm Cultivo ecológico, 62% for Nilda's farm, 55% for Benito's and 49% for Fabian's. Using an arbitrary scale proposed by the developers of TAPE, Cultivo ecológico is an agroecological farm (>70%), Nilda and Benito are in transition to agroecology (50–70%) and Fabian is not an agroecological farm (<50%). The score of each element is the average scoring of 3–5 indicators. For example, the score for the element 'Recycling' results from averaging the scores of 'Biomass and nutrient cycling', 'Water saving', 'Management of seeds and breeds', 'Renewable energy use and production'. The results are anecdotal, and may not say

[7] I do believe that diversity – also the diversity of views and opinions – is a positive attribute within the agroecological movement. However, it is often the case that people exacerbate differences in definitions or approaches to carve themselves a niche where they can fit and feel important, different. Exaggerating minor differences is often a strategy to protect such niches, in my view.

much to those who are unfamiliar with these farming systems, but we can clearly see that some of the elements get more variable scoring levels than others (Fig. 1.6b). The same can be said for individual farms, some of them score mostly >70% for all elements. The first row in the table of Fig. 1.6b scores the 'enabling conditions' for agroecology in the local context of each farm.

In the case described in Fig. 1.6, all the 10 elements received the same weight in the average score but there may be cases in which some may be weighed >1 and some <1 according to their local relative importance. To interpret TAPE results it is crucial to first characterise the farming context (Step 0), as systems that may be considered 'diverse' or 'efficient' or 'responsible' in a given context may not be so in another (El Mujtar et al. 2023). Applied within a certain context, through trained enumerators, TAPE allows (i) engaging in meaningful and thorough discussion about the multiple dimensions of agroecology with the local community, and (ii) identify areas for improvement.

But the aim of this section is not to discuss TAPE or whether the 10 elements are enough or not to assess agroecology. The tool was developed in a participatory way through an international online consultation, followed by a workshop that brought together representatives from 70 agroecology-related organizations from all continents, who gathered at FAO headquarters in Rome.[8] It was a real challenge, but the wide diversity of participants managed to reach an agreement within a week on the principles, methods, steps, dimensions, and indicators to be included. These 10 elements synthesize the views of many agroecology experts, practitioners, and activists from around the world. It is unfortunate that several organizations continue to propose new frameworks unilaterally, using 10, 13, or 20 principles, which adds to the confusion and ignores this international participatory process. The synthetic view of agroecology as we understand it today, represented by these 10 elements, is multidimensional, systemic, holistic, and transdisciplinary. These properties of agroecology will be addressed in the different chapters of this book.

1.5.2 The Last 100 Years of Agroecology

Strictly speaking, agroecology started 10,000 years ago with the onset of farming. A large share of the knowledge used in agroecology can be considered to be ancient or traditional, the result of domestication, breeding, invention and innovation by generations of farming families around the world. However, the first use of the term agroecology occurred 100 years ago, in a publication by Russian agronomist Basil Bensin (1930),[9] whereas the first use of the term in the title of a book was in

[8] Although it is often referred to TAPE as an FAO tool, it is actually a tool that was built collectively, through a process that was facilitated by the FAO.

[9] I will not dive much deeper into the history of agroecology: an informative recollection can be found in Gliessman (2007).

a

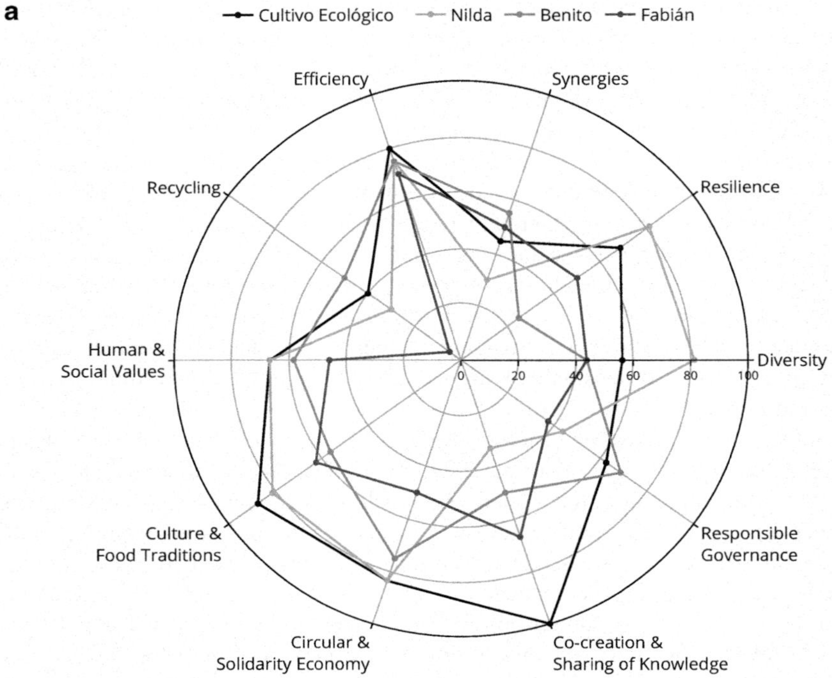

b

Evaluated Agricultural Systems

Elements of Agroecology	TA	Fa	HC	SC	AS	LS	NP	BT	SN	DC	ND	SR	MV	DM	TT	RC	MM	Va	CC	DH	OG	CE	AH	LL	FL	Element avarage
Enabling environment for agroecology	65	30	65	50	45	40	35	40	45	40	50	40	50	40	45	45	50	80	45	80	65	70	70	65	65	53
Recycling	65	5	55	60	65	60	65	50	50	75	30	70	25	85	65	50	50	25	55	40	50	40	50	75	55	53
Responsible Governance	44	38	63	50	50	31	50	69	56	56	44	56	50	56	63	31	63	81	31	88	63	63	63	56	63	55
Synergies	45	50	40	55	55	65	40	55	65	55	30	65	60	60	70	75	50	35	75	40	65	45	60	75	75	56
Diversity	69	44	56	63	31	56	56	44	75	31	81	50	94	63	50	75	44	44	81	44	75	56	69	75	81	60
Co-creation & Sharing of Knowledge	50	67	58	58	50	33	67	50	33	42	33	50	50	33	67	50	50	83	50	100	67	100	100	92	83	61
Culture & Food Traditions	13	63	13	25	63	63	63	56	75	69	81	56	58	81	50	81	81	63	69	75	69	88	75	69	69	63
Resilience	38	50	44	50	69	50	38	25	56	63	81	69	81	63	75	63	69	69	56	69	63	69	88	88	88	63
Homan & Social Values	38	46	58	67	58	50	63	58	67	71	67	58	58	71	46	71	71	79	75	63	88	67	46	71	92	64
Efficiency	55	70	75	65	60	65	65	75	55	70	75	60	70	70	70	70	90	75	80	85	65	80	70	50	80	70
Circular & Solidarity Economy	58	50	58	50	42	75	42	75	58	67	83	83	75	42	75	75	83	100	92	83	83	83	75	83	83	71
System avarage	48	48	52	54	54	55	55	56	59	60	61	62	62	62	63	64	65	65	66	69	69	69	70	73	77	

Fig. 1.6 Result of the first field testing of TAPE to assess the degree of transition to agroecology of north Patagonia farms in Argentina. (**a**) Comparison of the 10 elements across four farms in north Patagonia, a collective production unit called 'Cultivo ecológico' and three farm heads, Nilda, Benito and Fabián. (**b**) Average scores for the 10 elements of agroecology plus a criteria reflecting local enabling conditions attained by 25 farms from the same region, using a colour scale from red (low) to green (high) for both each element and the average score per farm system. (Modified from Álvarez et al. 2019)

German, Agrarökologie, by Prof. Wolfgang Tischler (1950) (Fig. 1.7). The early works of the organic farming pioneers around 1900s, A. Howard and F.H. King, inspired by traditional Indian and Chinese farmers, Rudolph Steiner's biodynamic farming, Eve Balfour's Haughley experiment, Masanobu Fukuoka's natural farming, and J.I. Rodale's organic gardening are certainly all precursors of agroecology. But also the ideas brought about by the emergence of the Liberation theology religious movement in Latin America, the pedagogy of Paulo Freire, the landless movements such as the MST in Brazil, the outcome of the Sandinist revolution and the US-backed *Contras* War in Nicaragua, the indigenous and peasant movements that resisted dictatorial governments along the Andes, the oil crisis affecting economically blockaded-Cuba after the fall of the Berlin wall, and other socio-political processes that ended up empowering peasant organisations throughout the continent, were important drivers for the creation of the agroecology movement in this region. Since then, agroecology has become part of the political agendas of progressive governments in Latin America and beyond. In 2012, Stéphane Le Foll, former French Minister of Agriculture, commissioned the *Institute National de la Recherche Agronomique* (INRA) to produce a report on Agroecology (AEF no 14773), which he defined as "a production model that is more economical in terms of inputs and energy, while sustainably ensuring their competitiveness". France's role as one of FAO member states was crucial at supporting – including financially – the concretion of the first International Symposium on Agroecology at FAO in 2013. "Agroecology made it to the Cathedral of the Green Revolution" pronounced the FAO Director-general of that time, José Graziano da Silva, at the closing of the symposium. The 10 elements that define agroecology emerged at that time. Nowadays, Brazil, Mexico, Uruguay, Argentina, Colombia, and Chile are all countries whose governments explicitly support agroecology through public policies and the creation of special offices, directions, laws and programs. The agroecology movement is growing rapidly in Europe – unfortunately faster than the number of practicing farmers – and several organisation are adopting and adapting the principles of agroecology in Africa and Asia (e.g. see the natural farming initiative of Andhra Pradesh State in India). Throughout these last 100 years, several works were published that greatly influenced our thinking and practice of agroecology. The selection of them illustrated in Fig. 1.7 is probably biased and incomplete, but hugely influential. And, as inspirational as so many agroecological farmers I had the pleasure to meet over the years.

1.6 Agroecology: Addressing the Food, Biodiversity and Environmental Crises

It is evident that increasing agricultural production is just one of the actions needed to solve the problem of world food insecurity. Rebalancing our diets, reducing food waste, and re-localising agricultural productivity are equally important. Currently, a

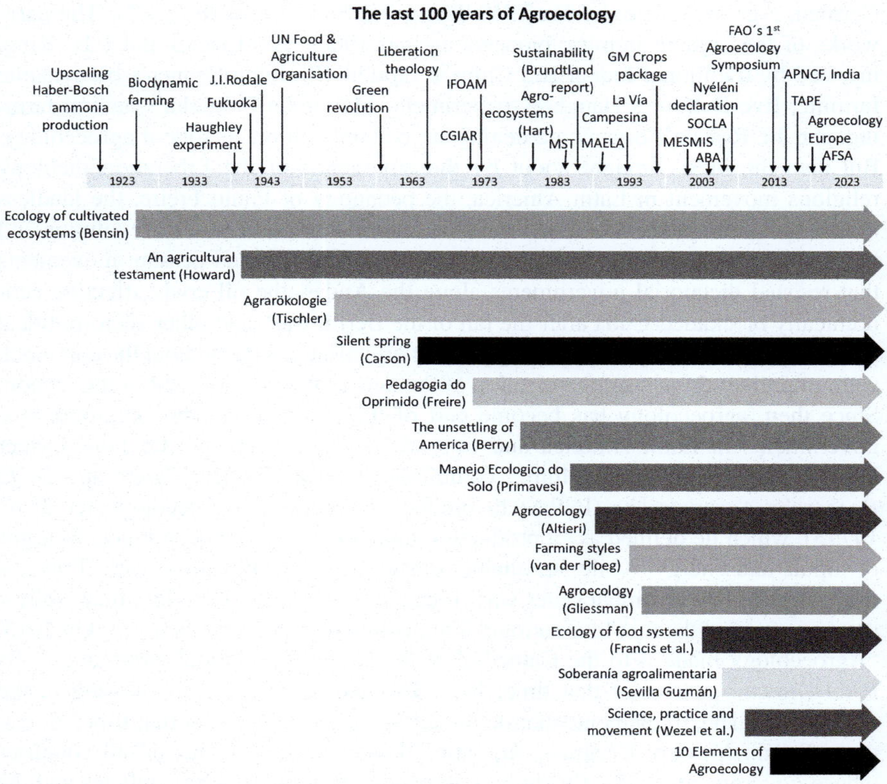

Fig. 1.7 A timeline of the last 100 years of agroecology, indicating on top the onset of key influencing events and the creation of relevant organisations, and below the publication of key influential works that shaped what agroecology is today. CGIAR: Consortium General for International Agricultural Research (1971); IFOAM: International Federation of Organic Agricultural Movements (1972); MST: Movimiento de los Trabajadores Rurales sin Tierra (1985); MAELA: Movimiento Agroecológico Latinoamericano (1989); GM: Genetically modified (1996); MESMIS: Marco para la Evaluación de Sistemas de Manejo incorporando Indicadores de la Sustantabilidad (1999); ABA: Associação Brasileira de Agroecologia (2004); SOCLA: Sociedad Científica Latinoamericana de Agroecología (2007); TAPE: Tool for Agroecological Performance Evaluation (2020); APNCF: Andhra Pradesh Natural Community Farming (2018); AFSA: Alliance for Food Sovereingty in Africa (2019)

significant portion of the world's food is not being produced in the areas where it is most urgently needed. Additionally, large amounts of food are wasted, used for animal feed, or converted into fuel. Given the low yields in many smallholder systems, enhancing the productivity of such farms would have a lower energy cost compared to further intensifying industrial farming. This approach would significantly impact global food security and contribute to the preservation of rural livelihoods and the global environment.

The model of agricultural intensification that emerged from the green revolution, while successful in increasing food production at the time, is no longer tenable. It is thermodynamically unsustainable, pollutes the environment, lacks economic viability, and perpetuates social inequality. While global food production needs to increase in the future, we must adopt a different approach. We require an ecological intensification of world agriculture that generates healthy and affordable food for both rural and urban consumers. It is crucial to decouple food production from fossil fuel energy and minimize environmental degradation. Additionally, agriculture should be compatible with biodiversity preservation and leverage natural ecosystem functionalities to replace chemical toxins. We need to develop affordable, biologically based technologies that work in marginal environments, enabling us to produce food while restoring degraded soils, mitigating climate change, and supporting sustainable rural livelihoods for the poor. Agroecology offers a model that can accomplish all of these goals, engaging different approaches to sustainable farming, guiding farmers and communities along transformative pathways, and embracing the broader landscape and food systems.

1.6.1 Agroecology and Other Approaches to Sustainable Farming

There are other practical approaches that share several principles with agroecology (refer to Chap. 10). These include organic and biodynamic farming, regenerative agriculture, permaculture, agroforestry, conservation and climate-smart agriculture, and circular farming, among others. While most organic and biodynamic farmers adhere to agroecological principles, there are instances of what is known as "conventional organics," where production systems prioritize organic inputs but neglect the social dimension, essentially following a conventional farming approach. Fortunately, such cases represent only a minority among organic farmers. In my opinion, organic farming and value chains may act as major drivers for a broad agroecological transition, particularly in the global North.

Agroecology encompasses most of the principles outlined in the definition of regenerative agriculture, holistic grazing, permaculture, conservation, and circular farming (see Fig. 1.8). However, regenerative agriculture has various definitions, and not all of them align with agroecology. Through a recent literature review, we have identified three primary forms of regenerative agriculture (RA) known as Philosophy RA, Development RA, and Corporate RA (Tittonell et al. 2022). Corporate RA refers to the regenerative agriculture model promoted by large multinational companies as a form of greenwashing, which deviates significantly from agroecology.

Agroecology adheres also to the three main principles of conservation agriculture, such as minimum soil disturbance, permanent soil cover, and crop diversification. However, it does not support the dependence of cropping systems on pesticides or

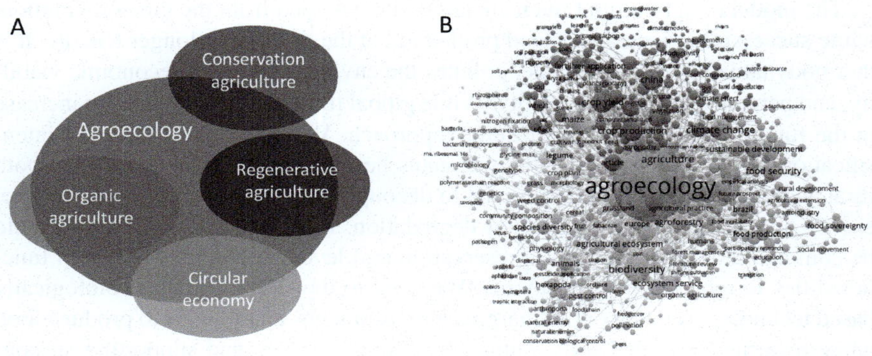

Fig. 1.8 (**a**) Agroecology shares principles and practices with other approaches to sustainable farming, such as organic, regenerative, conservation or circular agriculture. The degree of intersection between sets indicates how much agroecology and the other approaches have in common. (**b**) Co-occurrence analysis of the keywords of 4426 articles relating to agroecology in a 2021 literature review (Tittonell et al. 2022), showing the centrality of this concept

genetically modified crops. Circular and solidarity economy is indeed one of the ten elements that define agroecology, as mentioned in Barrios et al. (2020). However, it is important to note that not all practices labelled as "circular" agriculture today can be considered agroecological. Climate-smart agriculture, on the other hand, is defined based on three pillars: climate change adaptation, mitigation, and food security. It primarily represents an intention or goal rather than a farming approach that prescribes specific principles or practices. Agroforestry is often associated with agroecology. Agroforestry is certainly a technique that can be utilized within the framework of agroecology. It serves as a tool for designing a farm but should not be mistaken for a standalone movement or approach to sustainable farming.

1.6.2 Agroecology Transition Pathways

The transition from current to agroecological farming may follow different trajectories depending on several contextual and endogenous factors, and particularly on the starting point of the transition. For example, there are farmers who are referred to as "organic by default", indicating that they practice organic farming due to the inability to afford agrochemical inputs. Such farmers, predominantly smallholders, cannot be automatically considered to be agroecological either. In fact, it is incorrect to assume that all smallholder family farms are inherently agroecological. Some smallholder farms exhibit a high dependence on pesticides or suffer from degraded soils, lacking social equity, shared knowledge, or democratic governance within their contexts. The diagram presented in Fig. 1.9, adapted from Tittonell (2020), aims to illustrate the variations between different ecosystem states and transition

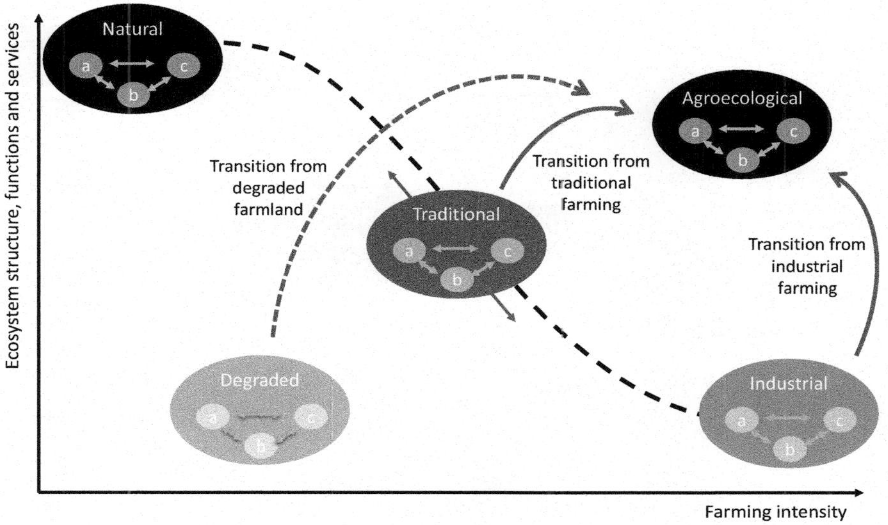

Fig. 1.9 A scheme representing the alternative pathways for the transition to agroecology, from traditional smallholder systems, industrial farming systems or degraded farmland. Each state of the agroecosystem (natural, traditional, industrial and degraded) comprises also different possible regimes, represented as three bubbles (a, b, c and their interactions) within each state. (Adapted from Tittonell 2020)

pathways towards an agroecological state using a conceptual state-and-transition model.

As natural ecosystems are converted into agroecosystems, they progressively lose certain functions and services in exchange for others. Traditional, multifunctional systems gradually give way to heavily transformed industrial farming systems that prioritize a single function: maximizing agricultural productivity per unit area. The transitions to agroecology from traditional or industrial systems represent distinct pathways characterized by different priorities, design principles, and time horizons. It is worth noting that the transition to agroecology can also commence from a degraded site, following a restorative or regenerative pathway. The diagram in Fig. 1.9 depicts agroecology as a multifunctional model of ecologically intensive farming that enables a high level of ecosystem services while also fulfilling the primary objective of food production.

1.6.3 *Landscape and Food Systems Approaches*

To effectively address the interconnected crises of food security, biodiversity loss, and environmental challenges, agroecology must adopt whole landscape and food systems approaches. While the transition of individual farms to agroecology is a

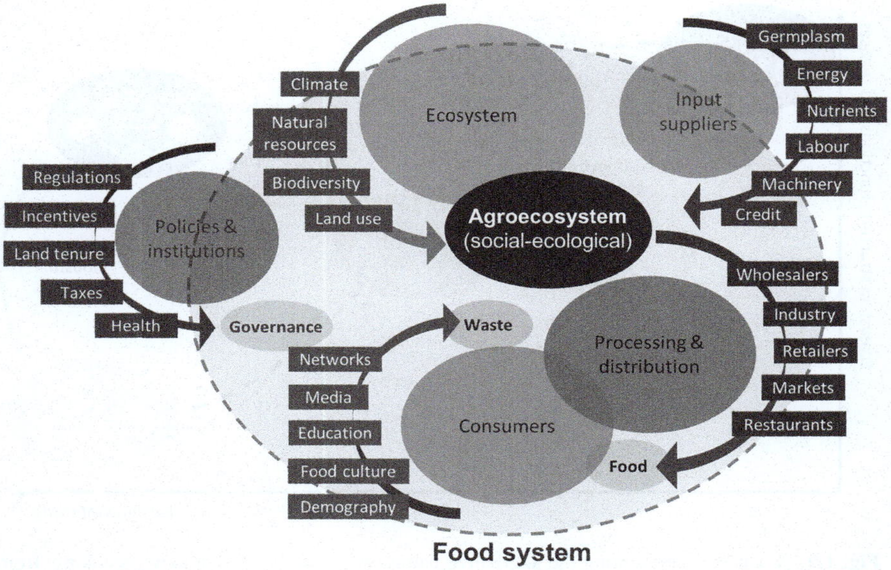

Fig. 1.10 A representation of the food system as conceptualised in this book, with the agroecosystem at its centre. The larger ecosystem, the input sector and policies are institutions are part of the food system but also broader than it

commendable starting point, reversing the decline of biodiversity necessitates the restoration of entire landscapes that encompass diverse habitats, trophic networks, and resources. Many essential functions of agroecosystems, such as pollination and natural pest control, operate at the landscape scale rather than the farm scale. Therefore, addressing these processes requires collective action, community involvement, ownership, and organization at the landscape level. Furthermore, tackling major environmental challenges, particularly climate change adaptation and mitigation, demands actions at the food system level (Fig. 1.10). This includes implementing changes in commercialization, consumption and waste management practices. Achieving global food security and sovereignty, as discussed earlier, goes beyond increasing agricultural production. Ensuring access to food, its utilization, stability, as well as promoting nutritional and cultural diversity, necessitates adopting a system perspective and a holistic approach to agroecology that considers the entire food system. The vertical integration of agroecosystems and the other components of the food system through value chains is addressed in Chap. 10.

1.7 The Need for a Systems Approach to Agroecology

Agroecology provides an ever-expanding knowledge base to inform the design of ecologically intensive agricultural practices across the world, both in the developing South and the developed North. Yet the ecological intensification of current agriculture requires stepping beyond agronomy, soil science or plant pathology, beyond the boundaries of the cultivated field or the animal herd, place human aspirations and decision making – individual or collective – at a central stage, and understand the complexity of ecological mechanisms and cross-scale feedbacks that operate at landscape and food system level. Lack of systems perspectives led to agricultural technologies being developed by reasoning at the scale of a single field or individual plant population. Such technologies – monocultures, fossil fuels, agrochemicals, genetically modified plants – form the basis of the current narrow model of agricultural production and are responsible for its vulnerability and social inequity. Transitioning to agroecology requires trans-disciplinary approaches (Fig. 1.11) that embrace the complexity, diversity and dynamics of the agroecosystem, the actor(s) managing it and the landscape and food systems in which they belong.

Taking a systems approach to agroecology, however, requires a solid understanding of what systems are and how to analyse them. The second chapter of this book provides key elementary concepts and definitions used in systems analysis, and links system properties to agroecosystem attributes that are central to the evaluation of their sustainability. The agroecosystem with its multiple dimensions and scales, is the core research object of the agroecology science. Understanding, analysing and designing agroecosystems requires a dialogue not just between scientific disciplines (as sociology, ecology, economics, history, soil science, anthropology, geography, human nutrition or agronomy), but also between different sources of non-academic knowledge (Fig. 1.11). Academic disciplines must interact among themselves and with the knowledge and innovation capacity of farmers, consumers, policy makers, development agents, social workers, private farm advisors, and traders. Integrating these sources of knowledge and disciplines requires not only attitudinal skills but also the methodological tools and approaches that we put together under the umbrella concept of *systems thinking*. This is a first reason to justify the need for taking a systems approach to agroecology, as this book proposes.

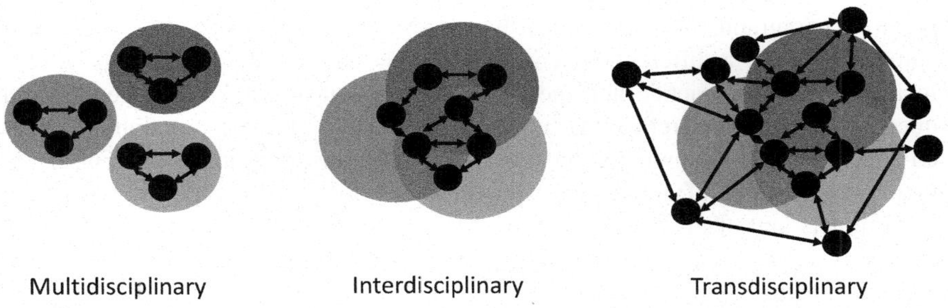

Fig. 1.11 The distinction between multidisciplinary (multiple academic disciplines operating simultaneously but in parallel), interdisciplinary (interacting academic disciplines) and transdisciplinary (academic disciplines interacting among themselves and with non-academic actors) approaches

Systems thinking is also necessary when dealing with complexity. The agroecosystem, as will be explained in the following chapters, is also a complex system that exhibits interactions and feedbacks within and across scales, non-linear dynamics, uncertainty and nested sub- and supra-systems. It may or may not have an associated spatial dimension and a temporal horizon. It is part of a landscape, with its ecological and cultural aspects, and at the same time it is part of a politico-institutional system. It is governed by nature and steered by humans; it has thus ecological, physical, socioeconomic and cultural dimensions. Understanding and managing social-ecological interactions in complex agroecosystems, at different spatial and temporal scales, and under growing uncertainty, calls also for a systems approach.

Agroecologists are often challenged by the complexity of the systems they need to design or manage, particularly when the design objectives or the problems to be addressed involve multiple 'nested' levels of integration (see Chap. 2). From cropping systems, to farm systems, landscapes, value chains, or food systems. Across these levels, often referred to as 'scales', the theoretical frameworks used to conceptualise and study them may differ widely. For example, cropping systems are typically studied as a 'sequence' of crops (plant populations) in time, often disregarding their spatial characteristics. Farming systems are conceptualised as interacting components (crops, livestock, infrastructure, etc.) and flows (financial, labour, nutrients, etc.) with clearly demarcated boundaries. Landscapes are typically described emphasising their spatial dimension, but also as a cultural object; they are rarely described as systems. Value chains are typically described as unidirectional flows. Food systems, on the other hand, are presented as interacting components, but with fuzzy boundaries and generally not associated with a spatial scale. This book attempts to highlight such differences between levels and approaches, but also to reconcile them, through a systemic approach.

Agroecology is often defined as consisting of three dimensions: science, practice and movement (Wezel et al. 2009). This expression constitutes a strong narrative that captures in a few words the complexity of the various dimensions of agroecology. However, it is not completely accurate in my view. The science of agroecology consists of the different disciplines and knowledge sources enumerated in the previous paragraphs. Thus, practical knowledge is actually *part* of the science of agroecology, not a different dimension of it. The science of agroecology already includes both the practice and the movement as sources of knowledge, research questions, experience, reflexion, guidance, etc. Because the construction, or rather, co-construction of knowledge in agroecology is transdisciplinary (cf. Fig. 1.11), there is no science of agroecology possible if it does not include the wisdom of practitioners and of the agents of change that we collectively call 'the movement'. Perhaps we should say, more accurately, that agroecology is *the* science of practice and movement.

1.8 About This Book

Systems approaches include both the phases of analysis and synthesis. In systems analysis we study the relationships between structures and functions, the processes they underpin, and use different methods to try to infer the purpose of such processes. Irrespective of the type of system we are analysing, a social system, a biological system, a political system, a physical system, we always follow approximately the same steps (i.e., structure-function-purpose), relying on different methods and disciplines, to arrive at conclusions and generate new knowledge. System synthesis, on the other hand, is what we do when we take the existing knowledge and use it to generate new realities. This is exactly what we do when designing or redesigning systems. In systems design, we know the purpose or goals we aim to achieve with our new design, and then propose a number of structures and functions that will deliver the necessary processes (i.e., purpose-function-structure). In systems design, we do not arrive at conclusions, we arrive at decisions. The methods and tools we use to support design may or may not be the same used in systems analysis.

The 10 chapters of this book are grouped into three parts based on these concepts. A first part in which the concepts of systems and system thinking are introduced (Chaps. 2 and 3). A second part in which methods for (agroeco-)systems analysis are described and illustrated with real life examples (Chaps. 4, 5, 6, and 7). This second part takes up most of the book, as it provides concepts, methods and tools necessary for taking a systems approach to agroecology. The last part introduces elements of design, and new tools to inform simultaneously systems analysis and design (Chaps. 8, 9, and 10). Design requires evaluation, and to evaluate systems we use indicators, participatory monitoring, multicriteria assessments, simulation models, etc. These methods are introduced in the last part of this book and linked with several concepts associated with the assessment of agroecological transitions.

Agroecology is not prescriptive, it does not deliver technological 'packages', and it questions narrow scientific assumptions and their supposed 'objectivity'. Yet it is scientifically rigorous, it requires methods that are often more complex than those used in disciplines such as agronomy, sociology, economics or ecology. Agroecological knowledge is co-constructed through dialogue, as explained earlier. In such a dialogue, we scientists and technicians have a responsibility: to train and inform ourselves on the most innovative methods that science can offer, and to develop new ones when gaps are identified, in order to enter the dialogue with farmers equipped with meaningful insights that can support and enhance the knowledge co-construction process. Farmers rely on us to get access to new knowledge that can be merged with their own.

1.8.1 What This Book Is Not About

Although it is difficult to discuss systems approaches in agroecology without touching upon design principles, this is not really a book about agroecological design. Writing such a book is my next project. Hence, this book will not prescribe methods to design agroecological farming systems and its components, such as agroforestry or integrated crop-livestock systems. It is also not a book about agroecological practices, such as composting or fermenting bio-fertilisers or biological pest control, or intercropping, etc. It is difficult for an enthusiastic agroecologist not to jump quickly forwards and start recommending practices, redesigns or bio-based solutions in general. This book is actually about the previous step, about all that needs to happen before we start recommending management practices, or informing operational decisions or systems redesign. It provides the necessary concepts, methods and tools to diagnose the problems at stake, their origin, and explore solutions together with different stakeholders. We cannot offer solutions without first understanding, categorising and explaining the problems at stake from a systems perspective.

Suggested Further Reading

The figures presented in the first part of this chapter were taken from a large diversity of international reports and scientific publications. The chapter provides a rather superficial glance at the various global trends that are relevant to agroecology. Each trend can be discussed much further and in much more depth. To keep the flow of the text, and not divert at any interesting crossroad, I decided to provide just an overview of the food security, environmental and biodiversity crises and list all the relevant sources about these here at the end of the chapter with their respective link. Online readers of this chapter can explore these sources further in search for more information.

- FAO State of Food Security and Nutrition in the World (2022) https://doi.org/10.4060/cc0639en
- FAOStat portal: https://www.fao.org/faostat/en/#data/QCL
- EDGAR: Emissions Database for Global Atmospheric Research, European Commission, 2022 https://edgar.jrc.ec.europa.eu/report_2022
- United Nations Environment Programme (2021). Food Waste Index Report 2021. Nairobi. https://www.unep.org/resources/report/unep-food-waste-index-report-2021
- Intergovernmental Panel on Biodiversity and Ecosystem Services (IPBES) Report. https://zenodo.org/record/6417333
- World Wildlife Fund – Living Planet Report (2022). https://livingplanet.panda.org/en-US/
- Tool for Agroecological Performance Evaluation (TAPE) https://www.fao.org/3/ca7407en/ca7407en.pdf

References

Alforja (ed) (1991) El campesino ve el ajuste estructural así. Reflexiones de Jorge Hernández (UPANACIONAL), Carlos Hernández (Consejo Justicia y Desarrollo), y Oscar Monge (UNAC). Centro de Estudios y Publicaciones Alforja, San José

Alomia-Hinojosa V, Speelman EN, Thapa A, Wei HE, McDonald AJ, Tittonell P, Groot JCJ (2018) Exploring farmer perceptions of agricultural innovations for maize-legume intensification in the mid-hills region of Nepal. Int J Agric Sustain. https://doi.org/10.1080/14735903.2018.1423723

Altieri MA (1989) Agroecology: a new research and development paradigm for world agriculture. Agric Ecosyst Environ 27:37–46

Álvarez V, De Pascuale J, Hara S, Cardozo A, Ocariz P, Furlan N et al (2019) Evaluación Multidimensional de la Agroecología. In: Sistemas Agropecuarios de Patagonia Norte. Primer Congreso Argentino De Agroecología. Universidad Nacional de Cuyo, Mendoza, pp 425–429

Andersson JA (2001) Reinterpreting the rural-urban connection: migration practices and socio-cultural dispositions of Buhera Workers in Harare. J Int Afr Inst 71:82–112

Barnosky A (2008) Colloquium paper: megafauna biomass tradeoff as a driver of quaternary and future extinctions. Proc Natl Acad Sci USA 105(Suppl 1):11543–11548. https://doi.org/10.1073/pnas.0801918105

Barrios E, Gemmill-Herren B, Bicksler A, Siliprandi E, Brathwaite R, Moller S, Batello C, Tittonell P (2020) The 10 elements of agroecology: enabling transitions towards sustainable agriculture and food systems through visual narratives. Ecosyst People 16(1):230–247. https://doi.org/10.1080/26395916.2020.1808705

BASIC (2020) An initial cost-benefit analysis of the pesticide sector at the European level. Bureau d'Analyse Sociétale d'Intérêt Collectif, France

Bensin BM (1930) Possibilities for international cooperation in agroecological investigations. Int Rev Agr Mo Bull Agr Sci Pract (Rome) 21:277–284

Cassidy ES, West PC, Gerber JS, Foley JA (2013) Redefining agricultural yields: from tonnes to people nourished per hectare. Environ Res Lett 8:034015. (8pp)

Dalgaard T, Hutchings NJ, Porter JR (2003) Agroecology, scaling and interdisciplinarity. Agric Ecosyst Environ 100:39–51

Dumont AM, Vanloqueren G, Stassart PM, Baret PV (2016) Clarifying the socioeconomic dimensions of agroecology: between principles and practices. Agroecol Sustain Food Syst 40(1): 24–47. https://doi.org/10.1080/21683565.2015.1089967

Edelman M (2014) Food sovereignty: forgotten genealogies and future regulatory challenges. J Peasant Stud 41(6):959–978. https://doi.org/10.1080/03066150.2013.876998

El Mujtar VA, Zamor R, Salmerón F, Guerrero AS, Laborda L, Tittonell P, Hogan R (2023) Lexical analysis improves the identification of contextual drivers and farm typologies in the assessment of transitions to agroecology through TAPE – a case study from rural Nicaragua. Agric Syst 209:103686. https://doi.org/10.1016/j.agsy.2023.103686

FAO (2011) Global food losses and food waste – extent, causes and prevention. Food and Agriculture Organization, Rome

FAO, IFAD, UNICEF, WFP and WHO (2022) The State of Food Security and Nutrition in the World 2022. Repurposing food and agricultural policies to make healthy diets more affordable. FAO, Rome. https://doi.org/10.4060/cc0639en

Figueroa BM, Tittonell PA, Ohiokpehai O, Giller KE (2009) The contribution of traditional vegetables to household food security in two communities of Vihiga and Migori districts, Kenya. Acta Hortic. https://doi.org/10.17660/ActaHortic.2009.806.4

Food and Agriculture Organization of the United Nations (FAO) (2018) The future of Food and Agriculture—Alternative pathways to 2050

Francis C, Lieblein G, Gliessman S, Breland TA, Creamer N, Harwood SL, Helenius J, Rickerl D, Salvador R, Wiedenhoeft M, Simmons S, Allen P, Altieri M, Flora C, Poincelot R (2003) Agroecology: the ecology of food systems. J Sustain Agri 22:99–118

Gliessman SR (1990) Agroecology: researching the basis for sustainable agriculture. Springer, New York

Gliessman SR (2007) Agroecology: the ecology of sustainable food systems. CRC Press/Taylor & Francis, New York, 384 p

Grau R, Kuemmerle T, Macchi L (2013) Beyond "land sparing versus land sharing": environmental heterogeneity, globalization and the balance between agricultural production and nature conservation. Curr Opin Environ Sustain 5:477–483

Heller C (2013) Food, farms & solidarity: French farmers challenge industrial agriculture and genetically modified crops. Duke University Press, Durham

Herrero M, Thornton PK, Power B et al (2017) Farming and the geography of nutrient production for human use: a transdisciplinary analysis. Lancet Planet Health 1:e33–e42

IFAD (2016) Rural development report: fostering inclusive rural transformation. International Fund for Agricultural Development, Rome

IFPRI (2016) Global nutrition report. International Food Policy Research Institute, Washington, DC

Instituto Brasileiro de Geografia e Estatística (IBGE) (2022) Divulgação do Censo 2022 segundo a idade e o sexo da população é adiada. https://www.ibge.gov.br/

IPBES (2019) Global assessment report on biodiversity and ecosystem services of the intergovernmental science-policy platform on biodiversity and ecosystem services. Brondizio ES, Settele J, Díaz S, Ngo HT (eds). IPBES secretariat, Bonn, Germany. 1148 pages. https://doi.org/10.5281/zenodo.3831673

Lowder SK, Skoet J, Raney T (2016) The number, size, and distribution of farms, smallholder farms, and family farms worldwide. World Dev 87:16–29

Michalscheck M, Groot JCJ, Fischer G, Tittonell P (2020) Land use decisions: by whom and to whose benefit? A serious game to uncover dynamics in farm land allocation at household level in Northern Ghana. Land Use Policy 91:104325. https://doi.org/10.1016/j.landusepol.2019.104325

Mottet A, Bicksler A, Lucantoni D, De Rosa F, Scherf B, Scopel E, López-Ridaura S, Gemmil-Herren B, Bezner KR, Sourisseau J-M, Petersen P, Chotte J-L, Loconto A, Tittonell P (2021) Assessing transitions to sustainable agricultural and food systems: a tool for agroecology performance evaluation (TAPE). Front Sustain Food Syst 4:252. https://doi.org/10.3389/fsufs.2020.579154

Murray C (2014) Low-risk diet vs. availability: a mismatch, Institute for Health Metrics and Evaluation 41. Washington University

Novotny IP, Fuentes-Ponce MH, Lopez-Ridaura S, Tittonell P, Rossing WAH (2021) Longitudinal analysis of household types and livelihood trajectories in Oaxaca, Mexico. J Rural Stud 81:170–181. https://doi.org/10.1016/j.jrurstud.2020.10.022

Reising CA, Lanari MR, Ladio AH (2022) The other sheep, resistant but forgotten: archetypal characterization of Linca sheep farmers in Argentine Patagonia. Ethnobiol Conserv 11. https://doi.org/10.15451/ec2022-10-11.25-1-21

Rietveld AM, Farnworth CR, Nawaz M, Groot JCJ, Timler CR, van der Burg M, Tittonell P (2023) Understanding diversity in gender norms in farming communities in Uganda: a Q-methodology approach, submitted

Samberg LH, Gerber JS, Ramankutty N, Herrero M, West PC (2016) Subnational distribution of average farm size and smallholder contributions to global food production. Environ Res Lett 11:124010

Silva Junior CHL et al (2021) The Brazilian Amazon deforestation rate in 2020 is the greatest of the decade. Nat Ecol Evol 5:144–145

Silva LC, Freitas KP, Oliveira JSL, Pinilla N, Jalil LM (2020) As Cadernetas Agroecológicas e os Quintais Produtivos: As agricultoras Familiares Agroecológicas do Nordeste Brasileiro Gerando Autonomia e Preservando a Agrobiodiversidade. Cadernos de Agroecologia – ISSN 2236-7934 -Anais do XI Congresso Brasileiro de Agroecologia, São Cristóvão, Sergipe – v. 15, 2

Soares JC, Santos CS, Carvalho SMP et al (2019) Preserving the nutritional quality of crop plants under a changing climate: importance and strategies. Plant Soil 443:1–26. https://doi.org/10.1007/s11104-019-04229-0

Steffen W, Richardson K, Rockström J, Cornell S, Fetzer I, Bennett E, Biggs R, Carpenter S, Vries W, de Wit C, Folke C, Gerten D, Heinke J, Persson L, Ramanathan V, Reyers B, Sörlin S (2015) Planetary boundaries: guiding human development on a changing planet. Science. https://doi.org/10.1126/science.1259855

Timmermann C, Félix GF, Tittonell P (2018) Food sovereignty and consumer sovereignty: two antagonistic goals?. Agroecology and Sustainable Food Systems, vol 42, pp 274–298. https://doi.org/10.1080/21683565.2017.1359807

Tischler W (1950) Ergebnisse und Probleme der Agrarökologie, Schrift. Landwirtschaft. Fakultät Kiel 3, 71–82

Tittonell (2014) Ecological intensification – sustainable by nature. Curr Opin Environ Sustain 8:53–61

Tittonell P (2020) Assessing resilience and adaptability in agroecological transitions. Agric Syst 184:102862. https://doi.org/10.1016/j.agsy.2020.102862

Tittonell P (2021) Beyond CO_2: multiple ecosystem services from ecologically intensive grazing landscapes of South America. Front Sustain Food Syst 5:190. https://doi.org/10.3389/fsufs.2021.664103

Tittonell P, Giller KE (2013) When yield gaps are poverty traps: the paradigm of ecological intensification in African smallholder agriculture. Field Crop Res 143:76–90

Tittonell P, Vanlauwe B, Leffelaar PA, Rowe E, Giller KE (2005) Exploring diversity in soil fertility management of smallholder farms in western Kenya. I. Heterogeneity at region and farm scale. Agriculture. Ecosyst Environ 110:149–165

Tittonell P, Vanlauwe B, de Ridder N, Giller KE (2007) Heterogeneity of crop productivity and resource use efficiency within smallholder Kenyan farms: soil fertility gradients or management intensity gradients? Agric Syst 94:376–390

Tittonell P, Muriuki AW, Shepherd KD, Mugendi D, Kaizzi KC, Okeyo J, Verchot L, Coe R, Vanlauwe B (2010) The diversity of rural livelihoods and their influence on soil fertility in agricultural systems of East Africa – a typology of smallholder farms. Agric Syst 103:83–97

Tittonell P, Klerkx L, Baudron F, Félix GF, Ruggia A, van Apeldoorn D, Dogliotti S, Mapfumo P, Rossing WAH (2016) Ecological intensification: local innovation to address global challenges. Sustain Agri Rev 19:1–34

Tittonell P, El Mujtar V, Felix G, Kebede Y, Laborda L, Luján Soto R, de Vente J (2022) Regenerative agriculture—agroecology without politics? Front Sustain Food Syst 6. https://doi.org/10.3389/fsufs.2022.844261

UNCCD (2017) Scientific conceptual framework for land degradation neutrality. A report of the Science-Policy interface, UNCCD-SPI Technical Series No. 01, SPI Report. UNCCD Publication

United Nations Environment Programme (2021) Food Waste Index Report 2021. Nairobi

van der Ploeg JD (1994) Styles of farming: an introductory note on concepts and methodology. In: van der Ploeg JD, Long A (eds) Born from within. Practice and perspectives of endogenous rural development. Van Gorcum, Assen, pp 7–31

van Dijk M, Morley T, Rau ML et al (2021) A meta-analysis of projected global food demand and population at risk of hunger for the period 2010–2050. Nat Food 2:494–501. https://doi.org/10.1038/s43016-021-00322-9

Votava EJ, Bosland PW (2002) Culivar by any other name: genetic variability in heirloom bell pepper California Wonder. Hort Sci 37:1100–1102

Wezel A, Bellon S, Doré T, Francis C, Vallod D, David C (2009) Agroecology as a science, a movement and a practice: a review. Agron Sustain Dev 29:503–515. https://doi.org/10.1051/agro/2009004

Willett W, Rockström J, Loken B, Springmann M, Lang T, Vermeulen S, Garnett T, Tilman D, DeClerck F, Wood A, Jonell M, Clark M, Gordon LJ, Fanzo J, Hawkes C, Zurayk R, Rivera JA, De Vries W, Majele Sibanda L, Afshin A, Chaudhary A, Herrero M, Agustina R, Branca F, Lartey A, Fan S, Crona B, Fox E, Bignet V, Troell M, Lindahl T, Singh S, Cornell SE, Srinath Reddy K, Narain S, Nishtar S, Murray CJL (2019) Food in the Anthropocene: the EAT-lancet commission on healthy diets from sustainable food systems. Lancet 393(10170):447–492. https://doi.org/10.1016/S0140-6736(18)31788-4

Chapter 2
Systems Approach: Analysis, Design and Modelling

Abstract The agroecosystem, or the ecosystem modified and steered by humans for the production of goods and services, is the central object of agroecology as a discipline. The concept of agroecosystem entails by definition a systemic approach. Yet its integration within classical approaches that study farming systems, livelihood systems or food systems is not always straightforward, and requires some attention. In this Chapter we recall some elements of systems thinking and theory to be able to provide a comprehensive definition of the agroecosystem and its properties. In the next chapter we describe the structure and diversity of agroecosystems in more detail. This chapter starts by defining systems, systemic approaches, and the agroecosystem as a special type of system. It briefly discusses the use of systems theory and modelling to inform the design and evaluation of sustainable agroecosystems.

2.1 Introduction: Elements of Systems Thinking and Theory

A system is a limited portion of reality, in which we can identify components (or sub-systems) interacting with each other and with the exterior environment through inputs and outputs. A model, on the other hand, is a simplified representation of a system that enables us to study its properties and behaviour. Almost every time we attempt to explain what a system is we automatically draw a few interconnected boxes and arrows: a model of the system. The system itself (i.e., the limited portion of reality) is known as the ontological system, while the model (its representation) is a semiotic system (Box 2.1). Moving from ontological to semiotic systems requires assuming variable degrees of reductionism, and explicit choices made regarding the level of detail in the model or the components and interactions that are represented. These choices depend on the objectives for which the model is built, the field of knowledge or discipline of the modeller, and his or her subjectivity. In other words, to represent a single ontological system we can build

P. Tittonell, *A Systems Approach to Agroecology*,
https://doi.org/10.1007/978-3-031-42939-2_2

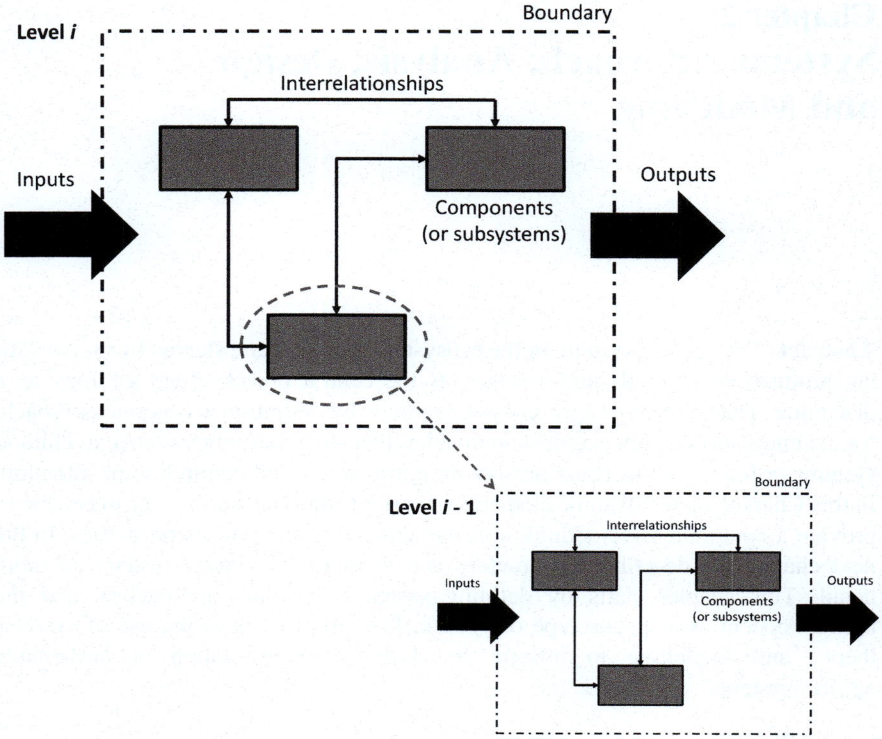

Fig. 2.1 A schematic representation of a system and its five basic elements: boundary, components, interrelationships, inputs and outputs. By zooming into one of the components (or sub-systems) of the system at level *i*, it is possible to identify a new system that can be defined as level $i - 1$. The interrelationships at level *i* become inputs and outputs at level $i - 1$

different semiotic systems. To represent a system, we use five basic elements: boundaries, components, interrelations, inputs, and outputs (Fig. 2.1). The interrelations within the system are also often referred to as 'interactions', but since this term has a specific meaning in statistics, it is preferable to use the term 'interrelations' which is more generic (i.e., they may represent interactions, but also flows, transactions, influences, etc).

Systems can be defined at different integration levels. If we zoom-in into any of the system components depicted in Fig. 2.1, we may find new sub-components operating within them. Thus, any component of the system *i* represented in Fig. 2.1 is in itself a system that we can denoted as $i - 1$ (or immediate sub-system). The internal relations between components defined within system *i* become inputs or outputs to and from system $i - 1$. The system represented in Fig. 2.1 may also be a component of a larger system $i + 1$ (or immediate supra-system). In agricultural systems, each integration level ($i - 1$; i; $i + 1$; etc.) often bears a relationship with a certain spatial scale, but this is not a straightforward relationship. For example, an agricultural field can be a sub-system of a farm system. But an agricultural field can also be a sub-system of an ecological network, or a sub-system of a watershed, a land tenure system or an administrative land unit. Similarly, an individual farm can be a component of a landscape. But it may also be a component of a green corridor, of a village territory, an irrigation scheme, a food supply chain or an economic sector. In some of these examples the spatial scales corresponding to integration levels $i - 1$, i, and $i + 1$ can be well defined, while in others not. Thus, for each particular system there are not only different possible semiotic systems but also different ways to aggregate or to scale up boundaries, components and interrelations both in space and time.

To clarify the concepts of ontological and semiotic systems, Box 2.1 presents an example of a farm system from the shores of Lake Hawassa, Ethiopia, and a representation of a farm system that was used to develop the dynamic simulation model FARMSIM (Tittonell et al. 2009a). In the picture, we can see the household, the enset fields, the maize fields, and the communal grazing area at the front. We cannot see the animals or the compost pit in the picture, as they are hidden by the trees that grow around the house, but they are there (I visited this farm several times). This is an ontological system. A portion of reality delimited, in this case, by the picture frame. The farm system is represented in the semiotic model as a delimited area of land that hosts a family and a homestead, or household component, an animal heard or livestock component, a manure and organic matter component and fields, which are represented as subsystems with two major components, soils and crops. These components and subsystems interact with each other in different ways. There are elements (driving variables, adjacent systems, contextual systems, etc.) that are represented as external to the farm system, such as the climate, the market or the communal resources.

Box 2.1 Ontological and Semiotic Systems

The picture below shows a portion of the agricultural landscape that dominates the western shores of lake Hawassa, in Ethiopia. Annual crops such as maize *(Zea mays)* and perennial crops such as enset *(Ensete* ventricosum) or Albyssinian banana are grown in a patchwork of small fields that alternate with areas of less quality land used for livestock grazing by the entire community. What we see in the picture is a limited portion of reality, an ontological system.

(continued)

Box 2.1 (continued)

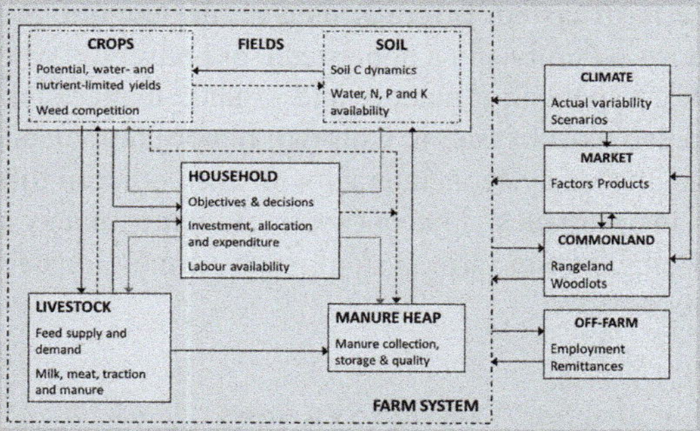

The various components of a farm system in this region can be represented in a generic way as in the diagram on the model FARMSIM (Tittonell et al, 2009a). This representation of reality, or semiotic system, is one possible model of the actual ontological system. A model is a simplification of a system for the purpose of studying it. This model of the farm system may be useful for the study of resource flows, such as nutrients, cash or labour. However, studying e.g. pest population dynamics or information flows between household members would require developing different models of the same system.

2.1.1 System Boundaries

Defining system boundaries is the first step towards defining integration levels and spatial scales. Choices are sometimes hard to make, as the selected system boundary may have a great influence on the result of our analysis. In general, system boundaries are chosen in such a way that the status of the system is influenced by external driving variables that behave independently from the system (although this is often hard to assert in biological systems). Let us use a semiotic system to clarify this point about system boundaries and external drivers. Let us assume then that a certain process can be described through the following equation:

$$Y = f(X, \theta) \tag{2.1}$$

Where, Y is a dependent variable, X is an independent variable, and θ is a parameter. To represent the system that is relevant to the process under consideration it is necessary to delineate system boundaries in such a way that the dependent variable Y (or state variable) becomes endogenous to the system, the independent variable X (or driving variable) becomes exogenous, and the parameter θ reflects a system attribute. Let us now translate this semiotic system into an ontological one, such as an annual crop. The endogenous dependent variable Y represents the total crop biomass produced in a season, X the seasonal amount of incident solar radiation (exogenous driver), and θ the radiation use efficiency by the crop canopy, or its ability to convert every unit of solar energy received into crop biomass (an attribute of the system). Identifying boundaries is not so straightforward as in this examples when systems are increasingly complex, but the same principle always holds: boundaries should be chosen in such a way that independent variables remain external to the system.

2.1.2 System Dynamics

Systems are also categorised in relation with the mutability of their components and interrelationships. We distinguish dynamic systems (diachronic) from static or immutable systems (synchronic). System dynamics reflects the 'behaviour' of a system, the way in which its configuration and functioning evolve with time, in terms of both endogenous changes or as a response to changes in their surroundings. The dynamics of a system can be represented as follows:

$$\begin{aligned} dY_1/dt &= f_1(X, Y, \theta) \\ dY_2/dt &= f_2(X, Y, \theta) \\ &\ldots \\ dY_n/dt &= f_n(X, Y, \theta) \end{aligned} \tag{2.2}$$

Where, $dY_{1\text{-}n}/dt$ are the rates of change in the state variables $Y_{1 \text{ to } n}$ (endogenous) over the time interval dt as a function of the driving variable X, the parameter θ and the status of the variable Y at time $t - dt$, or the previous system state. When we analyse the dynamics of a system, we aim to understand how the dependent variables $Y_{1 \text{ to } n}$ 'behave' as a function of the external driving variable X, of the parameter θ, and of time. This is represented as follows:

$$\begin{aligned} Y_1 &= f_1(X, \theta, t, Y_0) \\ Y_2 &= f_2(X, \theta, t, Y_0) \\ &\ldots \\ Y_n &= f_n(X, \theta, t, Y_0) \end{aligned} \tag{3.3}$$

The study of system dynamics originated with applications of electrical engineering control theory to the study of other kinds of systems (e.g., business systems) by J. Forrester in the 1950s. In agroecology, the study of system dynamics is essential to assess properties such as sustainability, vulnerability, robustness or resilience, all of which entail changes in the system (or not) over time. These ideas have been implemented by Masera et al. (1999) and López-Ridaura et al. (2002) in sustainability evaluation frameworks, building on the concept of dynamic system properties proposed by Gordon Conway (1987). Yet we must differentiate between short- and long-term system dynamics, and between cycles (e.g. seasons) and trajectories (e.g. land use history). Trajectories may be described through specific attributes such as continuity, reversibility or hysteresis. The choice of temporal scales to represent a system in agriculture is also associated with the time frame inherent to different types of agricultural management decisions, which may be operational (short-term), tactic (mid-term) and strategic (long-term). Besides, there is often also a close and logical association between spatial and temporal scales defined by the process that is being analysed. For example, a geological process cannot be studied in a time frame of seconds, or on a spatial extent of one squared meter; we must consider periods of millions of years and large tectonic blocks of several hundreds to thousands square kilometres. The temporal and spatial scales must be carefully chosen when representing a system, be coherent with the objectives and methods, and practicable.

2.1.3 System Organisation

The components of a system are connected through causal links that define the structure and functioning of the system. In a semiotic system, any given structure is necessarily associated with a specific function. Let us illustrate this with the example of an electronic device such as a mobile telephone. If we open up the case of a mobile phone and take a look inside we can see a number of components linked through electronic connections. Someone with sufficient knowledge of electronics may be able to identify sub-systems within the telephone, such as the wave transmission sub-system, the sound amplification sub-system, etc. If we dismantle the connections and separate all these components, their function as a whole – as a telephone – will be lost. If we throw all the components back into the case at random and close it, we may still have all the system components inside the case but the telephone will not work (Can we still call it a telephone?). A telephone, its function, is not only defined by its constituent components but also by the specific way in which they are connected, or organised.[1] In other words, a system is more than the sum of its parts. The analysis of biological systems or of social-ecological systems such as agroecosystems requires good understanding of the relationship between structure and function inherent to them. Sometimes, the structure of a biological system is not well known or cannot be clearly detected. In such cases, we study the relationship between *patterns* and functions, with the ultimate goal of eventually being able to characterise the underlying structures at a subsequent point in time.

Cybernetics is the study of the structures, functions and properties of regulatable systems irrespective of their nature: e.g., physical, biological, cognitive or social. In agroecology, we are primarily interested in an ontological system – oriented, causal and functional – constituted by material entities that relate to each other through physical flows of matter and energy. Human agency on the agroecosystem is essential to determine the causality at play between structures and functions. In such sense, agroecosystems are good examples of cybernetic systems, because they are regulatable, steered through human agency to pursue a number of goals. In other terms, their ontological determinism is causal, and the causal links can be described as transactions (direct exchange of energy and matter between entities) and relations (the indirect consequences of such exchanges). The semiotic systems we use to represent agroecosystems are also oriented, causal and functional, as defined by us as external observers. But the objects that constitute the system (the components) contain information based on *attributes* or behaviours associated with the real entities of the ontological system. In agriculture and ecology, the interrelations between these objects are often defined as *processes* by the external observer.

2.2 The Agroecosystem

In agroecology we deal with systems that are multi-dimensional, in the sense that we may distinguish minimally a biophysical and a socioeconomic dimension, or yet more dimensions, depending on the observer. Agroecosystems, which may be defined as ecosystems that were modified by humans for agricultural production, are governed by natural laws but steered by humans through design and management. In other words, they are cybernetic systems. The delineation of the

[1] It should be noticed in this example that even when the telephone is assembled correctly, it will not work unless it is connected to a telephone network, or supra-system, which is an excellent allegory to illustrate the concept of interdependency of systems across scales.

agroecosystem boundaries (space-time scale, and organisation) depends largely on the objectives of the external observer. Due to its two main interacting dimensions, agroecosystems can also be defined as social-ecological systems (cf. Tittonell 2020). As any system, an agroecosystem will have a certain organisation, a structure that sustains its functions, and exhibit different degrees of complexity, depending also on the level of detail at which they are assessed. Agroecosystems are mutable, they change over time and hence their behaviour can be understood by studying their dynamics. In other words, agroecosystems comply with all the elements associated with 'systems' presented in the previous section.

Towards the end of the 1980s, Fresco and Westphal (1988) proposed a conceptualisation of agriculture as a hierarchy of systems,[2] from individual crops or livestock herds to national agricultural systems (Fig. 2.2). This was useful at the

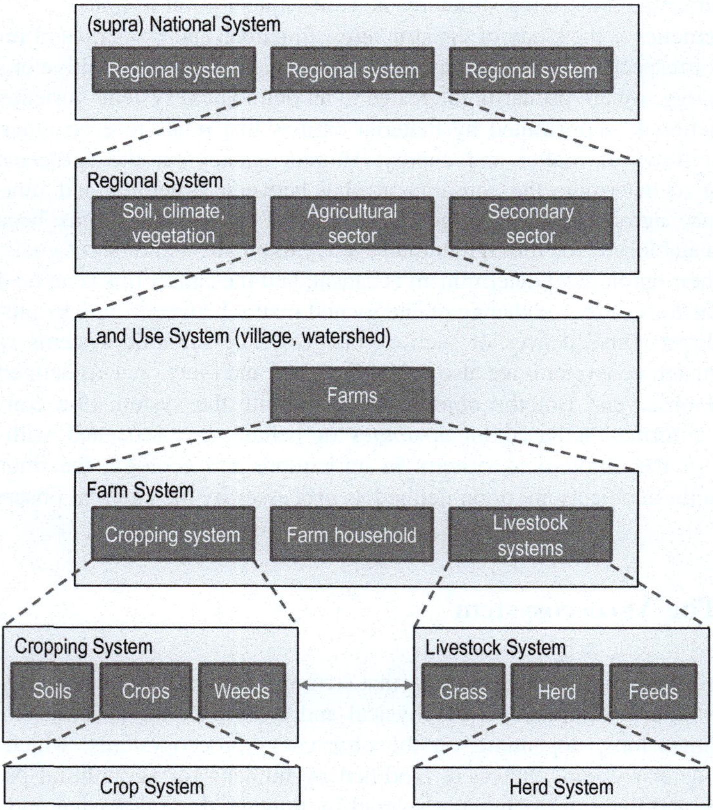

Fig. 2.2 Agriculture as a hierarchy of systems from Fresco and Westphal (1988). A similar scheme had been published earlier by Hart (1982), (1982)

[2]Agricultural systems as a hierarchy of nested systems were already described by Hart (1982) in Costa Rica.

time, and still today a good first approximation to understanding the complexity of agriculture, but we should not forget the old axioma inherent to systems analysis: a system is more than the sum of its parts. The conceptualisation used in Fig. 2.2, a nested hierarchy of sub-systems, is not always appropriate to represent the complexity of interactions, cross-scale feedbacks and emerging properties inherent to the dynamics of social-ecological systems as we understand them nowadays. In particular, the concept of *Panarchy* that emerged during the 2000s from the realm of resilience thinking has largely influenced the way in which we conceptualise agroecosystems and cross-scale feedbacks (e.g., Gunderson and Holling 2003). Cross-scale interactions in social-ecological systems may take two forms: hierarchical confinement, when supra-systems constrain the lower system (as in Fig. 2.2), or Panarchy, when both up- and downward interactions are simultaneously possible.

2.2.1 Agroecology vs. Classical Agronomy

The distinction between hierarchical confinement and panarchy discussed above may sound trivial or too abstract, but such differences constitute alternative theoretical underpinnings for the epistemological approaches used respectively in classical agronomy and agroecology (Table 2.1). The concept of nested subsystems as represented in Fig. 2.2 may be useful when studying the ecology of mono-specific populations (such as crops) or *autoecology*, which refers to the study of individual species in relation to their environment. This has been the approach followed by the strongly positivist school of Theoretical Production Ecology funded by C.T. de Wit in The Netherlands, since the end of the 1960s, which has influenced to a large extent the way in which we analyse agricultural systems worldwide (cf. van Ittersum and Rabbinge 1997). The concept of nested subsystems is less useful when dealing with, for instance, the ecology of insect pests and their natural enemies – who typically ignore the boundaries of single fields or farms – and their multi-trophic interactions, their genetic diversity, their niche and habitat preferences, etc. In such cases, we study agroecosystems by recurring to community ecology or *synecology*, the study of groups of organisms in relation to the environment. In addition, the nested hierarchy represented in Fig. 2.2 assumes that farmers are single decision-makers that operate within the boundaries of their farm. But human agency of the agroecosystem may be also dictated by collective decisions on communally owned resources in a territory, which may or may not coincide with the boundaries of a watershed or with an administrative unit.

When dealing with the ecological dimension of agroecosystems, synecology provides the theoretical backbone to agroecology. As the result of approaching systems analysis from the perspectives of either autoecology or synecology, classical agronomy and agroecology use respectively different indicators to assess agroecosystem performance (Table 2.2). For example, agronomy typically uses the land use efficiency or yield of a crop (kg per ha) as a primary indicator of performance. This indicator assumes that land is scarce, or perhaps the most limiting

Table 2.1 Approaches to agricultural systems analysis from the perspectives of autoecology (or population ecology) and synecology (or community ecology) illustrated through the assumptions made for three key criteria that are used to build semiotic models of the agroecosystem

Criteria	Autoecology perspective	Synecology perspective
Systems dynamics	Systems are assumed to have predictable outcomes, with an associated probability (risk), so that system feedbacks can be formalised; 'continuity' is often assumed to represent dynamics	Systems exhibit complex feedbacks, some of them yet unknown, randomness, and more complex systems dynamics such as hysteresis, non-linearity, irreversibility and discontinuity
(Bio)Diversity	Space-time diversity is generally seen as a burden; heterogeneity and asynchrony are seen as undesirable attributes in mechanised systems; species diversity (e.g., weeds, intercropping, crop-livestock integration) introduces complexity in systems operation; Systems design and management based on a theory of control (e.g., best agricultural practices)	Space-time and species diversity are seen as an attribute, as these lead to risk spreading and adaptability, and to natural regulation mechanisms through synergies, mutualism, facilitation or antagonisms (e.g., a multispecies rhizosphere enhances water and nutrient uptake) Systems design and management based on a theory of regulation (i.e., let nature do its job)
Moving across scales	Aggregation is done by assuming a nested hierarchy of systems from field to global scale; for example: Production at scale S (Ps) is calculated as: $Ps = Ye * Ae + \ldots + Yn * An$ Ye,n: yield in production environments e, n Ae,n: area of production environments e, n	New properties and interactions emerge as we move across scales (so that the whole is more than the sum of its parts), and there are cross-scale feedbacks (Panarchy). Production at scale S (Ps) could be calculated in different ways, for example like this: $Pe = (Y_1e + Y_2e + Y_ie + Y_1e * Y_2e * Y_ie) * Ae$ . . . $Pn = (Y_1n + Y_2n + Y_in + Y_1n * Y_2n * Y_in) * An$ $Ps = Pe + \ldots + Pn + Pe * \ldots * Pn$ $Y_{1,2,i}$e,n: yield of activities 1, 2, *i* in production environments e, n Ae,n: area of production environments e, n

While classical agronomy has largely built on population ecology, agroecology builds from the perspective of community ecology (modified from Tittonell 2014). This has implications for the way in which both disciplines address dynamics, diversity and upscaling

Table 2.2 Illustration of how different criteria and indicators can be used to assess the same system property

Property	Criterion	Indicator	Reference value	Method
Stability	Soil losses	Soil lost through water erosion	< 7 t/ha/year for grasslands	1. Estimation using RUSLE 2. Modelling sediment dynamics in a watershed 3. Runoff plots
Stability	Soil losses	Area affected by water erosion	< 30% of the area affected; >60% soil cover	1. Visual signs 2. Remote sensing 3. Soil cover assessments 4. Presence and distribution of trees
Stability	Economic performance	Indebtedness	<25% of total farm capital	Economic assessment and balance calculations
Stability	Economic risk	Income variability	CV < 50% < 20% negative exercises	Coefficient of variation in annual income Frequency of years 'in the reds'

A complete description of properties, criteria and indicators uses in agroecosystem sustainability evaluations can be found in the MESMIS framework: http://www.mesmis.unam.mx

factor, it can be calculated easily when dealing with monospecific crop stands, and it provides an image of the system in a single moment in time. With yield information, it is possible to calculate a value of production, deduct costs and assess a gross margin for the exercise (season, year, etc.). In agroecology, especially when polycultures are grown, the yield of a single crop is not as important as the total productivity of the multi-species system, calculated as the land equivalent ratio (the sum of the yields of all crops in a mixture relative to the sum of all yields when the same crops are grown individually).

In agroecology, productivity is not calculated for a single exercise but over the mid- to long term, aiming to capture both interannual variability and cumulative effects (positive or negative) over time. Similarly, while classical agronomy assesses 'yield gaps' of single crop species, agroecology places its emphasis on total farm and factor productivity gaps. While agronomic efficiency is calculated as a ratio of outputs to inputs, agroecology sees efficiency as an emerging property (calculated through matrices) at farm or landscape level. Instead of assessing nutrient flows and balances, agroecology considers nutrient networks within complex systems. And so on, agroecology uses different indicators to assess agroecosystem functioning and performance (cf. Table 2.1).

2.2.2 *The Farm System and the Farming System*

And there is a last distinction to be made: a farming system is not necessarily the same as the 'farm system'. The latter refers to the interacting components of an individual farm, its boundaries, components, inputs and outputs. A farm system is a concrete object, a farming system not. When we speak for example of 'smallholder farming systems' or 'maize-based farming systems' or 'organic farming systems', etc. we refer to a broad ensemble of agricultural systems that have a number of features in common; i.e., respectively their economic scale, the main crop they grow or the agricultural methods they employ, in these three examples. But we do not make, in these cases, any explicit reference to their extension in space. When we say 'smallholder farming systems' without further clarification, we refer, in principle, to the nearly 500 million small-scale family farms smaller than two hectares that exist in the world nowadays.

Thus, farming system is a rather vague concept that is used to mean different things in different contexts. Epistemologically, it has all the properties of an abstract object,[3] which poses problems for empirical enquiry because of its lack of physical references. An agroecosystem, by contrast, can be defined in the same way as an ecosystem at a specific spatial scale, location, boundaries, components and causal relations. In other words, agroecosystems qualify as concrete systems and can be thus subject to empiricism. When dealing with the analysis and design of agricultural systems, in agroecology but also in other disciplines, it is then preferable to define the agroecosystem as the *epistemological object*. An agroecosystem can be delineated in space and time by linking it with a physical management unit of interest (e.g., the farm, the landscape, the watershed, the region, etc.), and this allows empiricism. All these theoretical considerations may sound tedious, but they are important in systems analysis, design and modelling.

2.3 Analysis, Design and Modelling of Agroecosystems

Through systems analysis we aim to enhance our understanding of the relationship between structures and functions of systems, and ultimately infer their purpose (Fig. 2.3). In systems design we move in the opposite direction. The purpose is known, and through a process of knowledge synthesis we aim to identify the necessary functions to fulfil the specified purpose, as well as the structures needed to sustain such functions. As much as systems analysis is central to the agroecology discipline, design is central to the agroecological practice, including regenerative farming, agroforestry, permaculture, biodynamic agriculture or any other holistic form of agricultural management. Systems design requires evaluation, before (ex-ante) or after (ex-post) the implementation of the newly designed system, and continuous monitoring of system performance in time. In the case of agroecosystems, design and evaluation are two alternating phases in a continuous cycle, ideally an ascending spiral. Through the evaluation of a system – testing, monitoring – we may discover a need for re-design.

[3] Sets, numbers, values, etc. are examples of abstract objects; an abstract object does not exist at any particular time or space.

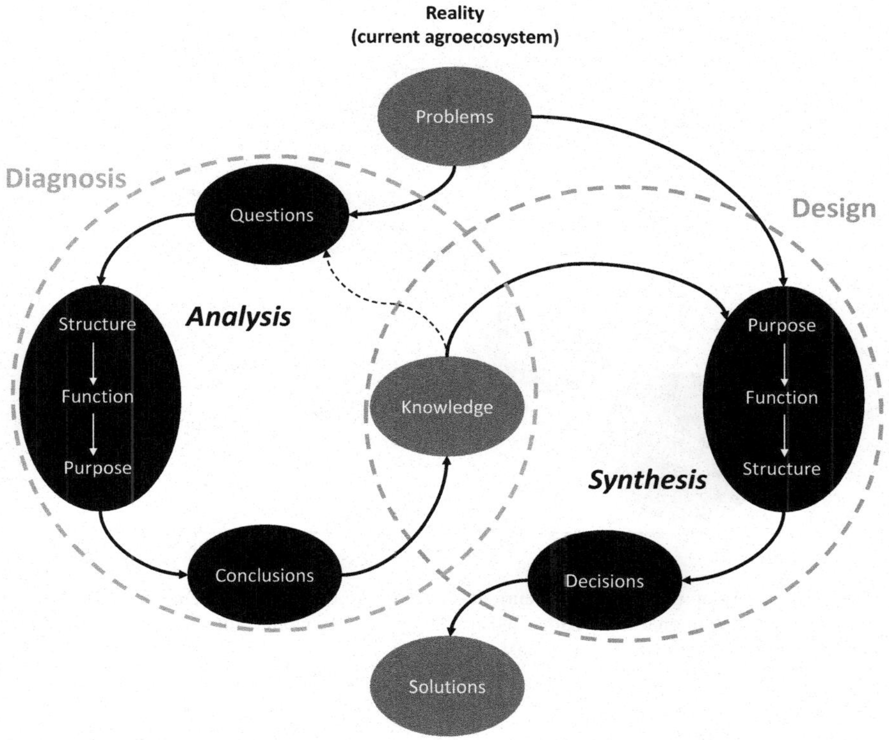

Fig. 2.3 Fundamental differences between research for diagnosing or understanding (analysis) versus research for design (synthesis) in the realm of agricultural systems, modified from Goewie (1997)

2.3.1 Systems Design and the Agroecological Transition

Design is a fundamental step in the transition towards agroecology. Gliessman (2010) speaks of four stages in the transition from current agricultural systems to sustainable ones. Namely, (i) increase efficiency of current systems, (ii) substitute toxic and/or non-renewable inputs and practices with alternative ones, (iii) redesign the system to propend towards agroecological principles, and (iv) reconnect food producers with consumers through new value chains and food systems. Achieving stage three in this transition, system redesign, requires innovation in terms of both technologies and practices, and institutional development (cf. Tittonell 2014). Institutions here means all forms of human organisation, from social norms to traditional by-laws, markets, or public and private organisations. This is illustrated in Fig. 2.4, where the trajectory from current systems to agroecological ones is depicted as the result of progress in both technical and institutional innovation. The key socioeconomic drivers that may push the system towards greater innovation in these fields are depicted as white arrows.

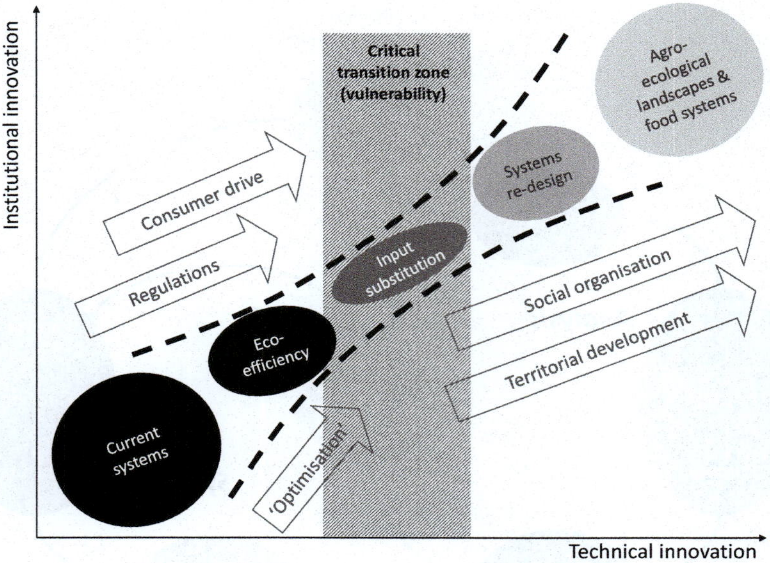

Fig. 2.4 Phases of transition from current towards agroecological systems as influenced by technical and institutional innovation, and driven by socioeconomic forces. See text for further explanation. (Adapted from Tittonell (2014))

Consumer demand for healthy food and government regulations, such as agri-environmental directives, as well as organic farming certifications, which function as a form of regulation, can serve as catalysts for initiating the transition towards agroecology (Fig. 2.4). However, it is important to recognize that these drivers alone are unlikely to bring about a comprehensive redesign of current agroecosystems. At best, they may result in a situation of input substitution, which refers to the process of replacing agrochemical inputs commonly used in conventional agriculture with organic or mineral inputs permitted in organic farming. It is worth noting that input substitution can occur without necessarily altering the overall configuration and functioning of the entire agricultural system, particularly in the case of large-scale organic farming operations.

Mainstream discourses in agricultural research for development emphasize the adoption of technologies aimed at increasing agricultural productivity while promoting greater eco-efficiencies. This pursuit is often referred to as sustainable intensification (Tittonell 2014, 2018). One category of technologies, known as precision agriculture, focuses on enhancing eco-efficiency by enabling farmers to produce more with fewer resources. These technologies are depicted in Fig. 2.4 as agricultural "optimization" technologies and practices. While optimization is an essential step, it alone is not sufficient to bring about a complete transition or redesign of the current agricultural system towards agroecological approaches.

The initial stages of transitioning, characterized by eco-efficiency gains and input substitution, correspond to the first and second stages in Gliessman's transition typology. However, farmers relying on input substitution can be considered a vulnerable population, particularly if the agroecosystem has not undergone significant transformations to provide robust regulation services, such as natural pest and disease control. To achieve further agroecosystem transformation through redesign, additional drivers are necessary to steer technical and institutional innovations. These drivers include various forms of social organization, such as consumer and/or farmer associations, interest groups, and socio-political movements. Moreover, governmental support through territorial development programs plays a crucial role in facilitating the desired changes.

2.3.2 Models and Experiments in Agroecosystem Design

The evaluation of agroecosystems requires considering multiple criteria simultaneously, reflecting the diverse objectives of various stakeholders involved in or concerned with the need for redesign. It also requires the ability to capture the interactions and feedback between different dimensions of the system across space and time, as well as the capacity to explore alternative futures through scenario analysis. Semiotic systems or models can be developed with the purpose of comprehending a system (analysis) or contributing to systems design (synthesis). Experimenting with entire agroecosystems is often impractical, especially when assessing alternative futures under uncertain contexts, such as climate change or market volatility. This can be achieved by building a semiotic system, a mathematical simulation model of the agroecosystem. These models can be of various natures or employ different algorithms to represent specific processes of interest (cf. Antle et al. 2016). Some models focus on representing particular dynamics of the system, while others optimize a set of objectives reflecting the preferences of different stakeholders. Additionally, there are models that account for emerging behaviours resulting from the interaction between individual agents following a set of rules, as well as spatially explicit models projecting land use trends based on past changes, among others. These models will be introduced in Chaps. 8, 9 and 10 of this book. Despite the potential benefits, the use of mathematical modelling in agroecology has not been widespread until recently, and one reason for this is the inherent complexity and uncertainty that agroecology deals with in attempting to understand reality. However, acknowledging the importance of embracing complexity and uncertainty, there is an increasing recognition of the value that mathematical modelling can bring to agroecological research and practice.

Models of complex social-ecological systems, such as agroecosystems, need permanent testing and ground-truthing through experimentation, field measurements or discussion with stakeholders to validate their outcomes. Although we defined agroecosystems as cybernetic systems that are steered by humans, a distinction should be made between systems that are purely mechanic from systems that depend on biological components such as microbes, or on stochastic drivers such as the weather. An engineer can easily design a machine on paper and, if the semiotic model developed is accurate enough, then the machine that could be built based on such design should work properly. A typical example of this was the design of the first vehicles for space travel that could not be tested before they were put in orbit – no experimentation could be made and yet in most cases the experience was successful (De Wit 1992). In biological systems, uncertainties are large and predictions are less reliable. We study structures to understand functions. We come up with ways of describing the causality at play between structures and functions and, thanks to experiments, with a probability of certitude associated with our statements. Experimentation is thus an essential step during the analysis and design of systems with important biological dimensions, such as agroecosystems.

Agroecology, low-input farming, and alternative practices such as conservation agriculture, organic farming, and traditional smallholder agriculture rely heavily on organic resources and biodiversity for their functioning. In these systems, the biological dimension holds greater significance compared to conventional systems, as they depend on processes like organic matter decomposition and biological nitrogen fixation for nutrient supply, soil-root feedbacks and rotational carry-over effects for disease suppression, crop-livestock interactions for nutrient cycling, and natural agents for pest control. Some of these biologically-mediated interactions may be lost or not fully captured in experiments conducted under controlled conditions, as they often simplify the complexity of actual agroecosystems (cf. Box 2.2). Scaling up the results of such experiments across diverse landscapes can be facilitated by using models, as will be demonstrated in subsequent chapters.

Models of biological systems are subject to high levels of uncertainty and require repeated testing through experimentation. Modelling and experimentation can be best used in combination to address complex agroecological questions. Currently, there is a lack of sufficiently comprehensive models that can capture the complexity of the social and ecological interactions within an agroecosystem. This may be one of the primary reasons for the relatively low adoption of mathematical modelling in agroecology. However, it is worth noting that complexity, diversity, and dynamics are inherent and often desirable characteristics of agroecosystems. While there are challenges in modelling the intricate nature of agroecosystems, ongoing research and advancements in modelling techniques offer potential to address these complexities and improve our understanding of the interacting social and ecological dynamics within agroecology.

Box 2.2. Models and Experiments

The schemes below represent key system components in an experimental field, a field on a farm and a portion of an agricultural landscape in a smallholder setting, next to examples of possible system models[(*)] delineated to represent abiotic interactions at field, farm and landscape level. An experiment is in itself

(continued)

Box 2.2 (continued)

a simplified model of the actual agroecosystem. The semiotic model shown next to it considers interactions between crop and soil components in the form of water, carbon and nitrogen flows. Moving up in scale certain processes become less relevant or more uncertain, while others emerge that dominate the behaviour of the system. Such is the case of the presence of trees and livestock, or the inherent spatial soil variability and land use system, in the two other examples illustrated here. At farm scale, household objectives and farm management decisions become central, as illustrated in the farm system model. The level of detail necessary to represent a certain system component, such as the soil, decreases as we move upscale, because uncertainties associated with error and space-time variability tend to overshadow the trends that could be observed or detected at lower scales. In smallholder systems, rural communities often share and manage a landscape or territory collectively. Collective decisions on natural resource management at community level tend to overrule individual farmer decisions.

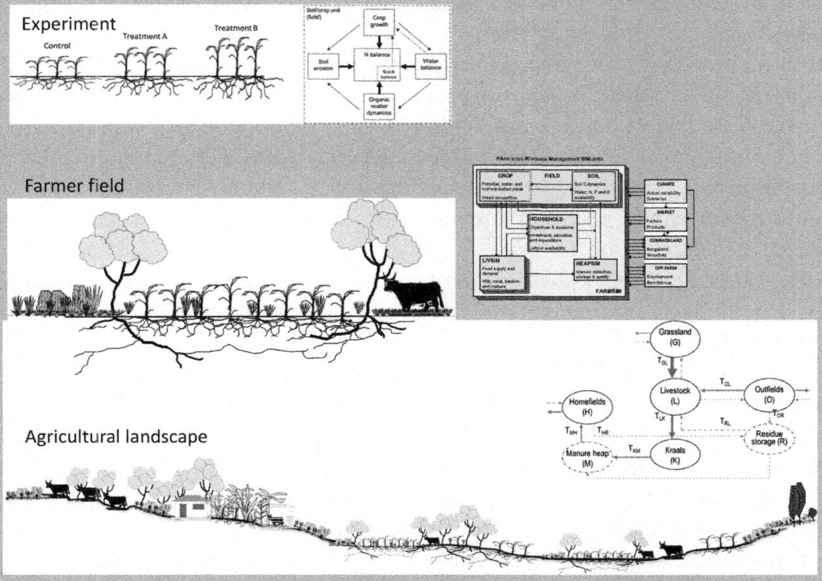

(*) The first two diagrams correspond to actual simulation models, published as Tittonell et al. (2006, 2009a, b)

2.4 Model Simulation and Scenarios in Agroecosystem Redesign

Simulation models are used in a diversity of applications in agriculture. From mere systems analysis, that is, to characterise the structure and functioning of agroecosystems at different scales, to the support of future-oriented (predictive or explorative) studies and scenario analysis, or as decision support tools in the realms of management practices and agroecosystems design. As defined earlier, a simulation model is a special type of semiotic model that uses mathematical language to represent the interactions that are observed in agroecosystems. Agroecosystems are complex and not all possible interactions can be simulated simultaneously. Complexity in a simulation model is defined by the number of components in the system, the nature of the interrelations among them, and the number of scales or integration levels considered. For example, crop productivity can be 'predicted' using a very simple model that considers rainfall and rainfall use efficiency by the crop species being simulated. Or, it can also be predicted by simulating every rainfall event, water dynamics into the soil profile, water uptake by roots driven by evapotranspiration, which in turn depends on species-specific crop leaf area, development stages and growth rates, etc., plus management factors such as sowing dates or plant densities. All these processes can be simulated with a crop growth model, which considers incoming solar radiation as a driver and the regulatory effects of air and soil temperatures, and of nutrient availability. The second option would require a much more complex model, with many interacting components and at least three levels of integration (e.g., the organ, the plant, and the soil-plant-atmosphere continuum). Yet, such a model would still be a simplification of the complexity of the ontological system, the actual agroecosystem.

The purpose of this section of the book is not to provide a detailed tutorial on model building, but rather to guide the reader in choosing an appropriate model. Selecting the right model requires several criteria to be considered. Firstly, it is essential to be aware of the range of modelling approaches used in agriculture. Secondly, a basic understanding of system processes is necessary to match the available models with the research questions or design objectives that will be addressed using a model. This involves considering factors such as scale, level of detail, and handling of errors. Furthermore, it is crucial to take into account the context of model application, the intended user, the expected results, and the availability of data for parameterization and testing. The choice of a model to support the redesign process will be illustrated through examples of model applications, with the aim of avoiding a prescriptive approach as much as possible. This is because simulation modelling in agroecology is a relatively new and evolving discipline. It is important to allow room for new modelers to experiment, innovate, and foster creativity within the field. Advancements in technology, particularly in dealing with large databases, such as cloud storage for real-time and location-specific simulations, as well as new technologies for generating, collecting, and sharing spatially-explicit data (e.g., crowd data sourcing), present significant opportunities for developing a new generation of agroecosystem models. These advancements open up possibilities for innovative approaches and support the integration of diverse data sources into modelling processes.

2.4.1 How Complex Should a Model Be?

Let us recap the basic concepts of systems analysis: a system represents a limited portion of reality (ontological system), and a model is a simplified representation of that system (semiotic system). The choice of model complexity depends primarily on the objectives of the study for which the model is being used, as well as subjective factors such as the user's disciplinary background and the availability of necessary data and information for model construction, testing, and execution. A quote often attributed to Albert Einstein states that "a model must be as simple as possible, but not simpler." This suggests the need to adopt a principle of parsimony, wherein the simplest approach or explanation is used to address a complex problem (known as Occam's razor approach). In other words, models should not be unnecessarily complex. However, they should also not be overly simplistic. The principle of parsimony in model development is not a trivial matter and has important implications for the amount and type of inherent error in a model. Striking the right balance in model complexity is crucial. Overly complex models can introduce unnecessary complexity and potentially obscure the underlying relationships or make the model difficult to understand and interpret. Conversely, overly simplistic models may overlook important interactions and dynamics within the system, leading to inaccurate results. Finding the appropriate level of complexity is a delicate task, considering the trade-off between capturing essential elements and avoiding excessive complexity. This also influences the level of error inherent in a model, as more complex models may introduce additional sources of uncertainty.

In a model with a large number of parameters, the cumulative error arises from the estimation of each parameter, in addition to the inherent uncertainty associated with understanding the process being modelled. On the other hand, a simple model with few parameters, often empirical in nature, as exemplified in the previous crop simulation example, implies a limited understanding or interest in the actual processes being modelled. It is important to note that there is no inherently right or wrong approach in this regard. The choice between a complex or simple model depends on the specific research objectives and needs, and it must be supported by credible justifications. As mentioned earlier, a general guideline in systems analysis is to construct models that consider one level of integration above and one level below the actual process being modelled. Let's use the same example of simulating crop productivity as a function of water availability discussed above. We may consider the crop-soil system as our reference integration level (i). To represent the rate of plant water transpiration, we may be interested in simulating leaf area expansion at plant level ($i - 1$). But we would not necessarily simulate, for example, the exchange of potassium through the membranes of the vacuoles in the palisade

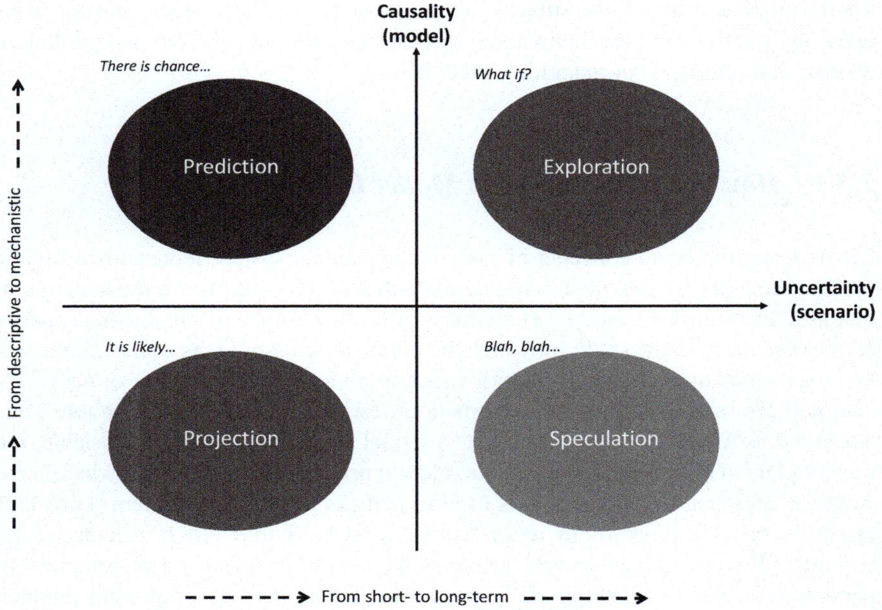

Fig. 2.5 A classification of future-oriented studies based on the level of causality inherent to the model used and the level of uncertainty associated with the scenarios analysed, based on Van Ittersum and Rabbinge (1997)

cells of the leaf tissue, which regulate stomata gas exchange. Such a process corresponds to integration levels $i-2$, $i-3$, and $i-4$. Many of the parameters that would be necessary to simulate such complexity have seldom been measured for most crops, let alone their variation under different environmental and management conditions.

Regardless of the complexity of a model, it is important to exercise caution when interpreting modelling results. To minimize the risk of overinterpretation, it is necessary to situate the model-based study within its specific context, considering two complementary aspects: the degree of causality inherent to the model and the degree of uncertainty inherent in the simulated scenarios (Fig. 2.5). Future-oriented or ex-ante studies used in agroecosystem design and evaluation can be classified based on these two criteria, as proposed by Van Ittersum and Rabbinge (1997). It is crucial to differentiate between the level of causality known regarding the studied processes and the level of uncertainty associated with the analysed scenarios. Uncertainty is often linked to the time horizon of the scenario, such as short-term projections versus long-term scenarios that consider major disruptions or changes in the future. For example, consider these two questions:

'what is likely to happen with food prices next year' → a low uncertainty scenario
'what if the gobal food market collapses in 20 years' → a highly uncertain scenario

When uncertainty is low but there is limited knowledge about the relevant causes and effects governing key phenomena (e.g., when an empirical function is used), the study is defined as a 'projection'. Under low uncertainty, but with an increase in causality (e.g., when employing a mechanistic, process-based model), the study can be defined as a 'prediction'. When a mechanistic, high-causality approach is used to investigate highly uncertain scenarios, the study is termed an 'exploration'. On the other hand, when analysing highly uncertain scenarios using empirical or descriptive models, the future-oriented study is simply labelled as a 'speculation'. Therefore, while it is common to hear terms like 'the model predicts' or 'predictive capacity', it is important to reserve the term 'prediction' specifically for models with high causality under scenarios of low uncertainty. Strictly speaking, when using a statistical model, even a complex one, we are not predicting but projecting a trend.

2.4.2 Types of Models that Can Be Used in Agroecology

In principle, any model that simulates a process or part of a process that is relevant to the functioning of an agroecosystem can be used in agroecology. Such processes can be biological, physical, social, institutional, transactional. However, most simulation models used in agricultural sciences in general tend to focus on representing biophysical processes, such as photosynthesis or light conversion into biomass, water or nutrient use and conversion, soil carbon dynamics, pest population dynamics, water flows, system productivity, etc. But most of these models are built following the principles that guide classical agronomy (cf. Table 2.1), such as linearity, predictability, uniformity, and focus on simulating individuals or populations and seldomly communities. And this limits their applicability in the realm of synecology, or community ecology, which is central for agroecological system design. The role of biodiversity, its functions and interactions, is poorly represented or not at all in most simulation models of soils, watersheds, cropping or livestock systems. Models that represent landscape-scale processes are usually spatially explicit and, strictly speaking, they do not represent processes but use empirical relations (generally regression models) to associate model outcomes to particular features (parameters) of patches or pixels in a map.

There are several possible ways of classifying models used in agriculture, ecology, landscape and rural studies. The simplest one is to classify them according to the object or primary system being modelled. This classification is not a strict one. For example, non-exhaustively, we may distinguish:

- Plant growth or crop eco-physiology models;
- Animal nutrition models;
- Cropping systems models;
- Livestock herd dynamics models;
- Farm system models;
- Household decision models;

- Landscape dynamics models;
- Pest or disease dispersal models;
- Land use change models;
- Agent-based multi-actor models;
- Community decision models.

Models may also be classified according to the type of algorithm they employ. For example, according to the way in which they deal with time in its calculations, such as static models using fix technical coefficients (or input-output tables in modelling jargon) versus dynamic simulation models using differential equations. When models are used to optimise an objective function or approximate a frontier they are termed optimisation models, such as linear goal programming models or Pareto optimality models. Classifications are many, as proposed in any good text book about simulation modelling. Yet, new algorithms appear almost every year that tend to blur the boundaries between the classical types of simulation models (e.g. as those outlined in Chap. 10). Yet, most of the complex processes that are important in agroecological design, such as multitrophic interactions, polyculture interactions, rhizosphere interactions, interacting population dynamics of pests and natural enemies, pest suppressive landscapes, etc., are poorly characterised as systems, let alone successfully modelled. Single component models such as crop growth models simulating the accumulation of biomass of monospecific populations (e.g. a wheat crop) as a function of incoming radiation, have limited use for agroecologists. Social and social-ecological interactions that are relevant to agroecological transitions at different scales are also poorly represented in models.

Figure 2.6 offers an overview of common models used in agriculture, ecology, rural and land use studies, classified according to the level of integration at which they are developed and applied, and the degree of determinism built in their structure, such as causal relationships and feedbacks. High causality models are often termed process-based models (Kanter et al. 2018). Dynamic component models such as crop or livestock simulation models (high degree of causality, explicit feedback loops) tend to summarise the existing knowledge about relevant biophysical processes into sets of differential equations and parameters, and use climatic data as independent input variables. The mere fact that they summarise the available knowledge in the form of functional relationships and feedbacks is already a great contribution of these models. They allow us to understand how systems work and identify knowledge gaps. However, as mentioned earlier, their use in agroecology is limited because they hardly internalise biotic interactions beyond their system boundaries (e.g. the monospecific crop field or livestock herd). Dynamic models at farm or landscape level use also functional relationships but they often need to sacrifice degrees of causality to represent relationships that are less well known or poorly parameterised (for example, farmer or household decisions).

At the lowest level of determinism are the models that use empirical equations obtained through statistical analysis. These are termed descriptive models. Examples of these are the econometric models, or the GIS-based analysis that use regression models. So-called land use models belong to this category. The various layers of

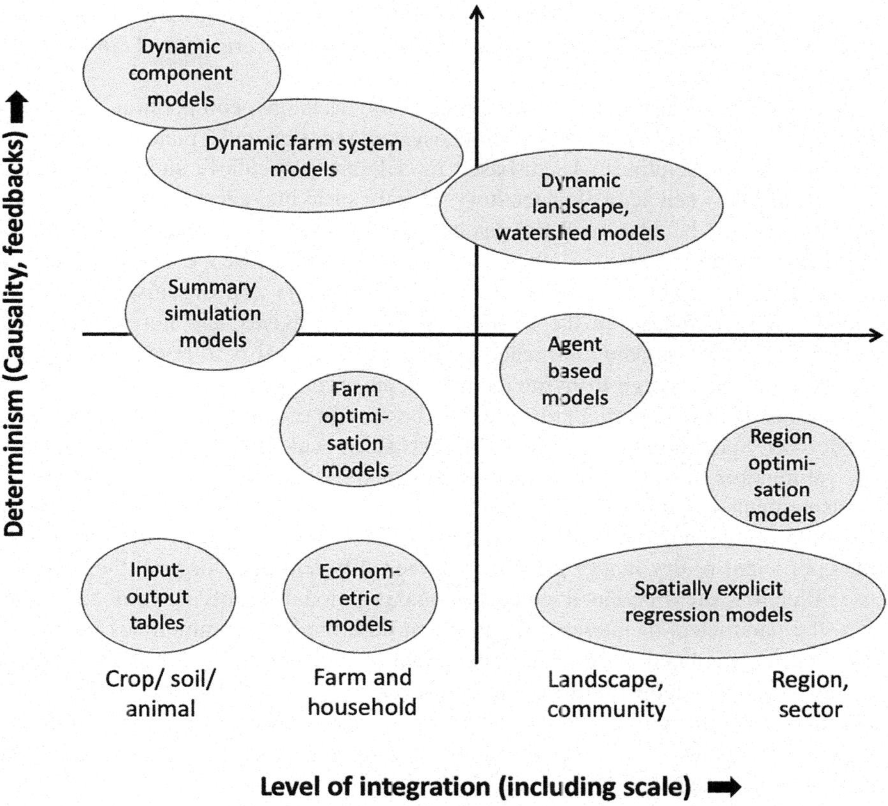

Fig. 2.6 A quick classification of common models used in agriculture, ecology, rural and land use studies, based on the level of integration at which they are developed/applied and the degree of determinism in their structure. (Adapted from Tittonell et al. 2015)

spatially explicit information are used as terms in a regression model, and simulated 'changes' in land use are actually projections, which are only valid when they are within the range of validity of the regression model (cf. Fig. 2.5). These models simulate no feedbacks, and their causality is at the lowest possible level, so that they function as a black box. This is not to say that they are not useful or reliable. Spatially explicit information is highly valuable. But it is important to avoid overinterpreting the outcomes of such type of models.

Optimisation models applied at farm or regional scale embrace a higher level of determinism because they rely on technical coefficients, which are normally derived from empirical data or even through the use of process-based simulation models. Yet their calculations are based on fixed input-output relations (technical coefficients) and consider no feedback, and no dynamics. Summary models are an intermediate case (Fig. 2.6), as they are built using process-based simulation models but retain only the most relevant feedbacks, according to the research question of interest, to reduce parameterisation efforts and associated parameter error (cf. Tittonell et al. 2009b).

Simulating social processes and interactions, decisions, competition, collaboration, etc. is also highly relevant in agroecosystem redesign and a major challenge for the modelling disciplines. Agent-based models are particularly suited to simulate interactions between actors in a territory. Strictly speaking, agents can be people but also plants, fields, insects, land patches, companies, microorganisms, etc., etc. Anything can be modelled as an agent. In land use and landscape studies, however, agents tend to represent individual or collective actors and organisations. These models place emphasis in the interactions between agents and not in the actual biophysical processes they influence or manage. The goal is to reveal the possible patterns that can emerge from interacting agents at higher scales.

Serious games are particularly suited to be used in combination with agent-based models (cf. Speelman et al. 2014a, b; Michalscheck et al. 2020). Agent based models are simulation models that require parameters, initialisation variables or actual measurements of output variables for their calibration. In a crop growth model, for example, common parameters are the specific leaf area of the crop at different stages, the coefficient of dry matter partition between different plant organs, the radiation use efficiency, etc. In collective decision-making models involving rural communities, the parameters of interest are those that describe how communities interact in the process of collective decision-making and the behavioural factors that influence it. For example, cooperation, communication, trust, leadership are all factors of interest or ‘parameters’ of the decision-making model that can be gathered through role playing games with stakeholders. Such parameters can be then factored in the agent-based model.

2.5 Agroecosystem Properties and Sustainability

Agriculture constitutes the most conspicuous form of interface between human societies and natural ecosystems. The *properties* that characterise agroecosystems are the result of interacting natural processes and human agency, while their *attributes* and *values* are defined largely by societal or cultural preferences. The properties of an agroecosystem can be used to evaluate its state and change over time, reflecting its characteristics and ‘behaviour’, as influenced by its structure and functions. When we design or manage agroecosystems we act upon their structures and functions, but such actions have an impact on their properties and attributes. From a systems perspective, understanding agroecosystem properties is essential to be able to evaluate their sustainability. When choosing indicators to monitor changes in the agroecosystem, as will be discussed in Chap. 8, we make sure that such indicators reflect variation in agroecosystem properties that reflect the consequences of our interventions. Ideally, the properties we choose to evaluate can tell us something about the sustainability of the agroecosystem. But before diving into a discussion on sustainability indicators, criteria, etc., let us first define what is meant by properties, attributes and values in the context of agroecosystems and sustainability evaluation.

2.5.1 *Properties, Attributes and Values*

In physics, a property is a characteristic used to describe the nature of a certain material, to differentiate between types of matter. In thermodynamics, more specifically, properties of substances are classified as intensive or bulk properties, when they do not depend on size or amount of material in the system, and extensive properties that are proportional to system size. Typically, density is an intensive property that results from the ratio of mass to volume. Both mass and volume are scale-dependent and therefore extensive properties. Properties can be also intrinsic or extrinsic, when they depend on the systems' own nature, independently of other things, or on their relationship with other things, respectively (Lewis 1983). Mass is an intrinsic property of an object, whereas weight is an extrinsic property that depends on the strength of the gravitational field where the object is placed. Properties can be isotropic, when they are independent from the direction of observation, or anisotropic otherwise. A supervenient property is one that depends on the underlying structures and functioning of a system. The colour of an object, for example, is supervenient on the underlying structure and reflective properties of its surface.

Although properties and attributes are often considered as synonyms, they may hold different meanings in certain disciplines. In mathematics, an attribute describes an object within a pattern; shape, colour or size, are typically attributes of an object. In philosophy, an attribute is considered to be ascribable while a property can be possessed (Cargile 1995). In informatics, an attribute is a characteristic that something or someone may have, for example:

<Grass palatability = "poor"/>

In this case, palatability is an attribute of grass (as defined by its physicochemical nature), while 'poor' is the value of this attribute (as perceived by a certain herbivore). Values can be operationalized in different ways by means of variables. In the example above, the attribute palatability could be defined as a dichotomy, such as 'good' or 'poor', or through an ordinal scale, such as 'poorly palatable', 'moderately palatable', 'palatable', or through a numerical scale 1, 2, 3, etc. Thus, the variable that is used to ascribe a value to the attribute can be binary, ordinal, be nominal or use rational numbers, discrete or continue, etc.

2.5.2 *Properties of Sustainable Agroecosystems*

In the various approaches to agroecosystems sustainability evaluation emerged around the turn of the twenty-first Century, both properties and attributes have often been used as synonyms (e.g., Smith and Dumanski 1994; Mitchell et al.

1995; Masera et al. 1999; Bossel 2000; López-Ridaura et al. 2002). However, the use of the term property is preferable over attribute when it comes to describing generic characteristics of agroecosystems. Properties can be defined generically in terms of being intensive and extensive, or intrinsic and extrinsic, etc., while attributes can be preferably used when values are being assigned to such characteristics. For example, Conway (1987) used the concept of intrinsic and extrinsic properties to define two types of agroecosystem properties: those that depend on the internal structure and function of the agroecosystem, such as productivity or stability, and those that reflect the response of the system to external driving variables, such as resilience or adaptability. When dealing with complex systems such as agroecosystems it is rather difficult to draw a line between intrinsic and extrinsic properties. Agroecosystem productivity depends not only on its structure but also on solar radiation and rainfall, two external driving variables. In open systems, everything is ultimately linked to external variables. Yet, the analogy may hold when we consider climatic variables to be 'inherent' to an agroecosystem.

Properties of dynamic systems, such as agroecosystems, are often evaluated with respect to time and using indicators that represent key variables of interest. Figure 2.7 illustrates agroecosystem properties as defined by the above-mentioned authors. The four figures illustrate properties that reflect the behaviour of a system over time. The 'target variable' in the vertical axis represents a certain indicator of agroecosystem performance, such as economic profitability, milk production, level of local employment, soil nutrient stocks or number of natural enemies of a certain insect pest. These variables, often termed target variables in the specialised literature, are able to reflect dynamic properties of agroecosystems that supervene their structures and functions. Let us consider the most widely used properties:

Stability
Stability or constancy refers to the capacity of the system to maintain the same level of the target variable, albeit normal oscillations, over the period of time considered (cf. Conway 1987; Masera et al. 1999; López-Ridaura et al. 2002). According to this definition, stability is an intrinsic property that depends on the systems' own structure and functioning. Loss of stability in an agroecosystem may occur due to soil nutrient depletion, for example, as illustrated with long term maize yield data from Togo in Fig. 2.8. Note that lack of stability in the case of agricultural productivity becomes evident over large periods of time, and should not be confused with inter-annual 'yield variability'. In Fig. 2.8, both systems exhibit yield variability from year to year, but only system B is stable.

Reliability
Reliability of a system can be defined with respect to the behaviour of an external driving variable, which means that it corresponds to the category of extrinsic or relational properties. In the face of normal oscillations in the external driving variable, an unreliable system will exhibit large variability in the level of the target variable over time. For example, in Fig. 2.8, the inter-annual oscillation of maize grain yields can be largely attributed to rainfall variability (external driving variable). If such oscillation remains within an expected range, then the system can be

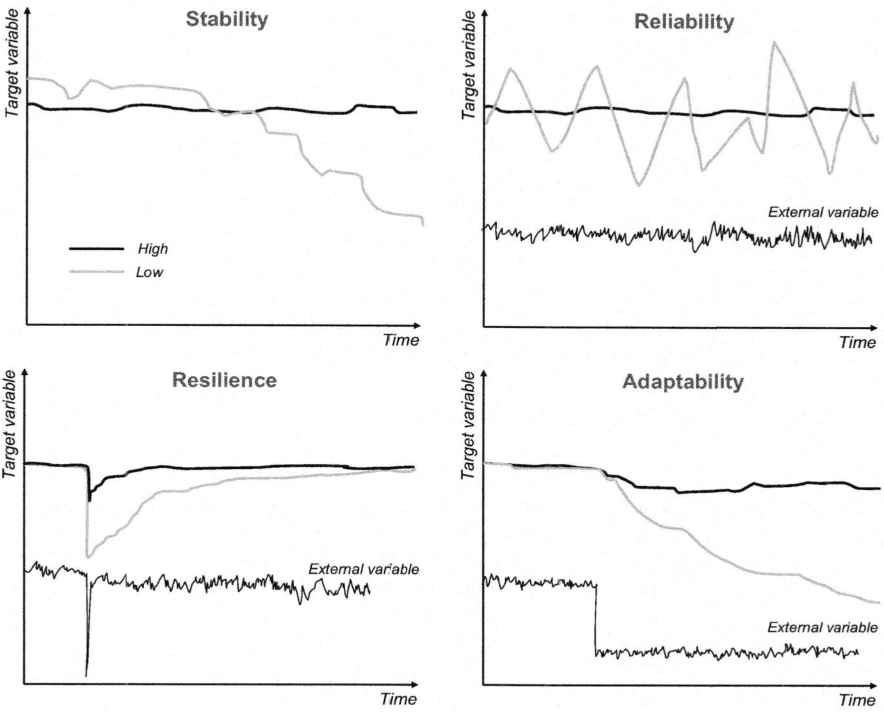

Fig. 2.7 The theoretical curves representing the evolution of a certain target variable (e.g., an indicator of agroecosystem productivity) in time. While stability is an intrinsic property that depends on the system's own structure and functions, Reliability, Resilience and Adaptability are responses of the system to changes in an external driving variable. Schemes adapted from Conway (1987) and Masera et al. (1999)

considered to be reliable. To illustrate this point further, let us examine the results of long-term cropping system trials at the Rusell Experimental Ranch of the University of California at Davis (1993–2012) (Fig. 2.9). Average tomato yields under conventional and organic management have been statistically similar over the almost two decades of experiment, respectively 66.7 ± 18.2 and 67.8 ± 9.0 t ha^{-1}. Yet it is evident from examining Fig. 2.9 that yields under conventional management are more volatile, less reliable, somewhat 'amplifying' the effects of climate, than yields under organic management, which are more reliable and contribute to 'buffer' climatic variability.

Resilience

Resilience is too an extrinsic or relational system property, defined with respect to the behaviour of an external driving variable. When there is a shock due to a sudden but temporary change in the level of the external driving variable, a resilient system will allow faster recovery in the level of the target variable than an un-resilient one (and may be also less affected by the initial shock, more resistant). This is illustrated

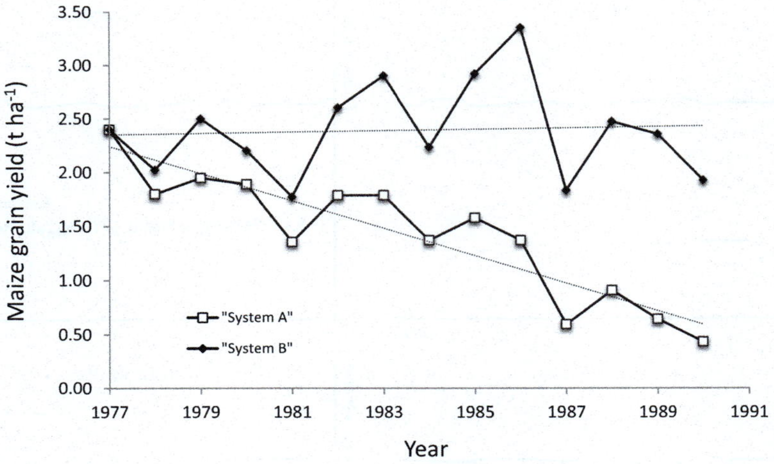

Fig. 2.8 Stability of maize grain yield in two types of agricultural management systems, generically termed as A and B. System B is stable over time, in spite of inter-annual variability. The data corresponds to a long-term agricultural trial (1977–1990) conducted in Togo and published as Kintché et al. (2016). Dotted lines in the background are trend lines

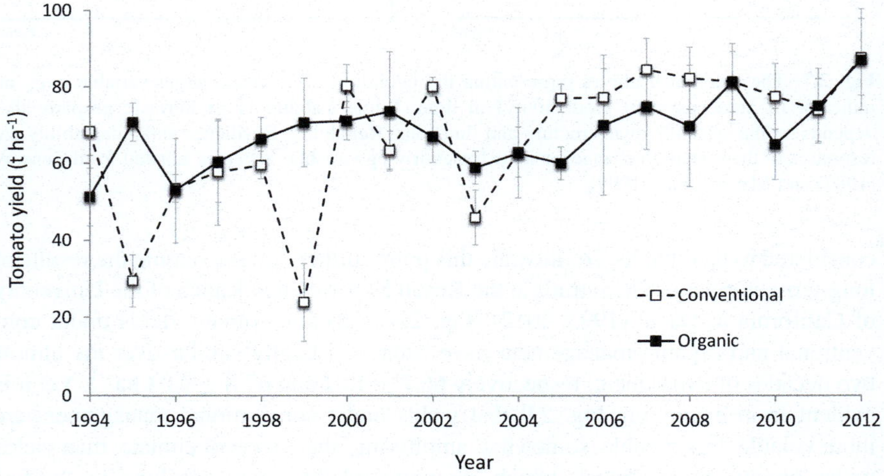

Fig. 2.9 Yield of field grown tomato in rotation with maize under conventional and organic management from 1993 to 2012 at the Rusell Experimental Ranch of the University of California at Davis. The data used in the analysis are publicly available here: http://asi.ucdavis.edu/rr

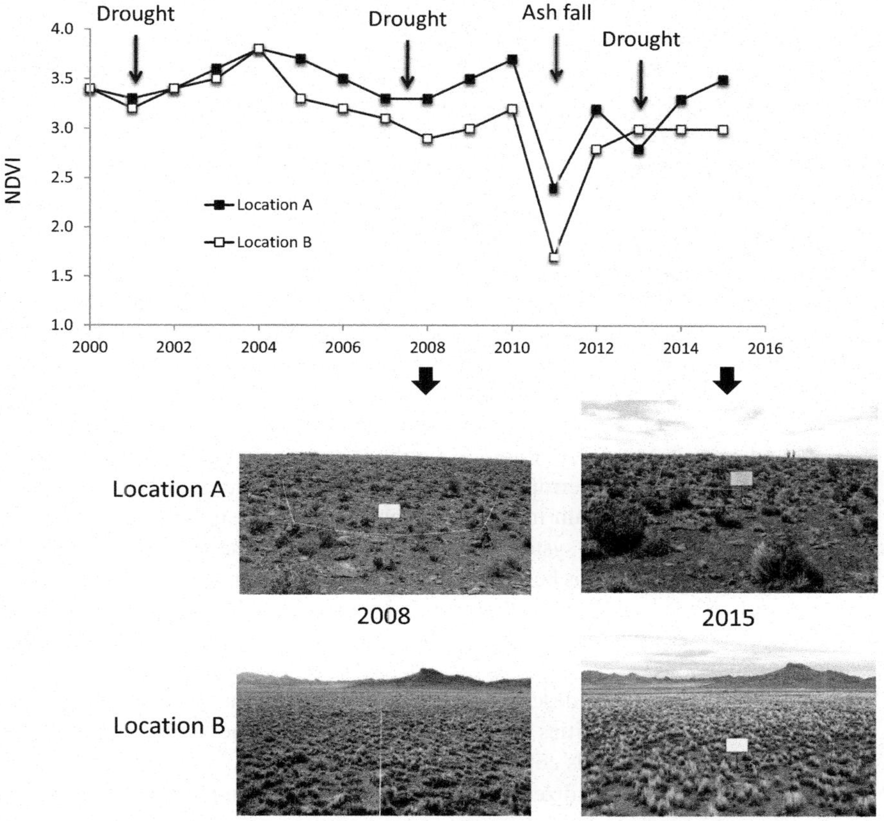

Fig. 2.10 Seasonal rangeland productivity from 2000 to 2016 at two locations in north Patagonia that exhibited high (location A) and low (location B) recovery rates under different grazing pressure; productivity is estimated through the normalised difference vegetation index (NDVI) using satellite imagery. Drought periods are indicated with arrows, as well as the heavy ash fall following the eruption of the Puyehue vocanic system in 2011. (Source: Bran et al. 2016)

by the primary productivity of dry rangelands in the Patagonia steppe, an area used for extensive livestock grazing, following a heavy ash fall after the eruption of a volcano in 2011 (Fig. 2.10). Grassland primary productivity, as estimated through the NDVI, recovered faster after the ash fall – the external shock – in location A as compared with location B. Moreover, due to the differential grazing pressure at both locations, rangeland productivity at location A recovers almost to its original level in the year 2000 after the heavy ash fall. The example illustrates how management decisions, such as stocking rates, can have an important influence on the resilience of the system to shocks.

The productivity of perennial vegetation in rangelands, grasslands or forests exhibit dynamics that can more easily illustrate resilience than in the case of annual cropping systems. A perennial vegetation ecosystem receives a shock at one point in time, and possesses a 'memory' that conditions its behaviour in subsequent years. Maize or tomato yields, as used in Figs. 2.8 and 2.9, represent fast-turning variables. A drought that takes place this year will not necessarily impact directly maize or tomato yields next year. The impact may be indirect when, for example, this year's drought affects the capacity of a farmer to afford land, labour or agricultural inputs for next year.

Adaptability
Adaptability, in this framework, is the capacity of a system to maintain acceptable levels of the target variable when there is a permanent change in the external driving variable(s), which may be abrupt (as in Fig. 2.7) or gradual. A non-adaptable system will exhibit an irreversible decline in the level of the target variable over time. Yet this definition, which follows principles of systems dynamics, falls short of describing what we nowadays understand by adaptability, particularly in relation with climate and global change. A simple example of abrupt change in external driving variables could be the interruption of a subsidy scheme, or a new regulation restricting the use of a certain input, or an abrupt change in input or output prices. Strictly speaking, adaptable systems are those able to continue performing at similar levels of the target variable as before such changes took place. In a broader definition of the term, however, adaptable systems may be also those that continue performing but perhaps delivering a different type of output, having adapted – changed – their production functions.

Speelman et al. (2014b) describe an example of this in their study of adaptation strategies of rural communities in Chiapas, Mexico after the signature of the free trade agreement between this country, the US and Canada. As maize prices dropped due to the imports of subsidized US maize, farmers were unable to compete and stopped growing maize. It was cheaper to buy it on the market than to produce it on the farm. Livestock keeping, coffee farming and non-timber forest products (collection of native ornamental species) replaced maize growing. Households reacted differently to this new reality, following strategies that were described as early diversification, late diversification and specialization. Adaptability, in such cases, cannot be simply assessed by examining the behaviour of a single target variable.

2.5.3 Selecting Evaluation Criteria and Indicators

Understanding system properties is essential to the identification of relevant criteria and indicators to diagnose, evaluate, compare and monitor agroecosystems. As will be seen in Chap. 8, agroecosystem design and evaluation are two alternating and co-depending stages in agroecological transitions that propend towards increasing sustainability (Fig. 2.11). Masera et al. (1999) proposed the use of system properties for the identification of indicators to evaluate agroecosystem sustainability across spatial scales, through what is known as the MESMIS framework (Indicator-based

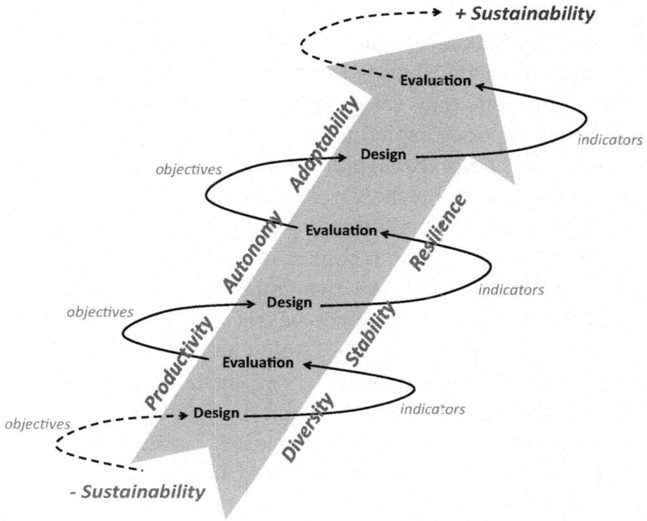

Fig. 2.11 A representation of how the phases of design and evaluation interact with each other during an agroecological transition. As the agroecosystem progresses towards greater sustainability, the various stages of (re-)design aim at the gradual improvement of key properties, such as productivity, diversity, stability, autonomy, resilience or adaptability

Framework for Evaluation of Natural Resource Management Systems, for its Spanish acronym). In this framework, it is assumed that a system is sustainable when it is productive, stable, reliable, resilient and adaptable. Expressed like this, these properties represent *attributes* of sustainable agroecosystems. Diagnostic criteria to evaluate agroecosystems are chosen to reflect one or more of these attributes.

It is evident that all these properties, stability, reliability, resilience and adaptability are interdependent and cannot be regarded in isolation from one another, that they are supervenient on both biophysical and socio-economic structures and functions, and that the definition of what is intrinsic or extrinsic to the agroecosystem remains largely arbitrary. Moreover, the examples presented to illustrate these properties focused on a single component of the agroecosystem, such as yields of individual crops. In diversified agroecosystems, as those proposed in agroecology, supervenient target variables that represent overall system productivity, returns to labour or aggregated gross margins are better able to reflect these properties at the scale of the entire agroecosystem. For example, even in an agroecosystem where maize grain yields may be non-reliable over time, other components of the system may be, compensating for the loss of maize yield and contributing to buffer income variability at the entire system level. In other words, a whole system, when diverse, may be reliable even when some of its individual components or subsystems are not.

More than one criterion can be used to assess the same attribute/ property (Table 2.2). For example, soil losses and long-term variability in the return to labour can be both proposed as criteria to reflect the stability of the agroecosystem. Then the next step is to identify indicators that reflect the selected criteria. An indicator is a variable that is used to assign values to a criterion. For example, when the criterion is economic profitability, one may select indicators such as the gross margin, or the return to land, to labour or to capital to evaluate it. In agroecosystems analysis an indicator is often – but not exclusively – quantifiable and it is most useful when meaningful benchmarks and threshold values can be assigned to it. Table 2.2 illustrates how the same property, stability in this case, can be assessed through different criteria; and how the same criteria, soil losses, can be evaluated through more than one indicator. Table 2.2 also shows two important elements of indicator-based sustainability assessments: their reference values and the methods for their assessment. An indicator is only useful when reference values (baselines, thresholds, benchmarks) are known. Also, more than one method can be used to assess the same indicator, as illustrated in Table 2.2 with the example of soil erosion.

In certain cases, different criteria may reflect the objectives of different stakeholders involved, and indicators and thresholds can be determined based on stakeholder preferences and weights. Once the indicators are selected to reflect the relevant criteria and properties of interest, and benchmark values and threshold levels are established for these indicators, the sustainability of an agroecosystem can be assessed by evaluating the extent to which the criteria are met or by scoring the corresponding indicators. Throughout the rest of the book, concepts such as properties, variables, values, and indicators will be revisited. The framework described here represents one approach for evaluating agroecosystem sustainability. The key takeaway message is that any variable or indicator chosen to monitor an agroecosystem should be closely linked to its fundamental properties. Through the redesign of an agroecosystem, the goal is to modify its structure and functioning in a way that positively influences its behaviour, ultimately enhancing properties such as productivity, stability, resilience, and adaptability.

The concept of sustainability is abstract and subjective. To make sustainability more tangible and measurable, various evaluation frameworks have been proposed, aiming to reduce subjectivity as much as possible. These frameworks attempt to transform the abstract notion of sustainability into concrete measurements that can be assessed and compared. They represent a significant advancement beyond the traditional Brundtland report definition of sustainability from 1987 (encompassing social, environmental, economic dimensions). The Brundtland report provided a fundamental understanding of sustainability by highlighting its broad dimensions, but it lacked specific quantifiable metrics to gauge sustainability in practical terms. The modern evaluation frameworks strive to operationalize sustainability, allowing for more concrete assessments and facilitating decision-making processes. Chap. 8 of the book delves deeper into these evaluation frameworks, illustrating how they can be applied to assess the sustainability of agroecosystems and agricultural practices. By incorporating measurable indicators and criteria, these frameworks provide a more comprehensive and pragmatic approach to evaluating and promoting sustainability in agriculture.

2.6 Summary and Concluding Remarks

An agroecosystem is a human-modified ecosystem designed for the production of goods and services. It operates based on natural laws and is influenced by human actors through design and management decisions. The term "agroecosystem" implies the concept of a system with all its theoretical and practical implications. A system is a distinct portion of reality that is isolated for study and becomes a cybernetic system when directed towards a specific goal. A model is a simplified representation of a system known as a semiotic system. The components, interrelations, boundaries, scales, and levels of integration of a system are defined based on the objectives for which the system is being studied. Multiple semiotic systems can be derived for the same ontological (real) system. In agricultural science, the analysis of systems has predominantly focused on autoecology, which is the ecology of single-species populations (e.g., crops). Agroecology, however, requires an approach to systems analysis from the perspective of synecology, which is community ecology (e.g., complex agroecosystems). It is important to distinguish between systems analysis (understanding how a system functions) and systems design (understanding how to make it work towards a specific objective), which is achieved through knowledge synthesis. Mathematical simulation models are invaluable tools for supporting design through systems synthesis, ex-ante evaluation, and scenario analysis. System properties and attributes can be used to aid design through sustainability evaluation and monitoring methods that are conceptually robust and applicable regardless of the type of agroecosystem or scale under consideration. A property is inherent to a system, while an attribute is assigned to it. Assigning attributes is accomplished through values, forming the basis for the development of indicator-based methods for sustainability evaluation. The most commonly used properties for assessing the sustainability of agroecosystems are productivity, stability, reliability, resilience, and adaptability. Diagnostic criteria are employed to assess the properties of agroecosystems, reflecting key issues related to their sustainability. Multiple indicators can be used to assess each criterion, and various methods can be employed for evaluating each indicator. To be useful, an indicator must be accompanied by reference values (such as thresholds, baselines, and benchmarks), units of measurement, and a specified assessment method. The redesign of agroecosystems involves modifying or creating structures to promote desired functions, which ultimately influence agroecosystem properties, ideally in a positive manner.

References

Antle JM, Basso B, Conant RT, Godfray HCJ, Jones JW, Herrero M, Howitt RE, Keating BA, Munoz-Carpena R, Rosenzweig C, Tittonell P, Wheeler TR (2016) Towards a new generation of agricultural system data, models and knowledge products: design and improvement. Agric Syst 155:255–268. https://doi.org/10.1016/j.agsy.2016.10.002

Bossel H (2000) Policy assessment and simulation of actor orientation for sustainable development. Ecol Econ 34:337–355

Bran D. et al. (2016) Marco conceptual y propuesta metodológica para el desarrollo de Observatorios de Sustentabilidad Rural. El Caso Jacobacci. Actas I Jornadas Nacionales de Investigación en Ciencias Sociales de la UNCuyo. Mesa 2: Complejidades y transformaciones territoriales abordaje desde los Observatorios. Mendoza, 25 de Agosto de 2016

Brundtland GH, Khalid M, Agnelli S, Al-Athel SA, Chidzero BJNY, Fadika LM et al (1987) Our common future. World Commission on Environment and Development

Cargile J (1995) Qualities. In: Honderich T (ed) The Oxford companion to philosophy, 2nd edn. Oxford, UK

Conway GR (1987) The properties of agroecosystems. Agric Syst 24:95–117

De Wit CT (1992) Resource use efficiency in agriculture. Agric Syst 40:125–151

Fresco LO, Westphal E (1988) A hierarchical classification of farm systems. Exp Agric 24:399–419

Gliessman SR (2010) The conversion to sustainable agriculture: principles, processes, and practices. In: Gliessman SR, Rosemeyer M (eds) Advances in agroecology. CRC Press/Taylor & Francis Group, Boca Raton

Goewie EA (1997) Designing methodologies for prototyping ecological production systems. Course reader MSc. Ecological Agriculture (F800-204). Department of Ecological Agriculture, Wageningen Agricultural University, Wageningen

Gunderson L, Holling C (2003) Panarchy: understanding transformations in human and natural systems, vol 114. Bibliovault OAI Repository, the University of Chicago Press. https://doi.org/10.1016/S0006-3207(03)00041-7

Hart RD (1982) One farm system in Honduras: a case study in farm systems research. In: Shaner WW, Phillip M, Schmehl M (eds) Readings in farming systems research, 59–73. Westview Press, Boulder

Hart RD, Pinchinat AM (1982) Integrative agricultural systems research. In Caribbean seminar on farming systems research methodology. IICA/INRA, Basse Terre

Kanter DR, Musumba M, Wood SLR, Palm C, Antle J, Balvanera P, Dale VH, Havlik P, Kline KL, Scholes RJ, Thornton P, Tittonell P, Andelman S (2018) Evaluating agricultural trade-offs in the age of sustainable development. Agric Syst 163:73–88. https://doi.org/10.1016/j.agsy.2016.09.010

Kintché K, Guibert H, Sogbedji JM, Levêquee J, Bonfoh B, Tittonell P (2016) Long-term mineral fertiliser use and maize residue incorporation do not compensate for carbon and nutrient losses from a Ferralsol under continuous maize–cotton cropping. Field Crop Res 184:192–200

Lewis D (1983) Extrinsic properties. Philos Stud 44:111–112

López-Ridaura S, Masera O, Astier M (2002) Evaluating the sustainability of complex socio-environmental systems. The MESMIS framework. Ecol Indic 2:135–148

Masera O, Astier M, López-Ridaura S (1999) Sustainability and natural resource management. The MESMIS evaluation framework. MundiPrensa-GIRA UNAM, México City

Michalscheck M, Groot JCJ, Fischer G, Tittonell P (2020) Land use decisions: by whom and to whose benefit? A serious game to uncover dynamics in farm land allocation at household level in Northern Ghana. Land Use Policy 91:104325. https://doi.org/10.1016/j.landusepol.2019.104325

Mitchell G, May A, McDonald A (1995) PICABUE: a methodological framework for the development of indicators of sustainable development. Int J Sust Dev World 2:104–123

Smith AJ, Dumanski J (1994) FESLM: an international framework for evaluating sustainable land management. In: World soil resources report 73. FAO, Rome

Speelman EN, García-Barrios LE, Groot JCJ, Tittonell P (2014a) Gaming for smallholder participation in the design of more sustainable agricultural landscapes. Agric Syst 126:62–75

Speelman EN, Groot JCJ, García-Barrios LE, Kok K, van Keulen H, Tittonell P (2014b) From coping to adaptation to economic and institutional change – trajectories of change in land-use management and social organization in a biosphere reserve community, Mexico. Land Use Policy 41:31–44

Tittonell P (2014) Ecological intensification – sustainable by nature. Curr Opin Environ Sustain 8: 53–61

Tittonell P (2018) Ecological intensification of agriculture, Ch. 13. In: Serraj R, Pingali P (eds) Agriculture & Food Systems to 2050. Global trends, challenges and opportunities. World Scientific, pp 437–472. https://doi.org/10.1142/11212

Tittonell P (2020) Assessing resilience and adaptability in agroecological transitions. Agric Syst 184:102862. https://doi.org/10.1016/j.agsy.2020.102862

Tittonell P, Leffelaar PA, Vanlauwe B, van Wijk MT, Giller KE (2006) Exploring diversity of crop and soil management within smallholder African farms: a dynamic model for simulation of N balances and use efficiencies at field scale. Agric Syst 91:71–101

Tittonell P, van Wijk MT, Herrero M, Rufino MC, de Ridder N, Giller KE (2009a) Beyond resource constraints – exploring the physical feasibility of options for the intensification of smallholder crop-livestock systems in Vihiga district, Kenya. Agric Syst 101:1–19

Tittonell P, Corbeels M, van Wijk MT, Giller KE (2009b) FIELD – a summary simulation model of the soil-crop system to analyse long-term resource interactions and use efficiencies at farm scale. Eur J Agron 32:10–21

Tittonell P, Gérard B, Erenstein O (2015) Tradeoffs around crop residue biomass in smallholder crop-livestock systems – What's next? Agric Syst 134:119–128

Van Ittersum MK, Rabbinge R (1997) Concepts in production ecology for analysis and quantification of agricultural input-output combinations. Field Crops Res 52:197–208

Chapter 3
Structure, Functions and Diversity of Agroecosystems

Abstract Systems thinking and theory are used in agroecology to support agroecosystems analysis and design, through unravelling the structures and functions that ultimately explain agroecosystem properties and attributes (cf. Chap. 2). In agroecosystems analysis, we move from describing and understanding structures to infer functions and properties, particularly of biologically-mediated processes. In agroecosystem design, we start from the objectives, identify the desired functions, and plan the necessary structures, including the more operational management practices. This way of thinking originated in systems engineering, and it does not always apply as such in the case of biologically-mediated systems for which some fundamental structures are still unknown. Yet, thinking in terms of structures allows us to categorise and describe the diversity of agroecosystems at different levels, to better understand their functions. Agricultural productivity and ecosystem services in general depend on such functions. Agroecosystems have both social and ecological dimensions that need to be considered in the characterisation of their structure. Across the world, different combinations of and interactions between these social and economic dimension resulted in a wide diversity of agroecosystems, which differ in the structure and function. This Chapter aims to contribute concepts, methods and examples for the characterisation of the structure, functions and diversity of agroecosystems through analysing its main components: the household, cropping, livestock and farm systems.

3.1 Structure and Functions of the Agroecosystem

Agroecosystems are complex and multidimensional systems that are governed by natural phenomena and processes, while also being influenced and guided by humans through their design and management (see Chap. 2 for more details). Similar to natural ecosystems, agroecosystems can be divided into two main components: the abiotic component, which includes the physical environment or habitat structure, and the biotic component, which encompasses the diversity of life forms or the biological community (Fig. 3.1). In addition to this ecological structure, agroecosystems also possess a sociocultural dimension, which involves livelihoods,

P. Tittonell, *A Systems Approach to Agroecology*,
https://doi.org/10.1007/978-3-031-42939-2_3

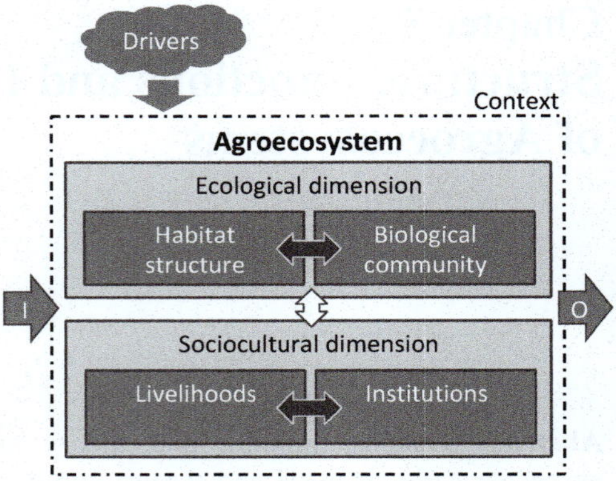

Fig. 3.1 Scheme representing the dimensions and components of agroecosystems. Block arrows indicate interactions between dimensions (white) and components (dark grey). Incoming and outgoing block arrows represent inputs (I) and outputs (O) to and from the agroecosystem, and the influence of external biophysical and socioeconomic drivers

economic activities, and institutions. Agroecosystems are considered open systems, meaning they receive inputs of energy, matter, capital, and information, and produce outputs of the same nature in the form of products, services, and externalities. They are influenced by external drivers, which can be either biophysical (such as rainfall) or socioeconomic (such as market prices), and operate within a specific context, which may include political, demographic, and other relevant factors. Agroecosystems often consist of nested sub-systems, such as cropping, livestock, and/or forest sub-systems that have *panarchical* effects on its functioning as a whole (cf. Chap. 2). The household is sometimes defined as a sub-system within the agroecosystem, encompassing human decision-making and components such as assets, labour, and knowledge. In other cases, authors may choose to define a separate decision-making or management sub-system as its own entity (e.g., Dogliotti et al. 2014).

3.1.1 Ecological Dimension

The components that define the habitat structure and biological community of the agroecosystem correspond respectively to the biotope and biocenosis, concepts that were used for a long time in classical ecology. Biocenosis is a term coined by the German zoologist K. Möbius (in 1877) to describe the community of interacting organisms cohabiting in a certain habitat, or biotope, within three categories: (i) Phytocenosis, (ii) Zoonecosis, and (iii) Microbiocenosis, corresponding to the plant, animal and microbial communities, respectively. The concept of biotope was introduced later to refer to the homogeneous environment that creates the physico-chemical conditions for the existence of a certain biocenosis.[1] The notion of

[1] Since the 1990, however, the term 'Biotope' has been used in the UK in the field of marine ecology to refer to both the habitat and the living communities of an ecosystem (Olenin S., Ducrotoy JP, 2006. The concept of biotope in marine ecology and coastal management. Mar Pollut Bull 53, 20–9)

Ecosystem = *Biotope* + *Biocenosis* (F. Dahl, 1908) has been more popular in continental European than in Anglophone science. In agroecosystems, the biocenosis consists of wild and domestic populations of animals, plants, fungi, protozoa and microbes. The use of the term biocenosis has declined over the years and virtually faded in the twenty-first Century. The biotope in an agroecosystem results from the interaction of its inherent soils and climatic conditions and the impact on these of human agency. Although biotope and habitat are used as synonyms in English, a distinction is made between both terms in other languages, where a habitat refers to the environment that hosts a population, while the biotope hosts a community. In this book, biotope and biocenosis will be further referred to as habitat structure and biological community, respectively.

Humans modify the biophysical environment in a diversity of forms, for example, through changes in vegetation structure (cover, density, height, vertical strata, connectivity, etc.), through soil tillage, levelling, ditching or terracing, through irrigation and drainage, through land fragmentation (parcelling, fencing, building of houses, lay out of cropping, grazing and forest areas, roads, etc.), by applying fertilisers and other agrochemicals, or through profound modifications of the microclimate in protected cultivation agriculture (greenhouses, plastic mulches, nets, artificial mist, etc.). But also, through choosing sowing dates of crops, or weaning dates of animals, for example. The biophysical environment is modified in these various ways to create favourable conditions for the domestic plants and animal species to prosper, and/or to enable or facilitate human life and activities. Both the cropping and the livestock subsystems of the agroecosystem consist of a habitat modified in a specific way and of an associated domestic and natural (or spontaneous) community.

3.1.2 Sociocultural Dimension

The sociocultural dimension of the agroecosystem encompasses two main aspects: the set of livelihood activities, which includes the households and their economic activities coexisting within it, and the institutions that play a role in enabling, motivating, and regulating these livelihoods (as depicted in Fig. 3.1). Additionally, the livelihood dimension also includes the decision-making or management component. Some studies refer to this component as the 'management' system, viewing it as central to the agroecosystem because it is responsible for steering the interactions among the biotic, abiotic, economic, and institutional dimensions. Institutions, on the other hand, are defined as the social order, mechanisms, or set of rules that govern the behaviour of a human community (as discussed by Chang and Evans 2005). This encompasses a wide range of entities, including public services, organizations, governments, and markets, along with the rules and norms associated with them. In rural settings, institutions often encompass traditional, both formal and informal, laws, customs, and behavioural patterns.

It is important to note that the distinctions made between livelihoods, decision-making, and institutions are not rigid boundaries; rather, they are arbitrary divisions that analysts use based on their objectives, interests, and background. Simplifications are made when defining a semiotic system to better understand the agroecosystem. For instance, when the objective is to analyse livelihood systems and strategies, as will be done in the next section, a different approach is taken. This involves considering the assets, contexts, activities, and outcomes that define livelihood strategies, as proposed by Ellis (1993).

In recent years, there has been a growing use of the terms ecological functions and ecosystem services to describe the processes occurring within (agro-)ecosystems and their value or utility to humans (Millennium Ecosystem Assessment, 2005). These functions and services encompass the interactions within agroecosystems (referred to as local ecosystem services) and the outputs that extend to regional and global scales (referred to as regional to global ecosystem services), as depicted by the block arrows in Fig. 3.1. Ecosystem services are typically classified into four categories: provisioning services, supporting services, regulating services, and cultural services, as outlined in Table 3.1. Prior to the publication of the extensive 800-page report by the Millennium Ecosystem Assessment, the literature from the 1990s often referred to agroecosystem *functions*, distinguishing between environmental service functions and productive functions.

One important implication of the increased popularity of the terminology and concepts related to ecosystem services is the recognition of the "landscape" as a fundamental unit of analysis. Many of the services listed in Table 3.1, such as watershed regulation, carbon sequestration, or biological pest control through natural enemies, depend on processes and functions that operate at scales larger than individual fields or farms. To properly assess these services, it becomes necessary to consider the spatial distribution of biotic and abiotic elements in the landscape and their interactions over time. This understanding has led to the adoption of "landscape approaches" to agriculture by the international research community. As an object of

Table 3.1 Ecosystem services as categorised in the Millennium Ecosystems Assessment framework

Category of service	Example
Provisioning	Food and biochemicals (incl. medicinal ingredients) Fresh water and air quality Wood and fibre Fuel and energy
Supporting	Soil formation Nutrient cycling Primary production Pollination
Regulating	Climate regulation (e.g., carbon sequestration) Flood regulation Pest and disease regulation Water purification
Cultural	Aesthetic Spiritual Educational Recreational

study, the landscape is more compatible with the notion of agroecosystems than with that of *farming systems*. An agroecosystem can be studied as a landscape, as long as it is clearly defined in terms of its spatial extent and scale (cf. Chap. 2).

Ecosystem services, at their core, are essentially ecological functions, and as per systems theory, we understand that functions are intricately tied to the underlying structures of the ecosystem. In simpler terms, ecological functions become services when viewed from a human perspective, with a utilitarian approach. This is why we treat ecosystem services as part of the sociocultural dimension of the agroecosystem. The remainder of this chapter will be dedicated to categorizing and describing the diverse array of structures and functions, encompassing both biophysical and socio-economic aspects that characterize agroecosystems.

3.2 Categories of Diversity of Agroecosystems

There are three major categories of diversity to be considered when characterising the structure and functions of agroecosystems. They comprise:

- The diversity of or between agroecosystems, as defined by agroecological conditions, but also largely by demography, culture, markets and infrastructure (e.g., remoteness, connectivity);
- Diversity within the agroecosystem in terms of land use, agricultural management, knowledge, and livelihoods (household or socio-economic diversity);
- Biophysical diversity within the agroecosystem, both in terms of habitat structure (abiotic) and biological communities (biotic);

This chapter provides an overview of different ways of characterising and describing these three major categories of diversity, using systems analysis concepts, and providing examples from around the world. Chap. 4 will provide methods to address the second category above, livelihoods and household diversity through farm typologies.

3.2.1 Farming System Classifications

Farming systems (i.e., as in 'ways of farming') and hence agroecosystems are diverse and their major forms around the world have been categorised on the basis of:

1. Their scale, *management system or production orientation*, such as smallholder versus large-scale systems, family versus entrepreneurial systems, subsistence versus (semi-) commercial systems or market-oriented, export systems, food- versus cash crops-oriented, etc.;

2. Their dominant *production activities* and means, or enterprise patterns and resource base (e.g., FAO), such as rainfed versus irrigated, cereal-based, tuber/root crop-based, crop-livestock or mixed-systems, perennial versus annual crop-based systems, or more specifically coffee-, rice- or maize-based systems, etc.;
3. The agricultural *methods, principles and/or philosophy* that guide management decisions, such as organic agriculture, permaculture, biodynamic farming, agro-ecological systems, versus industrial or 'conventional' farming, low versus high input systems, etc.;
4. Their *intensity of capital and labour* investments per unit of land area, such as intensive, semi-intensive or extensive systems, or the level of technology available, mechanisation, etc.;
5. Their *geographical location or context*, such as dryland, coastal or highland systems, urban agriculture, shifting or slash and burn systems, seasonal, transhumant, etc.

There are certainly other categorisations that escape these five groups. Rather than proposing yet a new way of categorising the diversity of farming systems around the world, as has been done by several authors before (e.g. Ruthenberg 1980; Dixon et al. 2001), this book focuses on ways (concepts, methods) to categorise and describe diversity within the ecological and sociocultural dimensions of the agroecosystem schematised in Fig. 3.1.

3.2.2 Agrobiodiversity

The biotic diversity in the agroecosystem is often referred to as agrobiodiversity. Definitions of this term differ between those that refer only to the genetic diversity of domesticated plants and animals, to those that include all living forms in an agroecosystem. These include, accordin to the UN Food and Agriculture Organization (FAO): "(i) harvested crop varieties, livestock breeds, fish species and non-domesticated (wild) resources within field, forest, rangeland including tree products, wild animals hunted for food and in aquatic ecosystems (e.g. wild fish); (ii) non-harvested species in production ecosystems that support food provision, including soil micro-biota, pollinators and other insects such as bees, butterflies, earthworms, greenflies; and (iii) non-harvested species in the wider environment that support food production ecosystems (agricultural, pastoral, forest and aquatic ecosystems)". The term "Biodiversity for Food and Agriculture" has been introduced more recently by FAO and encompasses the previous concept of agrobiodiversity, which includes genetic resources of plants, animals, and aquatic organisms, along with associated biodiversity and wild foods. Associated biodiversity refers to the biodiversity that coexists within agroecosystems, and thus it corresponds to the categories ii and iii of the FAO agrobiodiversity definition. The existence of multiple definitions, including those not mentioned here, can lead to confusion regarding these terms.

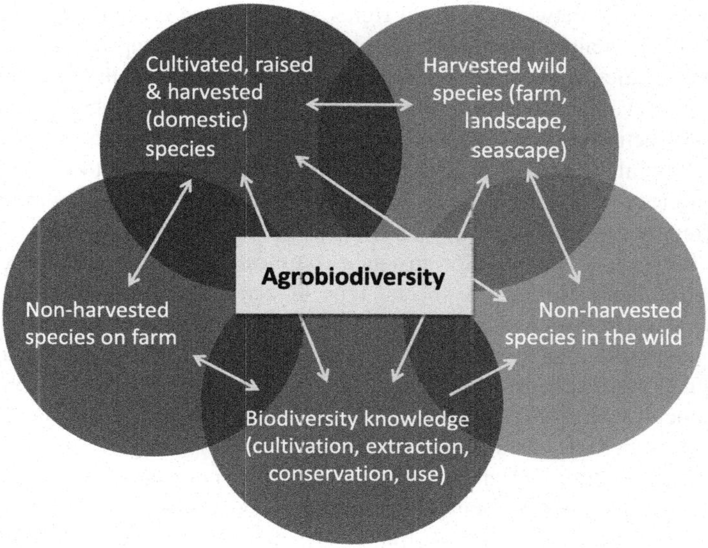

Fig. 3.2 Schematic representation of the definition of agrobiodiversity proposed in this book, consisting of (1) cultivated (or raised) and harvested species, (2) harvested wild species either on farm or in the broader landscape, (3) non-harvested species present on farm (e.g. hedgerows, soil biota, etc.), (4) non-harvested species present in the broader landscape, and (5) the knowledge necessary to cultivate, collect, extract, hunt, fish, domesticate, reproduce, conserve, process and use the other four categories of biodiversity. These categories are interdependent

Here, the concept of agrobiodiversity is expanded to include not only the genetic resources found in the biodiversity of animals, plants, and microorganisms but also the agroecological knowledge and practices associated with them, as well as the diversity of energy, biomass, and nutrient flows they generate or convey (Fig. 3.2). These five interdependent categories depicted in Fig. 3.2 mutually influence each other, meaning that the absence or reduction of one category can affect the others. The practice that plays a significant role in the existence and preservation of biological diversity in agroecosystems is the diversification of plant and animal species in space and time through various techniques such as polycultures, crop rotation, agroforestry, intercrops, and crop-livestock interactions. These techniques will be further defined and explored later on.

3.3 Livelihood Systems and Strategies

Livelihoods have been defined as the "*capabilities, assets (stores, resources, claims and access) and activities required for a means of living*" (Chambers 1994). This definition places emphasis on the means by which a 'living' can be sustained rather than on the economic results or incomes generated by the activities engaged in. In social sciences, the concept of livelihood is enriched by considering also cultural knowledge, social networks or legal rights as part of the capabilities of individual households to obtain a living. A rural livelihood system comprises all the components (capabilities, assets) and processes (activities) and their interrelations necessary to sustain a living in a given context, brought together by a rural family. Livelihood activities may take place on- or off-farm, and be agricultural or non-agricultural. The rural livelihood system may be equivalent to the farm system when families derive their income from agricultural activities that take place on their own farms. A rural livelihood system that is based on agriculture may still differ from a farm system when, for instance, families derive income from working off-farm for other farmers. A special type of activity that has been regarded in different ways by different authors is livestock keeping, sometimes considered to be a non-agricultural activity. In different parts of the world, livestock farmers may own a small piece of land where they live with their family and keep corrals and a few paddocks, while their livestock graze in communal land or concessions that may be several hundred hectares large. This led some authors to regard livestock keeping as an off-farm activity. Here, livestock farming is considered as an agricultural activity and hence a component of a rural livelihood system.

The sustainable rural livelihood framework of Scoones (1998), which has largely influenced our common thinking around livelihood strategies and resources over the last years, recognises four major livelihood strategies: agricultural intensification, extensification, livelihood diversification and migration. These strategies are the result of the ability of households, in a certain context and institutional frame, to access a range of livelihood resources classified as natural, economic, human and social capitals. The different livelihood strategies result in diverse livelihood outcomes, which are analysed through indicators. Livelihood outcomes and their trade-offs may cover aspects of livelihoods themselves, such as poverty reduction, job creation or general well-being, or aspects of sustainability pertaining to the household (adaptation, vulnerability, resilience) or the natural resource base. Here, resources and activities are characterised through structural and management variables and the corresponding livelihood outcomes through performance indicators, such as income, productivity, nutritional level, well-being, etc. (cf. definitions of variable types in Chaps. 4 and 8).

Almost simultaneously, Ellis (1993) proposed a different framework to analyse livelihoods, from a perspective of micro-policy analysis, and defined livelihoods as "*the assets (natural, physical, human, financial and social capital), the activities, and the access to these (mediated by institutions and social relations) that together determine the living gained by the individual or household*". Livelihood strategies are thus defined in terms of assets, access, context, activities and outcomes, as illustrated in Fig. 3.3. Livelihoods should not be seen as static entities but rather as evolving responses to changes in contexts, assets and institutions. For convenience, classical economics and policy analysis consider individuals as concerned with a single, main occupation or profession, and analyse their efficiency, their performance or their responsiveness to new technologies or policies on the basis of this assumption. Rural households in developing countries, however, must often engage in a diversity of activities to sustain their livelihoods. Such diversity, termed

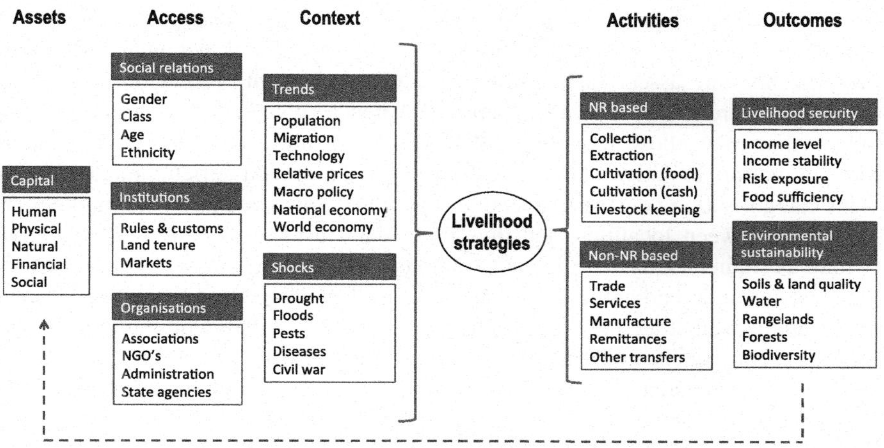

Fig. 3.3 A schematic representation of the concept of livelihood strategies based on assets, access and context, resulting in activities and outcomes, adapted from Ellis (2000). The dotted line at the bottom indicates a feedback between outcomes and capitals

Table 3.2 Livelihood strategies of the poor as defined by Dorward (2009)

Agroclimatic potential	Resource endowment	Local market opportunities	
		Low/stagnant	High/dynamic
Low	Poorer	Hang in (very difficult – subsistence livestock?)	Hang in (more local non-farm based)
	Wealthier	Step out (migrate)	Step out (local non-farm)
High	Poorer	Hang in (farm/ subsistence?)	Hang in (farm and non-farm)
	Wealthier	Step out (migrate) Step up (exports)	Step out (local non-farm) Step up (local markets)

portfolio, includes not only agricultural but also non-agricultural – often off-farm – activities. This diversity poses problems for economic and social analysis, and hence for policy development. Policies for diversification in rural areas typically aim to change the main occupation of rural people through non-farm employment or activities rather than promoting households to engage in diverse agricultural and non-agricultural activities. It is apparently easier – though not necessarily more effective – to design policies and evaluate their impact when the cross-sectorial nature of rural livelihood activities is ignored.

Explicitly or not, many of the existing categorisations of livelihood strategies consider the main determinants of such strategies as being external to the farm or household system. For example, Dorward (2009) differentiates households pursuing basically three main livelihood strategies (Table 3.2): 'Hang in', which takes place in situations of poor natural resource potential and market opportunities, and where households engage in activities to maintain their current livelihood (subsistence farming); (2) 'Step up', in situations of high agricultural potential and where investments in assets are made to expand current production activities (semi-commercial farming); (3) 'Step out', when activities are used to accumulate assets that may allow moving into different activities, not necessarily farming (i.e. migration to cities and/or local engagement in non-agricultural activities). This categorisation considers diversity in agroecological potential and market connectivity between localities, and thus more than just the diversity of livelihoods within a certain locality. The diversity of households is only considered through their level of resource endowment. Yet, households following hanging-in, stepping-up and stepping-out strategies can be recognised within a certain community, and not only as a consequence of resource endowment, as we will see later in Chap. 4.

Closely associated with the idea of diversity of livelihood strategies and their influence on farming practices is the concept of farming styles (van der Ploeg 1994). Farming styles are different expressions of the organised flow of activities that farming entails, when farming is understood as a social co-ordination of tasks oriented towards a goal (cf. Chap. 2: Agoecosystems as cybernetic systems). This concept was born from the observation of the dynamics of farm enterprises in western Europe, notably dairy farms in The Netherlands, and later on corroborated and/or adapted to different situations worldwide. The way in which the agricultural context shapes opportunities for farmers and the diversity of their livelihood strategies led Landais (1998) to propose typologies of agricultural contexts. In the following chapter, examples will be presented to illustrate different methods to delineate farm and rural household typologies, as well as some quick methods to characterise the agricultural context. Understanding the context is crucial to being able to interpret farm or rural household diversity, as these are always context specific and non-generalizable.

3.4 The Sub-systems of the Agroecosystem

The description of the household, cropping and livestock sub-systems is done following the simple and generic representation of the five elements of a system, namely boundaries, components, interrelations, inputs and outputs. Such representations could be graphical, numerical, mathematical or simply descriptive narratives. From now on, these sub-systems of the agroecosystem are treated as separate systems.

3.4.1 The Household System

Structurally speaking, a rural household system, or the household sub-system of a farm system, can be defined through four major types of components. The family,

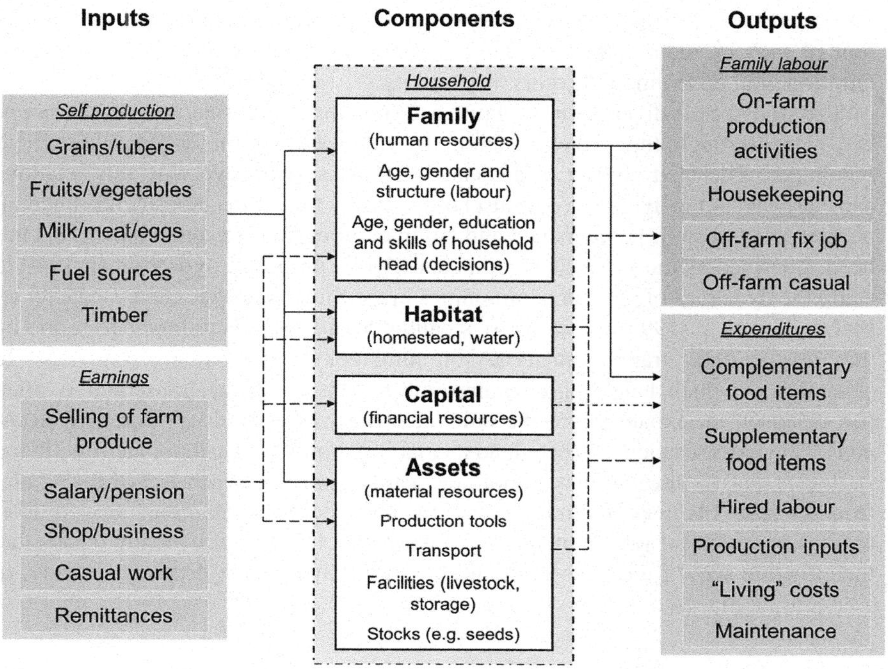

Fig. 3.4 Structural characterisation of a smallholder rural household system, with 'inputs' on the left, 'outputs' on the right and four system components at the centre

which includes their labour force, skills and capacities, then the habitat, the financial capital, and the assets they possess (Fig. 3.4). In some cases, financial capital and assets are considered as a single component. A functional characterisation of the rural household may include also the decision-making and/or agricultural management sub-system. In the example depicted in Fig. 3.4, decision-making is a function of the system that is influenced by family structure (age, gender and composition), livelihood objectives, preferences and production orientation, and conditioned by their capital, assets and context.

The flow of capital, information, energy, and matter to and from the household cannot always be easily categorized as "inputs" and "outputs" in a straightforward manner. In Fig. 3.4, the food produced on the farm and the earnings generated by the family through employment, business, or selling farm produce are depicted as "inputs" to the household system or income. Conversely, the labour provided by the family, both for on- and off-farm activities, as well as the expenditures and production and maintenance costs, are considered as "outputs" from the household system. However, this representation is arbitrary to some extent since some of these inputs and outputs are actually internal flows within the farm system. It's important to acknowledge that this definition of the household system serves as a guideline for quantitative household characterizations, assuming that these flows can be more or less quantified. Collecting data based on this framework would enable the calculation of various metrics such as capital, energy and material flows, efficiencies, nutrition, and economic indicators.

The study of the diversity in structure and functions of household systems can be best done through their categorisation, by delineating farm or rural household typologies. The next chapter is dedicated to that, so here, we will only examine some aspects of the dynamics of household systems. Over time, households undergo phases of expansion, maintenance and contraction of their resource base. This has been described as the Farm Developmental Cycle, proposed by Forbes in 1949 to study the dynamics of peasant communities in the Siberian steppes and used later on by Crowley et al. (1996) to categorise smallholder households in East Africa and by the FAO in their training materials on rural community studies (Fig. 3.5). In situations in which land is not limited, the expansion of the household is often paralleled by an expansion of the cultivated area, as a result of increased needs and/or increased capacity. Where land availability limits expansion, other livelihood strategies tend to develop that do not necessarily imply an expansion of the natural resource base. Likewise, the migration of the household head to urban areas or other regions in search for jobs, either seasonally, temporarily or permanently means that the effective size of the household and the farm labour force decrease. Finally, in

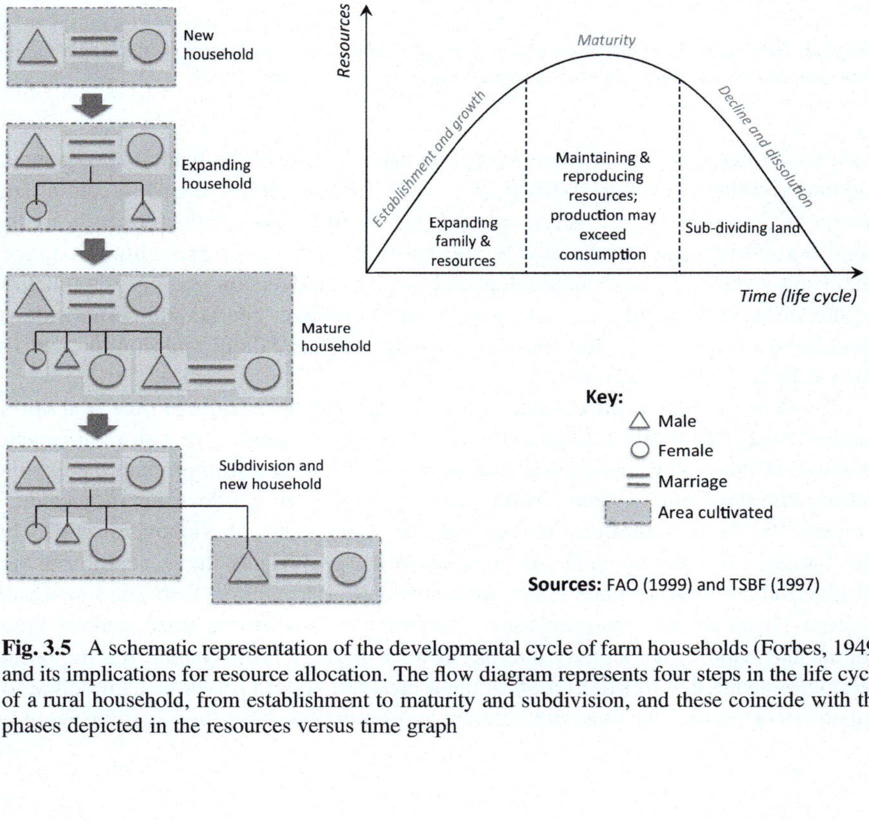

Fig. 3.5 A schematic representation of the developmental cycle of farm households (Forbes, 1949) and its implications for resource allocation. The flow diagram represents four steps in the life cycle of a rural household, from establishment to maturity and subdivision, and these coincide with the phases depicted in the resources versus time graph

situations in which wars, violence or the incidence of diseases such as HIV-AIDS claim the life of rural people, the basic structure of the family may be disrupted, as when e.g. young children are raised by their grandparents or other relatives, as commonly seen in parts of sub-Saharan Africa.

The structure of a rural household differs across cultures and regions, and sometimes within the same region when members of different ethnic groups inhabit the same or contiguous territories. Table 3.3 shows demographic, socioeconomic and structural indicators of households in two nearby districts of western Kenya, Vihiga and Migori, dominated by *Luhya* and *Luo* people, respectively. Although both districts may differ in their biophysical assets – yet both corresponding to humid tropical highlands with old lateritic soils – and main agricultural activities, the differences in terms of population density and farm sizes, women headed households or rural self-employment are considerably wide. Such differences are to a large extent cultural, as polygamous families are more common among the *Luo*. Notoriously, Migori district with greater infrastructure in terms of electricity and sanitation, with more land available per household and with a greater density of medical doctors does not exhibit significantly better socioeconomic or health indices. This example serves also to illustrate the importance of categorising agricultural contexts prior to any household categorisation or farm typology.

Continuing with the example from western Kenya, Table 3.4 shows the frequency of different crops grown on farms surveyed in these two districts. About 20 different crop species were found on these farms on average. It is evident that maize, and virtually also beans intercropped with maize, are grown in all the farms surveyed. Bananas are grown in more than 90% of the farms in Vihiga and only in 40% of them in Migori, where all households grow groundnuts (*Arachis hypogaea*) in contrast to none in Vihiga. About half of the farms surveyed in Vihiga grow yams (*Colocasia esculenta*) and cassava (*Manihot esculenta*) while none or a few of them in Migori, where more sorghum is grown in comparison with Vihiga. The differences in the biophysical background (soils, rainfall, altitude) at both localities are not wide enough to explain such differences in cropping patterns, except perhaps for groundnuts, which may be more easily grown on lighter soils. The differences observed in cropping patterns are thus largely cultural, reflecting different food habits and to some extent different market opportunities. Additionally, the work of Figueroa et al. (2008) in the region reveals that farmers in Migori tend to see themselves as 'livestock' farmers, as the *Luo* people traditionally were, despite the fact that the number of livestock heads they own does not differ much from those owned by Vihiga farmers, who see themselves as mainly agriculturalists.

3.4.2 The Cropping System

The yearly sequence of crops and fallows in a given area describes a cropping *pattern* (Plate 3.1). The cropping *system* can be then described as the integration of a cropping pattern with other farm resources and activities, at a given level of

Table 3.3 Indicators of demography, household size, structure and poverty levels in two districts of western Kenya, as published by the Kenya government and compiled by B. Figueroa et al. (2008)

Indicator	Vihiga district	Migori district
Demographic information		
Population (habitants in 2002)	550,800	565,080
District area (km^2)	563	2030
Population growth rate	3.3%	3.1%
Population density (inhabitants/ km^2)	1015	280
Sex ratio (female:male)	100:87	100:92
Male Life expectancy (years)	55	48
Female Life expectancy (years)	58	52
Infant mortality rate	100/1000	137/1000
Under 5 mortality rate	120/1000	213/1000
Socioeconomic indicators		
Number of households	105,701	113,930
Average household size (members)	4.5	4.5
Number of female headed households	37,691	5697
Average household income (KSh/month)	2000	–
Sectorial contribution to household income Agriculture Rural self-employment Wage employment Urban self-employment Other	60% 30% – 5% 5%	75% 5% 10% 5% 5%
Population below poverty line in rural areas	58%	48%
Contribution to national poverty	3.0%	2.2%
Main ethnic group	Luhya	Luo
Other ethnic groups	Luo, Gusii	Basuba, Luhya
Agriculture		
Average farm sizes (small-large, ha)	0.4–6	1.2–8
Main cash crops produced	Tea, coffee, sugarcane	Sugarcane, tobacco
Main food crops produced	Maize, beans, millet, potatoes	Maize, sorghum, cassava
Health indicators		
Doctor-to-patient ratio	1:50000	1:5280
Distance to nearest health facility (km)	5	8
Most prevalent diseases	Malaria, respiratory diseases, pneumonia	Malaria, respiratory diseases, diarrhea

(continued)

Table 3.3 (continued)

Indicator	Vihiga district	Migori district
Services ad facilities		
Households with electricity	16.7%	46.3%
Households with water infrastructures	29.6%	51.9%
Road type close to your household	"dirt road" 57.4%	"dirt road" 59.3%

KSh Kenya Shillings (75 KSh = 1 us$ at the time of survey)

Table 3.4 Agrobiodiversity of cultivated crop species grown on a sample of farms in Vihiga and Migori districts of western Kenya (n = 546); frequency (%) indicates the presence of these crops in the farms surveyed

Crop types	Frequency (%)	
	Vihiga district	Migori district
Cereals		
Maize	100	100
Sorghum	4	62
Finger Millet	8	15
Tubers		
Yams	48	0
Sweet potatoes	72	85
Cassava	52	8
Pulses		
Beans	96	92
Soya beans	48	46
Groundnuts	0	100
Cowpeas	84	38
Green grams	0	8
Fruits		
Bananas	92	38
Avocados	60	46
Mangos	48	46
Passion fruit	4	8
Paw paw	36	46
Guavas	72	62
Loquats	20	0
Pineapple	0	8
Lemon	0	8
Orange	0	15
Vegetables		
Tomatoes	8	8
Onions	8	0
Aubergines	0	15
Species richness	**18**	**21**

From: Figueroa et al. (2008)

Plate 3.1 Agricultural landscapes exhibiting different cropping patterns. **A**. Small parcels of winter cereals, potato and *quinoa* in the Ecuadorian highlands (4000 m.a.s.l.). **B**. Ancient terraces of cereal based cropping systems (wheat, *tef*) in the Ethiopian highlands. **C**. Grasslands and fodder maize in a mixed farm in Wisconsin, US. **D**. Trees and annual crops (coffee, moringa, banana, sweet potatoes, vegetables) combined on a slope in an agroecological farm in Minas Gerais, Brazil. (Photos: A and D, P. Tittonell; B and C, S. de Hek)

technology available. In the francophone literature, however, the *système de culture* is defined as the set of procedures used by humans to exploit the land in order to obtain plant products of interest. The French term is used in a more restricted fashion, as a theoretical representation of a certain way of farming, often referring to identical sets of techniques applied to a group of fields (cf. Dore et al. 2008). The principle of homogeneity suggests that within a specific cropping system, we can expect to find the same species, associations and/or sequences of crops, as well as similar cultural practices. It means that a cropping system can be shared among multiple farms or, within a single farm, different cropping systems can be identified. The francophone definition of a cropping system emphasizes it as a method or set of practices rather than a concrete system with fixed boundaries. This is why the term is also used to describe methods such as "no-till cropping systems," which refer to a specific way of cropping or a practice rather than a distinct system with clear boundaries. In quantitative agroecosystems analysis, it is more practical to define tangible objects rather than abstract concepts. Therefore, in this context, a cropping system is defined as the plant production sub-system of an agroecosystem. The scale

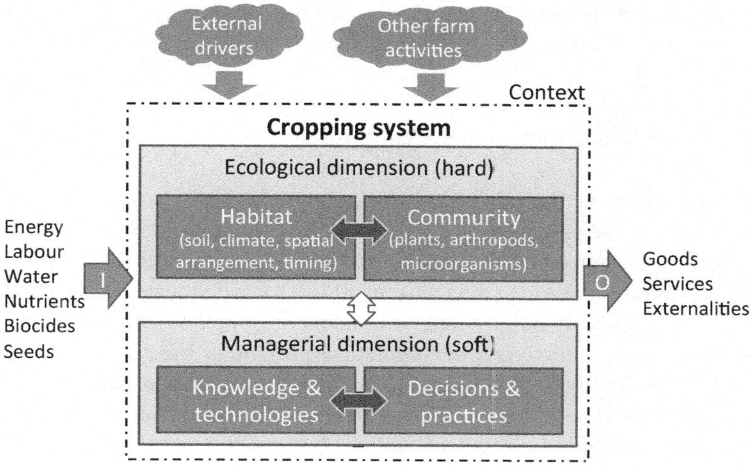

Fig. 3.6 Scheme representing the dimensions and components of cropping systems. Block arrows indicate interactions between dimensions (white) and components (dark grey). Incoming and outgoing block arrows represent inputs (I) and outputs (O) to and from the cropping system, and the influence of biophysical and socioeconomic drivers that may be external to the farm system or the result of other farm activities

at which the agroecosystem is defined determines how the cropping system is represented, whether it be a field, a group of fields, a farm, or a larger territory like a watershed. This allows for a more concrete and specific characterization of the cropping system for analytical purposes.

The cropping system is characterised by a sequence of crops organised in space and time (cropping pattern) and managed through a combination of techniques that aim to match the requirements of the various crops to the ecological niches of the agricultural landscape. It comprises concrete (hard) and abstract (soft) dimensions, as depicted in Fig. 3.6. As in the case of the agroecosystem, the ecological dimension of the cropping system consists of abiotic and biotic components. The abiotic component is defined by the habitat structure in which crops and other plants, arthropods and microorganisms live, grow and develop. Of particular interest is the habitat structure that crops experience, as this is shaped not only by the biophysical background (radiation, rainfall, altitude, soil type, etc.) but also by farmers' management decisions. A crop's habitat is modified by decisions such as planting date, plant density, planting depth, multispecies associations such as intercropping or agroforestry, by the presence of weeds, by nutrient management, or by infrastructures such as irrigation or greenhouse cultivation. The habitat determines the potential growth, production level and productivity of a crop. The biotic component of a cropping system consists of crop plants as well as non-crop plants, such as weeds growing within the crop area or other associated and service plants growing around the crop, such as windbreak trees of fencing shrubs or grasses. The biotic component comprises also animals (most of which correspond to arthropods) and all other organisms within the realm of the crop, particularly in the soil root zone or rhizosphere (bacteria, fungi, nematodes, etc.).

3.4.2.1 Physical Components of Cropping Systems

As the components of the ecological dimension of cropping systems are concrete elements, they can be associated with a scale of analysis, delimited in space and time. The state of the system can be measured and its dynamics monitored in time. In other words, this definition of a cropping system allows us to delineate it in space and time (boundaries), to identify sub-system components (biotic and abiotic) and to differentiate crops from other types of vegetation present in the landscape. Cropping systems can then be modelled, when state variables and rates of change are defined to represent them. The cropping system so defined is not a set of practices, but a physical entity, a limited portion of reality (system) that contains interrelated elements (crop and weed plants, soil, arthropods, micro-organisms), that receives inputs (energy, labour, water, nutrients, biocides), that experiences short- and long-term changes (dynamics) and that produces outputs (goods, services and externalities). For surveying purposes on the ground, the cropping system can be simplified

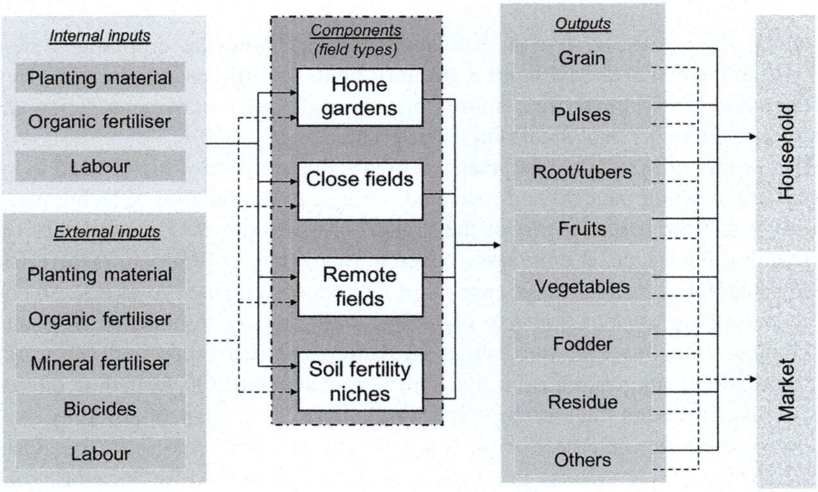

Fig. 3.7 A generic representation of a crop production sub-system of a smallholder farm system in East Africa, where the sub-system components are field types categorised according to their distance to the homesteads. Soil fertility niches refer to special parts of the farm where a natural or human-induced accumulation of soil fertility allows growing the most demanding and/or cash crops (e.g. old homestead o corral sites, valley bottom lands close to water sources for irrigation, etc.). Inputs to the crop sub-system may be from external sources (dotted line) or internal (solid line) to the farm system. Note that in this case all crops grown may be either consumed by the household (solid line) or sold on the market (dotted line). In some cases, there may be crops that exclusively grown for either the market or home consumption

according to the objective of the study to consist simply of physical entities that can be easily identified in the field. This is illustrated for a smallholder farm in Fig. 3.7 where the components of the cropping system are field types defined by their distance to the homesteads. This simplification is possible under the assumption that cropping patterns are comparable within these field type categories. Such typology of fields within farms may facilitate achieving several objectives including quick field surveys, targeting innovations and management practices, communicating with farmers about them, or even modelling complex farming systems (cf. Tittonell et al. 2009).

3.4.2.2 Cropping Patterns

Human agency through design and management governs the nature of the interactions between cropping system components through establishing cropping patterns. These cropping patterns correspond to the spatio-temporal arrangement of crops within a certain area (*NB*: note that this is how other authors define the cropping or farming system). The classical work of Ruthenberg (1980) classifies 'cultivation systems' according to:

(i) The type of rotation: natural fallow, ley system, field system, system with perennial crops;
(ii) The intensity of rotation or R: the proportion of the area under cultivation in relation to the total area available for arable farming[2];
(iii) Water supply: irrigated versus rainfed farming;
(iv) Cropping patterns and animal activities; and
(v) The degree of market integration: subsistence, partly commercialized farming (if >50% of the value of produce is for home consumption), and commercialized farming (if >50% of produce is for sale).

Such classification is still valuable for descriptive or mapping purposes at large scale but falls short from providing the sort of information that is needed in systems analysis. Particularly in agroecology, interest is placed mostly in cropping patterns, as these determine the underlying ecological structure (habitat + community) where ecological functions and services operate. The most common cropping patterns include:

Monoculture
It consists of growing the same crop on the same field year after year, and it is often practiced through sole cropping (one crop variety at the time, grown in pure stands). Monocultures may consist of annual or perennial crops. Conspicuous examples of monocultures are sugar cane or oil palm plantations, or the monoculture of cereals

[2] An R = 100 corresponds to a system where one crop is grown per year; 10 corresponds to a shifting system with two years of cropping and 18 of fallow; R = 300 for a system where three crops are grown per year.

and oilseed crops in vast areas (e.g., soybean in the South American Chaco), or the monoculture of single species of grasses for animal feeding (e.g. pure stands of *Brachiaria* following deforestation in the Amazon, or pure stands of *Lolium perenne* for dairy farming in western Europe). A common practice in intensive monocultures is ratooning, or keeping the stubbles of a crop to re-sprout after harvesting to produce a new crop, as in the case of sugar cane plantations. Monocultures are often also associated with uniform management practices over vast areas of land.

Rotation

Often referred to as crop rotation, although the alternation between cropping and fallow periods (ley farming) or between periods of cropping and pastures fall also within this category. Crop rotation consists of growing a series of different crops in the same field in consecutive seasons. In the francophone literature, a distinction is made between crop rotation and cropping sequence, assuming that a rotation should have start and an end that is planned *a priori*, whereas a crop sequence may be the consequence of season-to-season decisions. The consequence of a rotational plan is a diversification of crops, grassland and/or fallow areas in space. When the areas of the various fields in a rotation are identical or comparable, the ratio of the area of a certain crop relative to the total area cultivated indicates also the length of the rotation; i.e., a rotation of five years implies an area ratio of (1/5 =) 0.2 for any given crop. Other important parameters are the return time of a crop in a rotation and its frequency. Systematic crop rotations are more applicable on relatively homogeneous areas of land. In heterogeneous agricultural landscapes, farmers may design particular crop sequences (including monocultures) tailored to the particularities of each soil-landscape unit.

Multiple Cropping

It consists of growing two or more crops consecutively on the same field in the same year. This pattern is commonly practiced with short cycle crops in horticulture, in irrigated tropical areas that allow growing two to three crops per year (e.g., rice-wheat in the Indo-Gangetic plains of India, or rice-rice-rice in the Bangladesh delta region), in areas with bi-modal rainfall distribution that allow two cropping seasons per year, such as the highlands of East Africa or the humid savannahs of coastal West Africa, or through partial relay cropping in temperate regions (e.g., wheat-soybeans in the Pampas of Argentina, in which soybeans are sown before or during wheat harvesting), or soybean-late maize (*safrinha*) in the Brazilian Cerrado. A special case of multiple cropping is the use of catch crops, cover crops or green manures, which are sown before or after harvesting the main crop to either capture excess nutrients or fix atmospheric nitrogen, and incorporated as organic matter amendments or lain as mulch before planting the next main crop.

Polyculture

It consists of growing two or more crops simultaneously on the same field over a period of time. Polycultures may consist of associations of (i) two or more annual plants at random (mixed cropping) or following spatial patterns such as intercropping (distinct or same rows), companion planting or strip cropping (bands of different crop types alternating in space), (ii) of annual and perennial plants (alley cropping of annual plants in between rows of tree crops), or (iii) of several perennial (and annual) plants of different height (e.g., multi-story or multi-tier cropping of coconut, pepper, pineapple and grass). Sometimes annual crops are grown between rows of perennial crops for a number of years until the establishment of the productive phase of the latter. Agroforestry, permaculture or silvo-pastoral systems make use of one or more of these forms of polycultures. The intensive home gardens that can be seen in many tropical agroecosystems are good examples of centuries-old polycultures. A special case of polyculture is the association of different varieties of the same crop species with complementary characteristics, as is often done to reduce pest and disease or climatic risks in organic farming. According to some definitions, polycultures can also include animal production activities, or crop livestock integration, such as the ancient rice-duck-fish systems of Southeast Asia.

Although we categorise them as discrete entities, the various cropping patterns described above can co-exist on a single farm or landscape, or be used in combination (e.g., crop rotation and the use of green manures are also important components of a polyculture). In tropical hilly landscapes with a fluvial regime often the low lands or valley bottomlands are used to cultivate rice or vegetables. Such crops are restricted to those landscape units and therefore they do not rotate with upland crops grown on the slopes or the plateau soils. This means that while some rotations are systematic others are designed according to the biophysical properties of a field (or habitat in Fig. 3.6). For instance, national cotton boards and farmer cooperatives in parts of West Africa recommend farmers to systematically rotate cotton with cereals such as sorghum and maize every year. On fragile soils, however, farmers often allocate cotton to fields recently cleared from fallow vegetation, then grow cereals the following year to make use of the residual fertility, and a less demanding crop, often a legume such as cowpea, during the third season. After that the field is left as fallow for two or three years. There are a number of advantages to crop rotations and polycultures as compared to mono- and sole cropping, which relate to the maintenance of higher levels of diversity in space and time with the possibility to recycle carbon and nutrients, to outcompete weeds, or with the interruption of pest and disease cycles.

Smallholder families who farm small areas of land are often too restricted in their choices to be able to rotate crops or to keep the land as fallow. For example, when families farm less than one acre of land and produce enough food for just three months of self-sufficiency, as observed in densely populated areas of sub-Saharan Africa (Tittonell 2003). In such cases, the association of several crops in space is most common (Plate 3.2A). At the opposite end of the spectrum, in large scale commercial farming, mechanisation of large areas of cropland is easier when dealing with a single crop than with a mixture, preventing polycultures (Tittonell et al. 2020). Some large-scale farmers are incipiently adopting strip cropping in bands that are wide enough to be able to sow and harvest the crop with machinery (Plate 3.2B), particularly those producing for the organic market (Ditzler et al. 2021). Crop rotation and cover crops are of key importance in managing soil fertility. Cover

Plate 3.2 Examples of polycultures. **A**. Association of yams, cassava and rice in Benin; **B**. Strip cropping of annual crops (potato, cereals, grass and cabbage) in organic farming in The Netherlands. **C**. Dryland agroforestry in coastal Peru, showing locally adapted maize, fruit trees and *pisco* grapes in the background. **D**. Intensive rice, fish and duck systems in Indonesia. (Photos: P. Tittonell)

crops keep the soil covered during the time of the year in which the main crops are not grown, and contribute biomass to mulch the soil surface during the initial phases of the main crop, before canopy closure (Alliaume et al. 2014). Cover crops are often grown as intercrops with cereals and left in the field to expand in area once the cereal grain is harvested. This is the case of legume cover crops such as *Stylosanthes guyanensis*, *Mucuna pruriens*, *Dolichos lablab* or *Cajanus cajan*. The biomass produced by these cover crops is then used to mulch the soil during the subsequent crop.

3.4.3 The Livestock System

The livestock system is the animal rearing sub-system of an agroecosystem. It is restricted to domesticated animals kept in an agricultural setting to produce food, labour, fibre and manure, or to serve as savings or currency in certain cultures. The major animal groups or species considered are cattle, buffalo, sheep, goat, pig, poultry and fish. In the same way as was done for the cropping system, the definition of the livestock system followed here does not reflect simply a method or a set of practices, but rather a concrete system consisting of components (animals, pastures, facilities, etc.), inputs (feeds, labour, medicines, water, etc.) and outputs (meat, milk, eggs, wool, etc.). The nature of the interactions between components depends on both biophysical processes and human agency. A distinction must be made then between the livestock system of a farm or a rural village, defined by the various biotic and abiotic components organised in space and time, and the livestock *production* system, which describes a set of practices. Livestock production systems have been categorised by the FAO (http://www.fao.org/docrep/v8180t/v8180t0y.htm) using three major criteria: (i) whether livestock is the only activity of a farm or production system, (ii) whether livestock production is based on grasslands or landless, and in the latter whether the livestock species are ruminants or monogastric, and (iii) whether the mixed farming systems that combine livestock and crop production are irrigated or not. Further the classification considers the agro-climatic conditions under which production takes place, from arid to humid environments (Fig. 3.8).

Solely livestock production systems are those where more than 90% of the dry matter fed to animals originates from rangelands, pastures, annual forages or purchased feeds, and in which more than 90% of the total value of production comes from livestock activities. Mixed-farming systems are those where more than 10% of the dry matter fed to animals comes from crop by-products or stubble, or more than

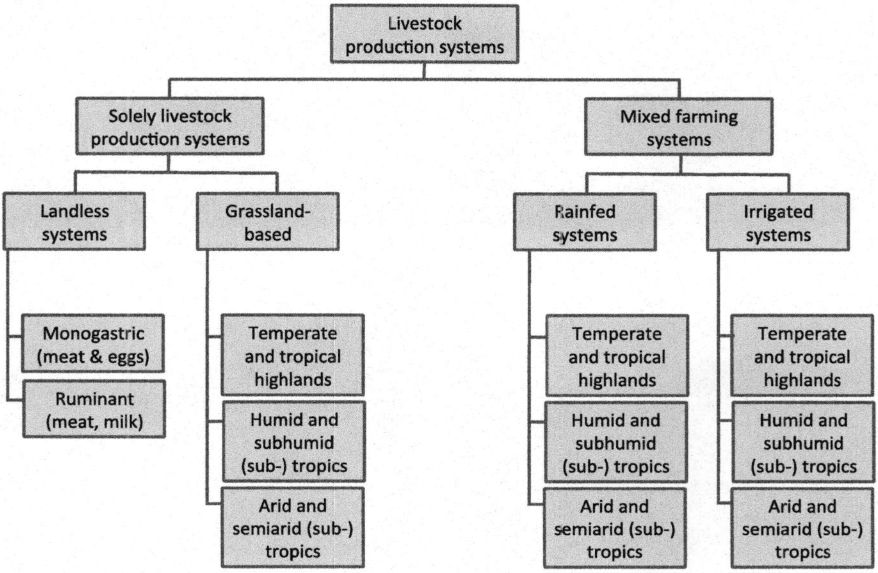

Fig. 3.8 Classification tree of livestock production systems proposed by FAO that has become something of a classic (Steinfeld et al. 2006). Here, 'milk' is incorporated in the landless production systems with ruminants, due to the proliferation of landless cow, sheep and goat dairies in the last decades. In the original version only meat (beef) was mentioned

10% of the total value of production originates from non-livestock farming activities (Fig. 3.8). According to this classification, solely livestock production systems can be landless or grassland-based, depending on whether less or more than 10% of the dry matter fed to livestock is produced on-farm, respectively, with annual average stocking rates above or below 10 livestock units per hectare of agricultural land. Landless systems can host monogastric animals such as pigs and chicken, or ruminants, especially cattle. In general, in landless systems decisions on feeding are decoupled from decisions on feed production, and nutrients are not recycled through the utilization of manure to fertilise fields (open systems). However, large-scale landless systems such as cattle feedlots are often integrated with parallel farming enterprises where cereals and legumes are produced to be processed into feed concentrates. Smallholder farmers, on the other hand, often opt for systems known as zero-grazing or cut-and-carry, in which animals are not allowed to graze and are fed roughages and concentrates produced on-farm or purchased, in different proportions (Plate 3.3A). Whether such systems can be classified as landless or mixed systems is hard to say. Grassland-based livestock systems and mixed farming systems (both rain-fed and irrigated) are further categorised following an agroecological zoning: (i) temperate zones and tropical highlands; (ii) humid and sub-humid tropics and sub-topics; (iii) arid and semi-arid tropics and sub-tropics. This

Plate 3.3 Diversity of cattle systems. **A**. A 'zero grazing' unit for dairy cows in western Kenya. **B**. Transhumant herds in a water point in southern Niger (showing fields of millet in the background). **C**. Extensive beef cattle ranching on native grasslands in Uruguay. **D**. Intensive pure graze beef cattle systems in the Basque Country, Spain. Photos: P. Tittonell

categorisation is, however, non exhaustive. Extensive sheep farming in the cold and arid, high latitude steppes of Patagonia (cf. Tittonell et al. 2021), for example, does not seem to fit in any of these categories.

This classification has nowadays a number of limitations that have been recognised by the authors earlier, not least because of the fact that when a classification exercise has been accomplished the systems classified may have already evolved and changed, rendering the classification gradually obsolete. Later classifications by FAO and ILRI teams improved on this initial one (Thornton et al. 2003; Kruska 2006; Robinson et al. 2011) by considering the livestock system within broader farming and livelihood systems, human population densities, using satellite imagery and the length of the growing period as determined by rainfall distribution. This classification, with its latest version published as Robinson et al. (2011), incorporates also elements of grassland utilization systems as described in the classical book of Ruthenberg (1980):

(a) Total nomadism: no permanent place of residence, no regular soil cultivation;
(b) Semi-nomadism: a permanent place of residence exists, supplementary soil cultivation is practised, but for long periods of time animal owners travel to distant grazing areas;
(c) Transhumance: a permanent place of residence exists, their herds are sent to distant grazing areas, usually on seasonal cycles;
(d) Partial nomadism: characterized by farmers who live continuously in permanent settlements and have herds at their disposal that graze in the vicinity;
(e) Stationary animal husbandry: animals remain on the holding or in the village throughout the year.

Global changes in demography, climate or markets and trade rules are reshaping livestock production systems worldwide. Category d, partial nomadism, known also as communal grazing, is becoming common as the village grazing lands degrade or shrink in their size relative to the size of the village herd. In areas of the Sahel, for example, nomadic and transhumant practices associated with certain ethnic groups are now gradually disappearing due to a combination of factors. On the one hand, increasing rural population densities mean that the areas available for grazing animals decrease, more intensive grazing affects the productivity of the remaining grasslands, watering points are more intensively used, and conflicts between livestock owners and non-livestock owners due to crop damage or for the utilization of crop harvest residues to feed livestock are increasingly common (cf. Tittonell et al. 2015). On the other hand, such traditionally pastoral ethnic groups are becoming increasingly sedentary, co-inhabiting the landscape with agriculturalists and growing crops by themselves (Diarisso et al. 2015). This process of sedentarisation is often voluntary and largely fuelled by people's demand for public services, such as schooling for children, electricity, telecommunications or access to public health. In other cases, sedentarisation has been enforced by law directly, or indirectly through regulations concerning land tenure and access to natural resources, country border regulations, etc. (Dossouhoui et al. 2023). Often sedentary pastoralists follow dual livelihood strategies, cultivating the soil where they settle and practicing semi-nomadic grazing, taking along the animals of other farmers in the village (Plate 3.3B).

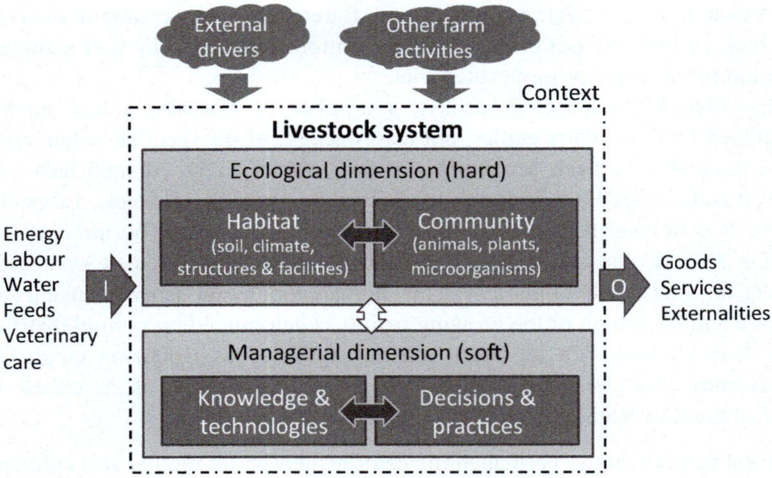

Fig. 3.9 Scheme representing the dimensions and components of livestock systems. Block arrows indicate interactions between dimensions (white) and components (dark grey). Incoming and outgoing block arrows represent inputs (I) and outputs (O) to and from the livestock system, and the influence of biophysical and socioeconomic drivers that may be external to the farm system or the result of other farm activities in the case of mixed farming systems

Zooming in to the livestock system, or more precisely in systems analysis terms, the livestock production sub-system of an agroecosystem, we distinguish the same dimensions and type of components as described earlier for cropping systems (Fig. 3.9). The habitat or abiotic component of the ecological dimension of a livestock system is represented by the climate and by the type of soil where livestock production takes place (except in landless systems), plus the human-made structures and facilities meant to alter (modify, regulate, improve) the natural habitat, such as parcelling, corralling, stalling, heating, cooling, automated feeding and watering systems, etc.. The biotic component of the livestock system includes the animal species kept for production purposes (cattle, sheep, goat, pigs, poultry, rabbits, etc.) plus those species that may affect livestock production positively (e.g., dung beetles, grassland birds) or negatively (e.g. ticks, moles, lice). The biotic components include also plants used as feeds, which can range from native through improved grasslands to cultivated fodder plants, and also plants used for medicinal purposes or shade.

It may be argued though that the structure of the vegetation where livestock graze (grasslands, shrublands, savannahs, forests) in spite of being a biotic component is also part of the habitat, as it regulates for example the microclimate where livestock graze, with direct effects on animal welfare (Plate 3.3C, D). The biotic component includes also the microorganisms and other life forms such as nematodes that may affect livestock production positively (biological N fixation) or negatively (parasites, diseases). A component that some authors consider as part of the livestock system while others not is the manure and waste management sub-system. This becomes increasingly important as we move from extensive to intensive, and landless livestock systems. Particularly in the latter, manure may be a sub-product that can be sold on the market, representing therefore a 'good'. In mixed systems, manure, feed refusals, bedding material and wastewater may become inputs to the crop production sub-system.

3.4.3.1 Grasslands

Grasslands are an important component of livestock systems. In general, grasslands are defined as an area covered with grass and grass-like vegetation, which according to its ecological structure may be referred to as steppe, prairie, pampa, meadow, veld, campo, savannah, etc. According to the international terminology for grazing land and grazing animals of Allen et al. (2011), grazing lands are any vegetated areas where animals graze or has the potential to be grazed by animals (domestic and wild), and may include pasture lands, croplands, forestlands, grasslands and rangelands. Each of these terms covers also a diversity of situations that can be categorised. It is not the intention in this chapter to categorise systems, but for the sake of consistency, the terminology adopted here is the one depicted in Fig. 3.10.

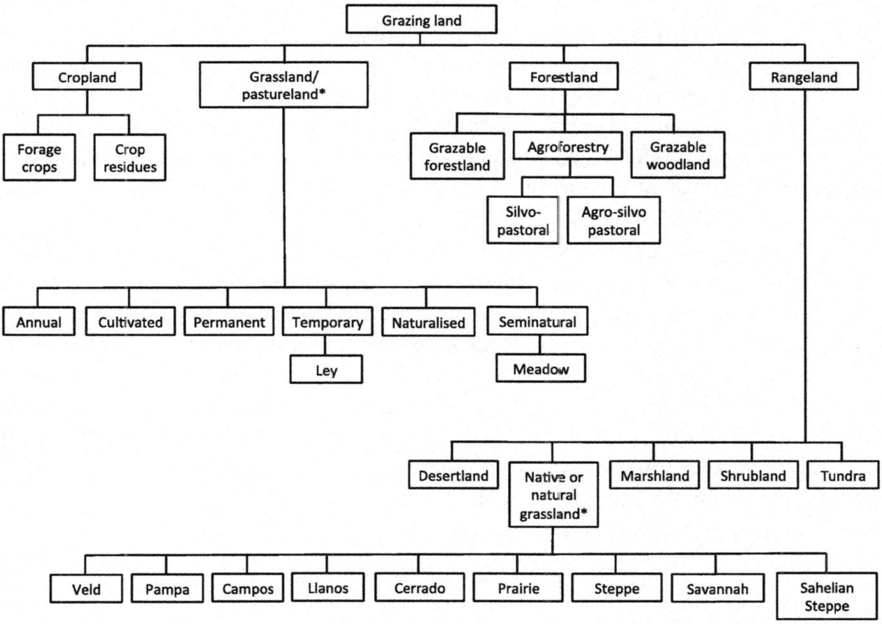

Fig. 3.10 Classification of grazing lands following Allen et al. (2011) and as proposed by FAO. (*) The term grassland is synonymous with pastureland when referring to an imposed grazing-land ecosystem. The vegetation of grassland in this context is broadly interpreted to include grasses, legumes and other forbs. When these grasslands are incorporated in a crop rotation they are termed leys. Meadows are often associated with the conservation of hay or silage, although they may exist as a result of discontinuous features of hydrology, landscape position or soil characteristics that differ from the surrounding landscape and vegetation (e.g., mountain meadow, wet meadow, etc.). Note that grassland is also a term used for native grasslands, which are a special type of rangeland

Confusion often emerges with the use of the term grassland. Grassland is used in some cases as synonymous with pastureland, referring to grazing ecosystems dominated by introduced herbaceous species, from fields cultivated annually to permanent and/or semi-natural swords. Grassland is also used to refer to a special type of rangeland consisting of naturally occurring or native herbaceous vegetation dominated by grasses, ranging from prairies to steppes, and including also savannahs with scattered woody species (*NB*: such confusion does not exist for example in Spanish, since two different terms are used: *Pasturas* to refer to pastureland-grasslands and *Pastizales* to refer to native grasslands).

Rangelands are areas in which the indigenous or spontaneous vegetation is predominantly grasses, grass-like plants, forbs or shrubs that are grazed or have the potential to be grazed, used as a natural ecosystem for the production of grazing livestock and wildlife. In some cases, grasslands can be the result of deforestation or removal of the original vegetation, and they may comprise native, naturalised and/or introduced species, so that the classification becomes rather blurry. Leys are a special type of pastureland that alternates with periods of crop production following a rotational scheme. Ley periods are very important to restore soil quality and break the cycle of crop pests and diseases (as much as the cropping period may contribute to 'clean' the pastureland from livestock parasites or unpalatable and toxic plants) in intensive mixed farming systems. The proportion of years of cropping and years of ley in the rotation depends on the type of soil and climatic conditions as well as the type of crop/livestock activities considered. Intensive dairy systems typically include crop-ley rotations with the objective of producing fodder crops such as silage maize or grains used to produce feed concentrates during the cropping period, making use of carbon and nutrient transfers that are possible through the application of animal manure to the crop/ley fields.

Such crop livestock interactions can be also classified according to the level of farming intensity, largely determined by agro-climatic conditions, and considering the degree to which crops products or sub-products participate in the main livestock diet (Fig. 3.11). Crop livestock interactions and trade-offs on the multiple uses of crop residues (e.g. for livestock feeding vs. soil mulching, composting, use as fuel, etc.) are important in smallholder farming systems (cf. Andrieu et al. 2015; Alomia-Hinojosa et al. 2020) specially in situations of intermediate agroclimatic potential or farming intensity. In the extensive pastoral systems livestock feed mostly on rangelands (Plate 3.3C), whereas in the most intensive systems fodder and pastures are cultivated exclusively for livestock feeding.

3.4.3.2 Livestock System Components

For surveying purposes on the ground, the livestock system can be simplified according to the objective of the study to consist simply of 'herds' or animal categories that can be easily identified in the field, as illustrated for a smallholder farm in Fig. 3.12. This type of simplification allows building field surveying protocols that generate valuable data for economic analysis, energy flows, nutrient and feed balances, environmental externalities, risks, etc. For more sophisticated analysis

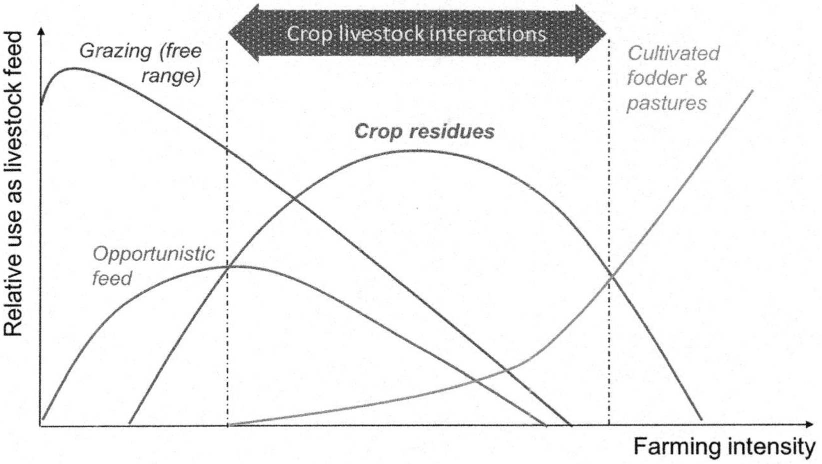

Fig. 3.11 Hypothesised relationship between farming intensity or agroclimatic potential and the importance of crop residue biomass as a source of feed for livestock. The relative importance of various biomass sources as feed is represented considering only roughages, not concentrates, along a gradient of farm intensity defined by the number of cropping seasons per year, population density, market connectivity, as developed by the former System-wide Livestock Program-SLP by B. Gérard. (Adapted from Paul et al. 2013)

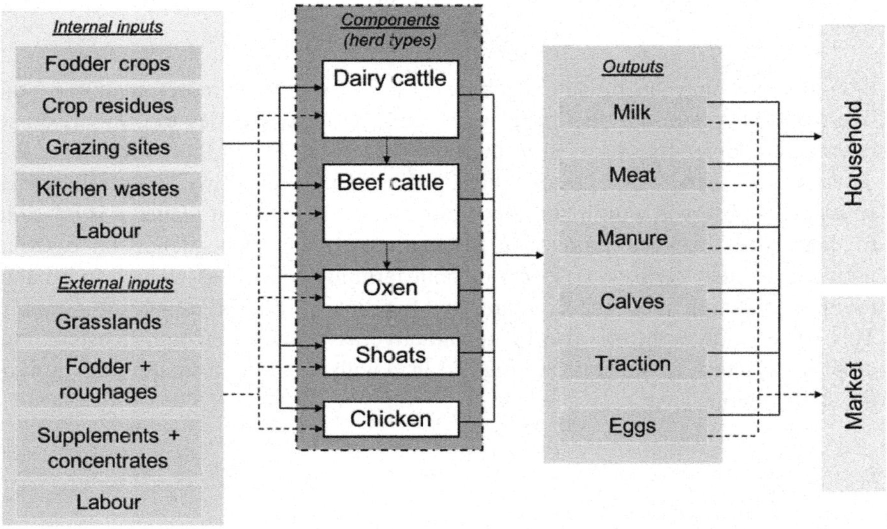

Fig. 3.12 A generic representation of an animal production sub-system of a smallholder farm system in East Africa, where the sub-system components are herd types or animal categories. Shoats refer to sheep and goats which are often managed as a single lot. Inputs to the livestock sub-system may be from external sources (dotted line) or internal (solid line) to the farm system. Note that in this case all livestock products and services may be either consumed by the household (solid line) or sold on the market (dotted line). There may be animals that exclusively kept for either the market or home consumption

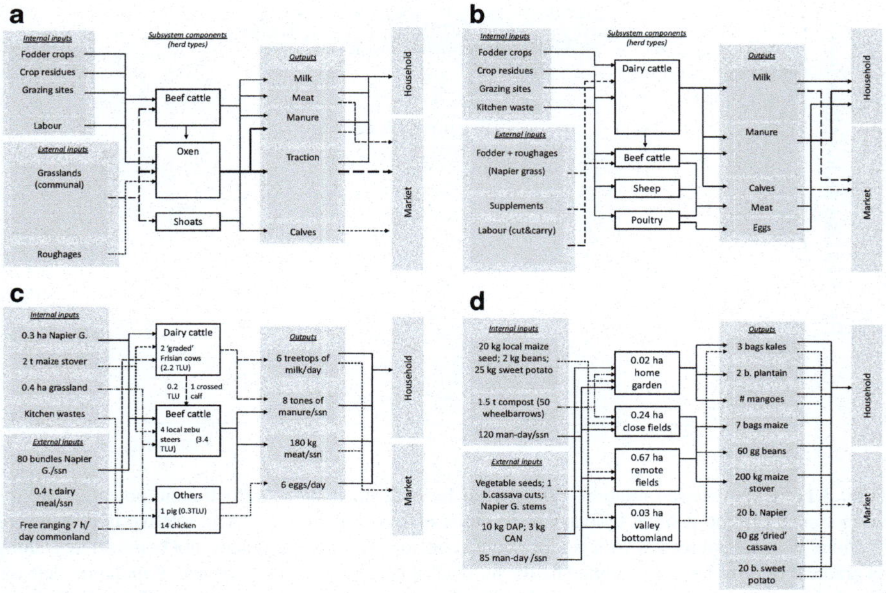

Fig. 3.13 Generic representation of subsystems of the farm system. (**a**) and (**b**) depict two distinct livestock production subsystems co-existing in a village of western Kenya. (**c**) and (**d**) are quantitative applications of this graphical method to describe respectively the livestock and crop sub-systems of a smallholder farm in the same village

this approach may be limited. Such generic and graphical representation of the livestock subsystem may be useful when delineating typologies to categorise and describe the diversity observed in the field (cf. Chap. 4) and to communicate this diversity to others. For example, Figs. 3.13a, b represent two distinct livestock subsystems observed within the same village in Kenya. At first glance it is possible to distinguish between production orientations, assets, dependence on external inputs and other livelihood aspects of both farming systems as well as to be able to infer something about the structures and functions of both livestock sub-systems. This representation allows also for quantitative assessments of both livestock and crop sub-systems, as illustrated in Fig. 3.13c, d with data from a smallholder farm of the same region.

3.4.4 The Farm System

The farm system may also be considered as a sub-system of the agroecosystem, which integrates the household, cropping and livestock subsystems, and any other farm activity (e.g. bee keeping, aquaculture, forest, service provision). According to the scale at which it is defined, an agroecosystem (e.g. a landscape, a territory, a region) may comprise several farm systems within its boundaries, or just one, when the agroecosystem is defined at farm scale. The farm system should not be confused with the farming system, a term that should be preferably avoided for the reasons already explained in Chap. 2. The farm system represents an individual farm, with a single identifiable node for decision-making that may or may not delegate to others the decisions and responsibilities concerning individual farm sub-systems. In a small-scale family farm, the decision node is generally the household. A distinction is often made between family farms and enterprise farms, with all kinds of possible intermediate situations in between. This distinction varies with the definition adopted in every country. In the USA, for example, 97% of all the farms are considered to be family farms, irrespective of their scale, when a member of the family that owns the farm is in charge of its operation. In Argentina, the same criterion is used but in combination with the size of the farm and with the main source of labour force. Family farms are considered to be those that are run by a family member and that are small in scale, and where the family uses mostly its own labour, sometimes complemented with temporal or permanently hired labour. The farm area below which farms are considered small-scale varies across the various regions of Argentina depending on agro-environmental, climatic, social and economic conditions (up to 500 ha in the humid Northeast; up to 1000 ha in the Pampas, Cuyo and Gran Chaco regions; up to 2500 ha in the Northwest and up to 5000 ha in Patagonia). Figure 3.14 illustrates a mixed family farm system in central Argentina (252 ha), located in a transition zone between the Pampas and the Chaco regions.

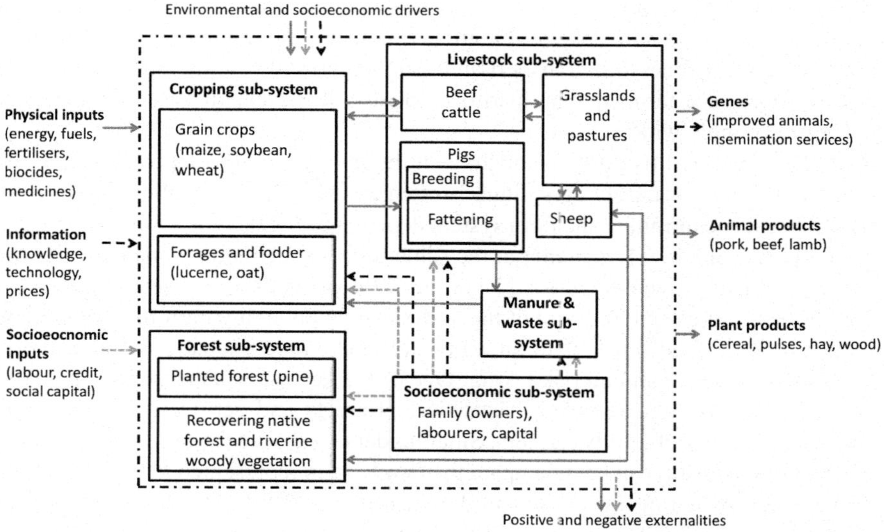

Fig. 3.14 Simplified graphic representation of a crop-livestock family farm (252 ha) in the foothills of Sierras de Comechingones, Cordoba, Argentina. The arrows indicate the main flows of physical resources (energy, biomass, water, nutrients) in solid dark grey lines, information in dotted black lines, and labour and other socioeconomic resources in dotter light grey lines. Note that the term Socioeconomic sub-system is used instead of Household sub-system, as this is more appropriate in this context. Positive externalities include also ecosystem services, while negative externalities include air and water pollution

3.4.5 *Entrepreneurial vs. Family Farm Systems*

In enterprise farming, the central decision node may be the person or family that owns or usufructs the farm land, a technical team, an executive manager and/or even a steering board. Shareholders usually own the large farm operations that are common in several countries of South America, and which may own or rent land in different regions or countries. In large mixed farming operations several sub-systems may be identified and each of them have its own decision node, responsible team or person. For example, large-scale dairy or beef systems may have one or more animal subsystems (e.g. dairy operations that incorporate also pig production), a crop production sub-system, a sub-system that deals with animal feeds or even two of them, one that deals with pastures, forages and silage and another one that takes care of milling and preparation of feed concentrates, pellets, etc. In large commercial farms there may be also a sub-system dedicated to accounting, buying and selling inputs and products, and all sorts of administrative tasks.

Farm systems may also incorporate processing activities such as making cheese, wine, oils, preserves, or have sub-systems dedicated to select and pack fruits and vegetables, flowers, wool and other fibres. In large vegetable greenhouse operations, for example, there is often one team dedicated to operating the ferti-irrigation sub-system, which may become quite a complex operation when having to deal with several crops at different stages of development throughout the year. Moreover, the fields and other facilities or production units that integrate a farm system may be located in a single portion of land or scattered in a landscape or region. Sometimes the farm system is a sub-system of a greater industrial production system, and thus associated with processing, transport and commercial subsystems. Such is often the case of sugar cane farms that are part of a sugar mill, or several vineyards that may belong to a single winery.

In the scientific literature, as much as in the policy and development world, the term family farm or family agriculture is often associated with smallholder, or even with subsistence farming. The farm system depicted in Fig. 3.14 is a family farm, run by more than one family member at the same time, of medium to small size for the region, and that provides jobs for 11 families. It is of small scale when compared to other farms in the same sector, but the scale is large enough within its own context for the owning family to obtain an acceptable income in exchange of their work, knowledge, investment and risk taking. One of the key aspects that render this farm system viable is its diversity of activities, products and services, a typical risk spreading strategy of family farmers. Smallholder or subsistence farms are generally family farms as well. But let's compare the family farm represented in Fig. 3.14, with the one represented in Fig. 3.13c and d, which illustrate the livestock and cropping sub-systems of a family farm in western Kenya. This particular household (schematized in Fig. 3.4) owns 2 crossbred dairy cows, 4 zebu steers, one pig and 14 chicken, farms 0.96 ha of land, barely manages to earn an income to pay for school fees, and produces enough food per year to cover about six months of the family self-consumption needs. Terms such as family farms or family agriculture may be informative about certain aspects of the functioning of the farm system (goals, preferences, production orientation, etc.) but are too imprecise to be used in systems analysis, as illustrated by the disparity between these two cases of family farming that were just described.

3.4.6 Semiotic Models of the Farm System

The quantitative analysis of farm systems often relies on the use of models, which can take various forms, such as graphical conceptual models, graphical quantitative models, or mathematical models based on differential equations, spreadsheets, or optimization routines. When building a model to represent a real system, essentially converting ontological systems into semiotic systems, certain assumptions and decisions must be made to simplify the complexity of the real system. These choices are made with the intention of creating a model that is informative and useful for achieving specific objectives. However, the process of simplification is not solely determined by the objectives of the analysis; it is also influenced by the background of the observer or modeler. The observer's knowledge, professional training, cultural background, and previous experiences all come into play when deciding what aspects of the real system to include or exclude in the model. This inherent subjectivity can affect the way the model represents and interprets the farm system. Figure 3.15 illustrates how different observers might create different models of the same system based on their backgrounds and perspectives.

We visited the mixed farm presented in Fig. 3.14, located at the foot slopes of the Sierra de Comechingones, during a postgraduate course on Ecological Systems Analysis at the National University of Río Cuarto in Cordoba, Argentina. Students were asked to collect information in the field to model the farm system, starting from graphical conceptual models to subsequently develop them into mathematical models of the dynamics of the farm system using differential equations. The objective of the modelling exercise was to be able to quantify relevant physical flows to study environmental impacts and ecosystem services associated with the activities of this farm. Figure 3.15 shows two distinct graphical models built by two different groups of students to represent the same farm system, both with similar objectives. Although the model in Fig. 3.15b is more detailed than the one in Fig. 3.15a, this does not necessarily imply that the former is better than the latter. The graphical model has to be good enough to understand the structure and the functioning of the system with the level of detailed deemed relevant (or practicable) and to communicate about these to others. This just illustrates how different semiotic systems may be delineated to represent the same ontological farm system. Each model may emphasize different elements of the farm, leading to diverse understandings of the system and potentially different outcomes when analysing its behaviour or response to changes. Awareness of this subjectivity is crucial when using models for quantitative analysis. It reminds us that models are tools for understanding and decision-making, but they are not perfect representations of reality. Acknowledging

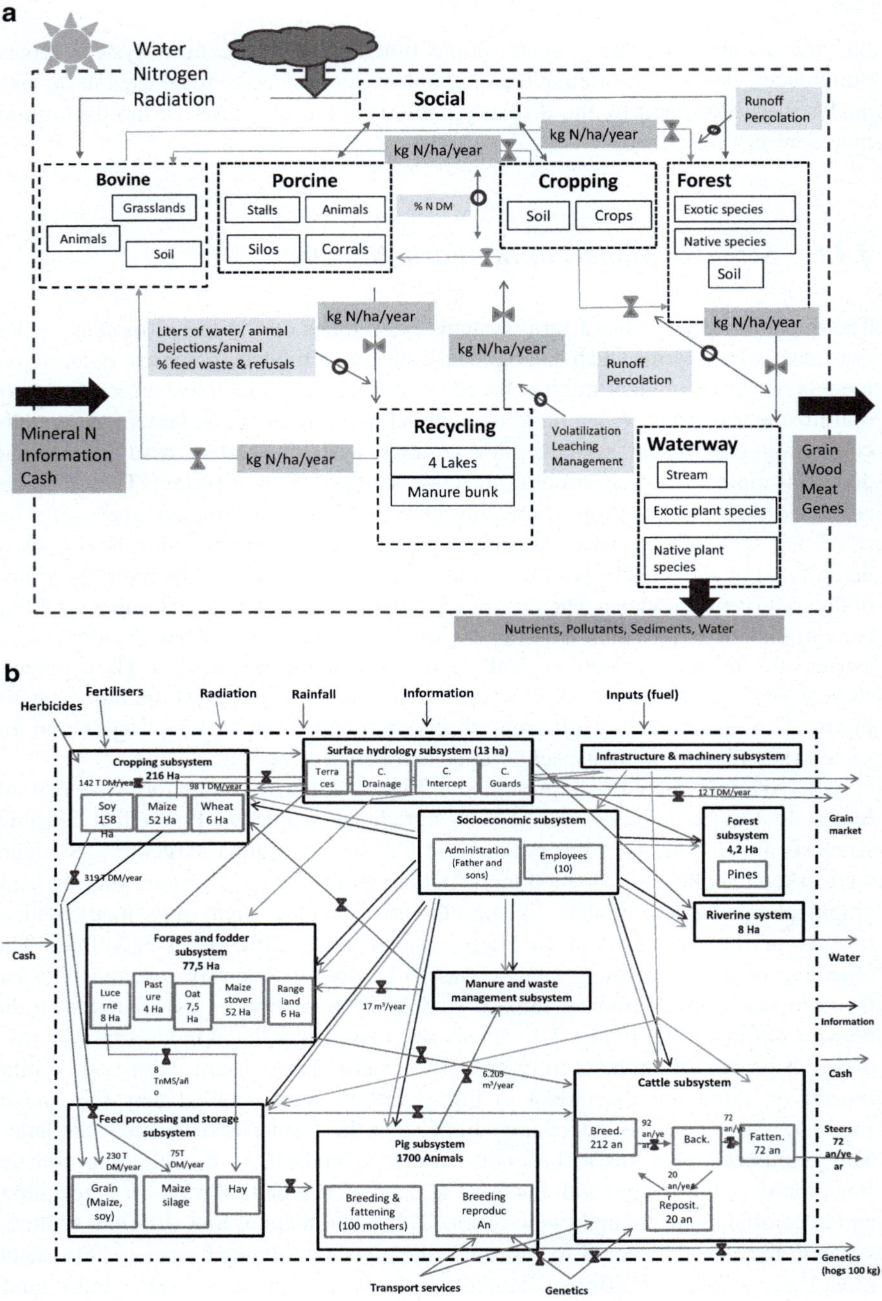

Fig. 3.15 Graphical models of the mixed farm system presented in Fig. 3.14 as delineated by two different groups of postgraduate students during a modelling exercise. Students were given total freedom to choose the symbols and graphical language they preferred so long as they collected enough information and in an orderly way to build a dynamic model of the farm using differential equations

the limitations and potential biases of models can help ensure that their application is appropriate and meaningful for the specific context and objectives at hand.

3.5 Summary and Concluding Remarks

This chapter presented basic systems analysis concepts to characterise and describe the diversity of agroecosystems, such as livelihood systems, the household, the cropping, the livestock and the farm system, and illustrated them with examples from around the world.

- The agroecosystem is the social-ecological entity that can be delineated on the ground at a given scale of analysis and that can be used to 'materialise' more abstract concepts such as farming system, landscape or region.
- Agroecosystems are concrete systems with ecological and socioeconomic dimensions, and where tangible and non-tangible elements of reality can be observed, analysed, measured, quantified, monitored.
- The livelihood system consists of the capacities, assets and activities engaged by a rural household to obtain a living, within a context of social relations and institutions that mediate their access to them.
- Rural household systems consist of a family, their habitat, their capitals and their assets, and they undergo phases of expansion and contraction of their resource base, affecting their livelihood decisions – within a socio-cultural context – throughout their life cycle.
- The cropping system as defined here (i.e., the crop production sub-system of an agroecosystem) differs from the concept of cropping system used often in the literature to refer to cropping methods or cropping patterns. A cropping system has ecological and socioeconomic dimensions. Its ecological dimension (habitat + community) is concrete, measurable, quantifiable and can be delimited in space and time.
- The livestock system as defined here (i.e., the animal production sub-system of an agroecosystem) differs from the concept of livestock *production* system used often in the literature to refer to animal rearing methods or grassland utilisation patterns. A livestock system has ecological and socioeconomic dimensions. Its ecological dimension (habitat + community) is concrete, measurable, quantifiable and can be delimited in space and time.
- The farm system is a sub-system of the agroecosystem that integrates the household, cropping, livestock and other sub-systems corresponding to farm activities (e.g., apiculture, forest, aquaculture, processing, etc.), through a central decision-making unit and one or more management sub-units.
- A distinction is made between family and enterprise farming as regards to the level of engagement of the family members in decision-making. Definitions of family farming vary from country to country and consider criteria such as farm size and main sources of labour.

These definitions are not meant to propose yet another norm or rigid conceptual framework. They are just working definitions that are useful to interpret the concepts that will be discussed in the rest of the book. The term ´system´ is use in many different ways, often to refer to methods, techniques or practices. Here we use the term system to refer to a limited portion of reality with interconnected elements and a purpose. Hence, the definitions proposed here for the various components of the agroecosystem differ from other definitions in the literature in one particular aspect: they follow a systems approach that allow for qualitative as well as quantitative analyses.

References

Allen VG, Batello C, Berretta EJ, Hodgson J, Kothmann M, Li X, McIvor J, Milne J, Morris C, Peeters A, Sanderson M (2011) An international terminology for grazing lands and grazing animals. Grass Forage Sci 66:2–28

Alliaume F, Rossing WAH, Tittonell P, Jorge G, Dogliotti S (2014) Reduced tillage and cover crops improve water capture and reduce erosion of fine textured soils in raised bed tomato systems. Agric Ecosyst Env 183:127–137

Alomia-Hinojosa V, Groot JCJ, Speelman EN, Bettinelli C, McDonald AJ, Alvarez S, Tittonell P (2020) Operationalizing the concept of robustness of nitrogen networks in mixed smallholder systems: a pilot study in the mid-hills and lowlands of Nepal. Ecol Indic 110:105883

Andrieu N, Vayssieres J, Corbeels M, Blanchard M, Vall E, Tittonell P (2015) From farm scale synergies to village scale trade-offs: cereal crop residues use in an agro-pastoral system of the Sudanian zone of Burkina Faso. Agric Syst 134:84–96

Chambers R (1994) Participatory rural appraisal: challenges, potentials, and paradigm. World Dev 22(10):1437–1451

Chang H-J, Evans P (2005) The role of institutions in economic change. In: Reimagining growth: towards a renewal of development theory. Zed Books

Crowley EL, Soule MJ, Carter SE (1996) Off-farm income and farming in western Kenya. Report to USAID/Kenya, Nairobi

Dahl F (1908) Grundsaetze und grundbegriffe der biocoenotischen forshung. Zool Anz T 33:349–353

Diarisso T, Corbeels M, Andrieu N, Djamen P, Tittonell P (2015) Biomass transfers and nutrient budgets of the agro-pastoral systems in a village territory in South-Western Burkina Faso. Nutr Cycl Agroecosyst 101:295–315

Ditzler L, van Apeldoorn DF, Schulte RPO, Tittonell P, Rossing WAH (2021) Redefining the field to mobilize three-dimensional diversity and ecosystem services on the arable farm. Eur J Agron 122:126197. https://doi.org/10.1016/j.eja.2020.126197

Dixon J, Gulliver A, Gibbon D (2001) Farming systems and poverty: livelihoods in a changing world. FAO and World Bank, Rome, Italy, p 410

Dogliotti S, García MC, Peluffo S, Dieste JP, Pedemonte AJ, Bacigalupe GF, Scarlato M, Alliaume F, Alvarez J, Chiappe M, Rossing WAH (2014) Co-innovation of family farm systems: a systems approach to sustainable agriculture. Agric Syst 126:76–86

Dore T, Clermont-Dauphin C, Crozat Y, David C, Jeuffroy MH, Loyce C, Makowski D, Malezieux E, Meynard JM, Valantin-Morison M (2008) Methodological progress in on-farm regional agronomic diagnosis. A Rev Agron Sustain Dev 28:151–161

Dorward A (2009) Integrating contested aspirations, processes and policy: development as hanging in, stepping up and stepping out. Dev Pol Rev 27:131–146

Dossouhoui GIA, Yemadje PL, Diogo RVC, Balarabe O, Tittonell P (2023) "Sedentarisation" of transhumant pastoralists results in privatization of resources and soil fertility decline in West Africa's cotton belt. Front Sustain Food Syst 7:1120315. https://doi.org/10.3389/fsufs.2023.1120315

Ellis F (1993) Peasant economics, Cambridge books. Cambridge University Press, number 9780521457118, December

Figueroa BM, Tittonell P, Ohiokpehai O, Giller KE (2008) The contribution of traditional vegetables to household food security in two communities of Vihiga and Migori Districts, Kenya. International symposium: underutilized plants for food, nutrition, income and sustainable development, 3–7 March, Arusha, Tanzania

Forbes M (1949) Time and social structure. An Ashanti case study. In: Fortes M (ed) Social structure: studies presented to A.R. Radcliffe-Brown. Clarendon, Oxford, pp 54–84

Kruska RL (2006) Seré and Steinfeld version 3. ILRI data set. International Livestock Research Institute (ILRI), Nairobi

Landais E (1998) Modelling farm diversity: new approaches to typology building in France. Agric Syst 58:505–527

Paul BK, Herrero M, Notenbaert M, Groot J, Peters M, Tittonell P (2013) A feeding systems typology for East Africa. In: Tielkes et al. (eds) Agricultural development within the rural-urban continuum. Proceedings of the Tropentag conference, 17–19 September 2013, Stutgart, 587 p

Robinson TP, Thornton PK, Franceschini G, Kruska RL, Chiozza F, Notenbaert A, Cecchi G, Herrero M, Epprecht M, Fritz S, You L, Conchedda G, See L (2011) Global livestock production systems. Rome, Food and Agriculture Organization of the United Nations (FAO) and International Livestock Research Institute (ILRI), 152 pp

Ruthenberg H (1980) Farming systems in the tropics, 3rd edn. Oxford, Clarendon Press

Scoones I (1998) Sustainable rural livelihoods: a framework for analysis, IDS working paper 72. IDS, Brighton

Steinfeld H, Wassenaar T, Jutzi S (2006) Livestock production systems in developing countries: status, drivers, trends. In World Organization for Animal Health (OIE). Scientific and Technical Review 25:505–516

Thornton PK, Kruska RL, Henninger N, Kristjanson PM, Reid RS, Robinson TP (2003) Locating poor livestock keepers at the global level for research and development targeting. Land Use Pol 20:311–322

Tittonell P (2003) Soil fertility gradients in smallholder farms of western Kenya. Their origin, magnitude and importance. Quantitative approaches in systems analysis no. 25, ISBN 90-6754-713-1, Wageningen, The Netherlands, pp. 233

Tittonell P, van Wijk MT, Herrero M, Rufino MC, de Ridder N, Giller KE (2009) Beyond resource constraints – exploring the physical feasibility of options for the intensification of smallholder crop-livestock systems in Vihiga district, Kenya. Agric Syst 101:1–19

Tittonell P, Gérard B, Erenstein O (2015) Tradeoffs around crop residue biomass in smallholder crop-livestock systems - What's next? Agric Syst 134:119–128

Tittonell P, Piñeiro G, Garibaldi LA, Dogliotti S, Olff H, Jobbagy EG (2020) Agroecology in large scale farming: a research agenda. Front Sustain Food Syst 4:584605. https://doi.org/10.3389/fsufs.2020.584605

Tittonell P, Hara SM, Álvarez VE, Aramayo VM, Bruzzone OA, Easdale MH, Enriquez AS, Laborda L, Trinco FD, Villagra SE, El Mujtar V (2021) Ecosystem services and disservices associated with pastoral systems from Patagonia, Argentina – A review. Cah Agric 30:43. https://doi.org/10.1051/cagri/2021029

van der Ploeg JD (1994) Styles of farming: an introductory note on concepts and methodology. In: de Haan HJ, van der Ploeg JD (eds) Endogenous regional development in Europe. Vila Real, Portugal 1991. Luxembourg (1994) 7–31. Ook in: Born from within. Practice and perspectives of endogenous rural development, J.D. van der Ploeg, A. Long (eds). Van Gorcum, Assen, p. 7–31

Part II
Analysis-Oriented Approaches

Chapter 4
Categorising Diversity Through Rural Household Typologies

Abstract Agroecology sees opportunity in diversity. This comprises the diversity of structures and functions that characterizes soil life, for example, but also the diversity of activities, knowledge systems, management practices and land use that coexist within a rural territory. Within any given agricultural context, a region, a watershed, a landscape or a village, individual farms or rural households may differ from one another on the basis of their assets, activities, location, family structure, livelihood strategies, decision-making and preferences, opportunities, production orientation, marketing strategies, integration in the food chain, etc. The resulting diversity of households and their strategies can be categorised by means of farm or rural household typologies. The distinction is made here between farm and rural household typologies because in some cases we can find rural households that are too constrained in terms of land availability and thus cannot be considered to be farms; in the case of commercial farming, on the other hand, farms may be entreprises in which no family lives, and thus they cannot be considered to be 'households'. This Chapter presents concepts, definitions and methods for the structural and functional categorisation of farm and rural household diversity, necessary for better targeting agroecology practices and policies.

4.1 What Use for Typologies?

Typologies are used in a diversity of applications in research and development, in policy making, in risk analysis, in subsidy schemes, for monitoring and evaluation, in econometric studies, etc., and often taken as the basis for scaling-up studies or for implementing scaling-out actions. Because the objectives for which typologies are delineated may differ widely, there is no universal set of criteria to categorise farm diversity. In other words, farm typologies should always respond to the objectives for which the categorisation of farms is needed. Generically, such objectives may include:

P. Tittonell, *A Systems Approach to Agroecology*,
https://doi.org/10.1007/978-3-031-42939-2_4

1. Targeting of policies, development actions or agricultural technologies to different types of farms or rural households, on the basis of their dominant enterprises, land use, resource management, wealth class or objectives;
2. Characterising the socio-economic structure and trends in a farm or rural household population, such as equity and wealth distribution;
3. Analysing the dynamics of household diversity as affected by interventions or external drivers, in terms of households shifting through categories, widening poverty traps, resilience and transformability;
4. Model-aided (ex-ante) analysis of policy or development actions and scaling up of results to the level of territories or regions on the basis of the frequency distribution of different farm types.

Different methods are used to categorise diversity, including statistical clustering, participatory rankings, expert knowledge, etc. A few examples of typologies resulting from these methods will be presented here. Often in the literature we find categorisations of farms or rural households based on their resource endowment. These typologies are usually drawn from the analysis of census or survey data conveying information on farm resources such as land, livestock, production facilities or machinery, household characteristics such as family structure (age, gender, education), land tenure, or farm activities such as land use types, market orientation, etc.

When such data are gathered at one point in time they offer an overview of the diversity of households in the form of a static picture of otherwise dynamic systems. Such diversity is often termed transversal diversity. Longitudinal surveys, on the other hand, are those that are repeated at several points in time, whether at a fix interval or not. They offer insight in the dynamics of the farm or household system. In many cases, governmental agencies collect regular (often annual) data on farms through census or through activity records when subsidy mechanisms are in place, and this information is of great value to study farm dynamics. Yet, from a purely static picture of farm diversity, by asking the right questions during a survey, and by complementing all this with historical information about the major drivers of farm diversity, it is also possible to infer aspects of the dynamics of the farm systems. Examples of this type of studies will be presented here as well.

4.2 Structural and Functional Farm Typologies

Structural farm or household typologies are typically delineated through categorising farms based on their current resource endowment, asset levels, land use type or family composition. Since they reflect diversity in the structure of the system, they are often used to make hypotheses about their functioning (cf. Chap. 2: Structure-function relationships in systems analysis), often resulting in diverse management practices and farm performance. For example, in smallholder farming contexts, it may be assumed that farmers owning livestock are more likely to use

Table 4.1 Number of livestock heads (tropical livestock units) owned by household differing in resource endowment, and average soil organic carbon, available phosphorus and maize yields at farm scale (ranges indicate average minimum and maximum values recorded; i.e. best vs. worst fields)

Household resource endowment	Tropical livestock units per farm	Soil organic carbon (g kg^{-1})	Available phosphorus (mg kg^{-1})	Maize grain yields (t ha^{-1})
Low	0–1.0	0.6–1.5	3–12	0.3–1.4
Medium	1.0–2.5	1.1–1.9	2–30	0.8–1.9
High	2.1–7.7	1.2–2.6	6–65	0.9–3.7

Source: Tittonell (2003)

animal manure to fertilise their soils, and thus levels of soil organic matter may be higher on such farms (Table 4.1). Further, it is often observed that farmers owning livestock are also the wealthier ones and may have access to certain agricultural inputs (e.g. hybrid seeds, mineral fertilisers) or services (hired labour, animal traction) that influence their farm management and productivity. Thus, management decisions concerning soil fertility in this case, and their crop yields and economic revenue, are highly influenced by structural properties of the farm system (resources, assets). Indicators of performance, such as crop and livestock productivity, are in this case a *consequence* not only of environmental factors but also of the choices and possibilities associated with the structure of the farm system. In a study of smallholder rural households in coastal Peru done by Cabrera et al. (2005), the authors used age classes to study how household welfare may be affected by family composition, which is also typically a structural characteristic of a household that may reflect wealth differences.

Functional farm typologies, on the other hand, must reflect not only the diversity in the structure of farm systems but also in their dynamics, in their behaviour and functioning. The definition of functional typologies proposed here, for the particular context of family agriculture, assumes that the diversity of the livelihood strategies followed by the farm household determines how a farm system functions. The livelihood strategy determines the objectives, production orientations and resource and labour allocation priorities within the farm. By so doing, livelihood strategies have a direct impact on the functioning, dynamic properties and long-term trajectory of the farm system.

For example, rural livelihoods may rely largely on earning an income by employing one or more of the family members off-farm. This may impact on labour availability to complete activities that require labour on-farm, such as land preparation for early planting or weeding the crop fields on time, thereby impacting on crop productivity. Households that trade off their own farm's productivity for the acquisition of externally owned income are typically those that have either insufficient land to achieve self subsistence, insufficient labour to tend to their own crops, or insufficient capital to access agricultural inputs; or a combination of all of these. In other cases, depending on skills, on chances or on education level, household members may engage in more remunerative activities off-farm that can generate

Table 4.2 Key elements of a functional typology for household categorisation applied in western Kenya by Tittonell et al. (2005a)

Farm type	Resource endowment[a] and production orientation	Main characteristics[b]
1	Predominantly high to medium resource endowment, mainly self-subsistence oriented	Variable age of the household head, small families, mostly constrained by land availability (lack of family labour compensated by hiring-in). Permanent sources of off-farm income (e.g. salary, pension, etc.)
2	High resource endowment, market-oriented	Older household head, numerous family (starting land subdivision), mostly constrained by labour (hired-in) due to large farm areas; cash crops and other farm produce are the main source of income
3	Medium resource endowment, self subsistence and (low-input) market-oriented	Young to mid-aged household head, young families of variable size in expansion, mostly constrained by capital and sometimes labour, farm produce and marketable surpluses plus complementary non-farm enterprises
4	Predominantly low to medium resource endowment, self-subsistence oriented	Young to mid-aged household head, variable family size, constrained by availability of land and capital, deriving income from non-farm activities (e.g. ox-plough service, handicrafts)
5	Low resource endowment, self-subsistence oriented	Variable age of household head, variable family size, often women-headed farms constrained by land and capital, selling their labour locally for agricultural practices (thus becoming labour-constrained)

[a]Referring to assets representing wealth indicators (i.e. land size, livestock ownership, type of homestead, etc.)

[b]These refer to the family structure (age of the household head) in relation to the position of the household in the 'farm development cycle' (Crowley et al. 1996), to the main constraints to agricultural production faced by the household, and to the main source of income

enough income to buy food and/or to invest in agricultural assets, inputs and labour to increase farm productivity. Table 4.2 presents an example of a functional typology based on these criteria.

Building a functional farm typology often implies starting from a structural typology, either when using some form of statistical clustering or even when typologies are obtained through participatory wealth ranking by members of a rural community. Other definitions of functional farm typologies include categorisations of farm households based purely on their behaviour, irrespective – at least explicitly – of their livelihood strategy or resource endowment. For example, Ruiz Meza (2014) categorised smallholder coffee farmers in Chiapas, Mexico, based on their response to external shocks and stressors imposed by climate change and by the international coffee market. Obviously, the response of different farmers was determined by the structural characteristics of the farm and by the livelihood strategies of the households, but such relationships were not made explicit in the categorisation. Adelhart Torop and Gosselink (2013) categorised organic dairy farmers in The Netherlands in terms of their incentives for farming, their goals, the challenges they foresee, the success factors they identified, and their criteria for excellence. Based on these functional criteria, they depicted five farm type portraits described in Box 4.1. Further, farms or rural households may be categorised functionally based on their propensity to adopt a certain technology, on their attitude towards risks, on their level of agricultural productivity, on their ability to innovate, etc.

Box 4.1 Portraits of Dutch Organic Dairy Farms Through a Functional Typology

In a study of positive deviants among organic dairy farmers in The Netherlands, Adelhart Torop and Gosselink (2013) developed a farmer functional typology building on the concept of farming styles (van der Ploeg 2012), on the basis of criteria such as: incentives for farming, farmers´ goals, the challenges towards the future, factors for success in framing, and criteria for excellence. They came up with the following portraits:

1. *Fine tuners.* These farmers are close to the average in terms of scale and productivity but deviate from the rest through their overall fine-tuned management. High quality milk is produced for exclusive dairy products. The farmers are strong in system thinking and are optimizing at the farm level as well as on regional level in terms of environmental impact. Pure breeds are kept to guarantee consistent milk quality. No antibiotics are used on these farms, which can be seen as proof that the system is in a sophisticated balance. The success of this type of farmer is a combination of perfectionism and well-balanced decisions concerning herd size.
2. *Intuitive dairy farmers.* The system is well established. Animals are adjusted to the local circumstances. Production is high, replacement rate is low, animal health is good. Antibiotics are given when needed. More people are working on the farm and everybody can express their own quality. Extra value is added through the production of dairy products. The intuitive farmer is strongly value-driven and less mechanistic and rule bound. Experience, intuition and enthusiasm are an important part of the success of this type of farmer.
3. *Steady farmer.* The daily routine with a dairy herd has been like this already for a long time. Challenges are for example in nature conservation or diversifying the farm. This farmer deviates because of its soberness and efficient time management. Its success lies in setting priorities and diversification.
4. *Entrepreneurial farmer.* Breeds are robust and resilient and there is a technical view on farming. Attention to the livestock is at herd level.

(continued)

Box 4.1 (continued)

There are more people involved than the family only. The farm is managed as a company with sophisticated, technical, mechanistic management practices. The success of this farm lies in the management capacity to stay tuned with the large herds they own.

5. *Integrated farmer*. Keeps cows as part of the system, milk is a by-product of manure: a deep litter stable provides good quality manure which is used for the arable crops of the farm. Calves and heifers free grazing to reduce the workload and to make them robust and resilient. People are an important part of the system, the farm is often also used with a social purpose. The system needs to be transparent and easy in order to work. There are a lot of people involved in the farm enterprise. This style of farming is rough and resilient. The system view and a strong ideology are characteristic for the success of this style.

Reference: *Adelhart Torop, R., Gosselink, K., 2013. Analysis of positive deviants among organic dairy farmers in The Netherlands. MSc thesis Farming Systems Ecology, Wageningen University, 53 p.*

In all these examples, the causality at play between the structure and the functioning of the farm system is either implicit or hypothesised. Implicit means that it is assumed that households that exhibit different levels of resource endowment or types of livelihood strategies may be more or less likely to adopt a certain technology, more or less risk averse, run a more or less productive farm or be more or less innovative. Hypothesised means that the analysis is run by reasoning in the opposite sense, as done in econometric studies: given the trends in adoption, risk management, productivity or innovation among households observed through extensive surveying, the analysis aims to unravel the characteristic of household types that exhibit high or low levels, frequencies or probabilities of these behaviours.

It is evident that the functioning of a farm cannot be explained solely by its structure, in isolation from contextual drivers such as the climate or the market, and particularly not when the dynamics of the system are concerned. As the relationship between farm resource endowment, external drivers, farm performance (past and present) and livelihood strategies may lead to chicken-and-egg type of enigmas, it is convenient at this point to outline some basic definitions:

External Drivers In the light of the discussion on system boundaries from Chap. 2, external drivers are variables that influence the system and that are not directly influenced by the system. Climatic, demographic and market variables fall in this category. Depending on the scale of analysis, policies, resource management regulations (e.g. water use in irrigated areas), local land use systems (e.g. communal grazing rights) or inherent land productivity may also be considered external drivers. Institutional, historical and organisational drivers are also external drivers when the unit of analysis is the farm or household system.

Structural Variables They define the structure of the system (cf. Chaps. 2 and 3) in terms of system boundaries and components, their size, diversity and organisation. In delineating farm typologies, structural variables of prime interest are those that describe agricultural assets, activities and family structure. Typical structural variables include total land area owned and cultivated, land quality (e.g., upland, hillside, lowland), land use variables such as area of cropland, grassland, forest, orchards or gardens, livestock variables such as numbers, units, type (species or breeds), farm facilities such as stalls or greenhouses, machinery, and household characteristics such as quality of family housing, education, age and gender composition, family labour, etc. In some cases, some structural variables may be expressed in terms of their monetary value.

Functional Variables From the perspective of systems analysis (cf. Chap. 2), functional variables refer to inputs, outputs and interrelations, or to variables that represent dynamic system properties or trajectories. Variables that reflect livelihood strategies such as percentage of labour force or land or capital investments allocated to different on-, non- or off-farm activities are often employed as functional variables. Since typologies are delineated in response to specific objectives, it is hard to generalise or select *a priori* which variables will reflect the functions of interest. For example, if the objective is to categorise households based on their dependence on external nutrients, then the total system throughput of a certain nutrient may be used as a functional variable. If the objective is to categorise farms based on market orientation, then the area share of a certain cash crop (which may be also a structural variable) may be used. Sometimes ratios calculated from structural variables, such as land-to-labour ratios, cattle densities or food self sufficiency may be used as variables that reflect system functioning.

Management Variables They reflect decision-making on farm production (combination of factors) and husbandry within operational, tactic and strategic time horizons for planning. Rates and methods of fertiliser application, sowing dates and densities, cultivar choice, stocking rates, feed ration composition, crop rotation, irrigation frequency, pruning method, soil tillage method, crop residue management, frequency and opportunity of weeding, vaccination, pesticide spraying or harvesting method and timing are all examples of agricultural management variables. These variables are largely determined by structural and functional variables, and strongly influenced by tradition, by technology dissemination and by practical recommendations (context).

Performance or Target Variables They must reflect the result of the interaction between management decisions and external driving variables such as the weather or market prices for inputs and outputs, and are often the variables of greatest interest for the farmer and for most other stakeholders. Performance variables typically reflect factor productivity, such as cereal or milk yields per year (land productivity), benefit-cost ratios (capital productivity), or economic return to labour (labour productivity). They reflect resource use efficiency (e.g. water or nutrient productivity) and externalities, which may be positive (e.g., preservation of habitats and biodiversity, water capture, carbon storage, pest regulation through hosting natural enemies, etc.) or negative (e.g. green house gas emissions, nitrate load in ground water, active principles of pesticides released to the environment, tonnes of soil erosion per ha per year). Often performance or target variables are used as indicators of system sustainability, regarded as diagnosis criteria for dynamic system properties such as stability, resilience or adaptability.

Although we often see farm typologies delineated on the basis of management variables (e.g. use of mineral fertiliser) or performance indicators (average yields), in reality these variables are just surrogates of the actual structural and functional variables that explain how the system behaves. In other words, they are not the cause, but a consequence of how the system works. Using surrogate variables to delineate farm typologies is equivalent to taking a shortcut. In some cases, this procedure may be enough to arrive at a satisfactory working typology. In systems analysis, however, it is always useful to respect the hierarchy in the logical chain of causalities that explain how a system works. A typology is always a hypothesis, a model, a simplification of reality. One way of expressing farm typologies, management and performance as conceptual models could be the following:

Typology $= f$ (structural variables, functional variables, external drivers, ε)
Management $= f$ (Typology, external drivers, ε)
Performance $= f$ (Management, external drivers, ε)

The external drivers that appear in each of the three models may differ in their nature. The term ε represents the extent of unexplained variability. For example, the average productivity of a certain crop is determined by past and current soil management, which is different in different farm types, according to their resource endowment and production orientation.

This way of describing farm or rural household diversity, which is just a proposal, may allow unravelling complicated puzzles that emerge out of interactions that take place across time scales, or the memory of the system. For example, current management is conditioned by past performances through learning, through soil degradation or through current financial health. Also, repeated performance or management outcomes over time may make a farm system shift across categories in the typology. Since the livelihood strategies of rural households determine to a large extent their structure and function, and since the context explains the diversity and magnitude of external drivers, let us examine these two elements first before embarking on methods to delineate typologies.

The way in which the agricultural context shapes opportunities for farmers and the diversity of their strategies led Landais (1998) to propose typologies of agricultural contexts. Later in this Chapter, examples will be presented to illustrate different methods to delineate typologies, without making much explicit reference to the context. Yet, understanding the context is crucial to being able to interpret farm or rural household typologies, as these are always context specific and non-generalizable. The next section summarises some quick methods to characterise the agricultural context.

4.3 Characterising the Local Agricultural Context

The first question we need to address with regards to the agricultural context is how broad we need to go in terms of scales to unravel the relevant aspects of the context of agroecosystems. Obviously, the world economic trends or the global changes in terms of climate, technologies or societal preferences are always elements of the context. But these are common to all agroecosystems in the world. National level policies or regional development plans are also crucial aspects of the context. Demographic trends such as age, ethnicity and gender structure in rural areas, and in the particular community under study, may be of prime importance. For example, Fig. 4.1 illustrates the age composition of rural communities across Zimbabwe. The data from the 2012 census shows that sex ratios (male/female) are equal or just above one among children below 5 years, and they fluctuate around one until the age of 20. Between 20 and 60 years of age, sex ratios are strongly biased towards females, likely due to outmigration of males from rural areas in search for jobs.

Any survey of rural households must be designed and interpreted in the light of this background information. For example, activity calendars, the calculation of labour productivities or the characterisation of decision-making processes on-farm should be designed to capture the actual – and not the assumed – role of women, as there is likely to be a majority of effectively female headed households in most rural

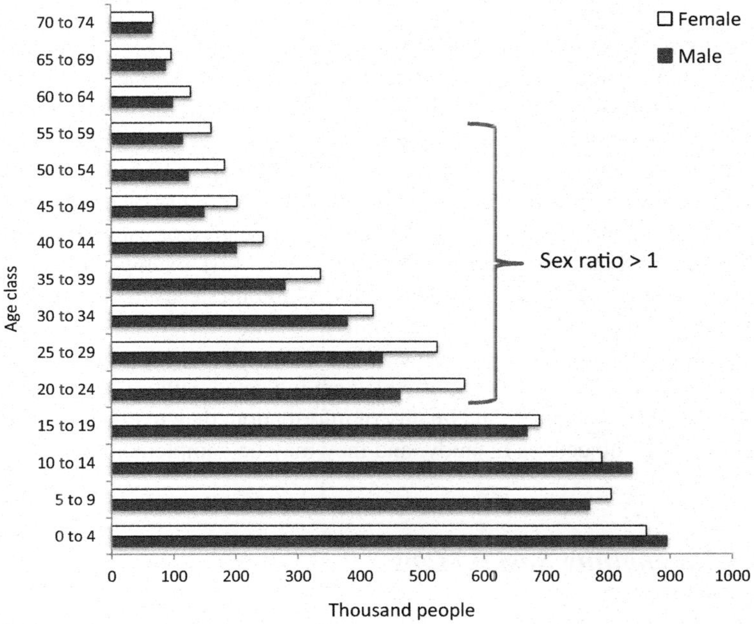

Fig. 4.1 Age classes and sex ratios in rural Zimbabwe, according to the 2012 census

territories. The same analysis would apply to any other demographic characteristic that describes the context.

Having illustrated the effect of the broader context, let us examine now some methods and examples that relate to the 'local' context, understood in different places as either the 'community', the 'village', the 'micro catchment' or the 'territory'. In systems analysis terms, this corresponds to the scale at which we define the agroecosystem (cf. Chap. 2). Some methods used to characterise the local agricultural context include:

(i) Community resource maps
(ii) Community social map
(iii) Village or territory transects
(iv) Venn diagrams and local institutional landscapes
(v) Trend lines and community story lines
(vi) Observational techniques

Most of these methods are well described in manuals on participatory research, for example, in the series of Socio Economic and Gender Analysis manuals produced by the Food and Agriculture Organisation (FAO) of the United Nations (http://www.fao.org/gender/seaga/en).

4.3.1 Community Resource Maps

Community resource maps can be drawn in many different ways, depending on the objectives of the study, and involving the community. Ideally, one or more members of the community should draw the map, in consultation with the representative group of community members selected for the exercise. Sometimes a provisory map can be drawn on the ground, on a dust or sand surface, and then copied by the facilitator to a piece of paper and confirmed and or improved through discussions with the community. Figure 4.2 shows two community resource maps corresponding to village territories in Burkina Faso and in Zimbabwe. The first example emphasises the geographical aspects of the territory; the location of homesteads, roads, water sources, forests, grazing areas, cultivated fields, hills, lowlands, etc. The second one is less spatially explicit and focuses on resource flows within the territory, specially between crops and livestock, the magnitude of which are indicated with groundnut pods. Both maps provide different nature and quality of information, they may be adequate or not depending on the objectives of study, and they may be complementary.

4.3.2 Community Social Maps

Closely related to the resource map is also the social map of the community, in which the different households of the village are depicted on the map and information is

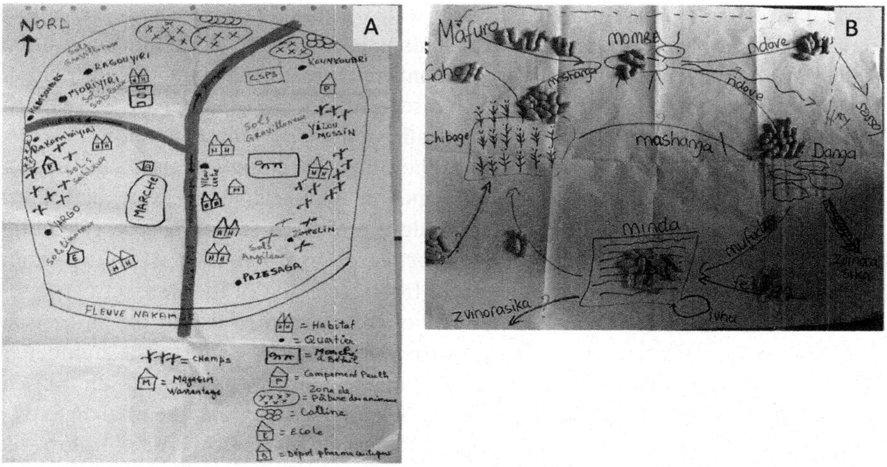

Fig. 4.2 Examples of community resource maps drawn in participatory workshops in (**a**) Burkina Faso (Diarisso et al. 2015) and (**b**) Zimbabwe

assigned to each of them in terms of age, gender, ethnicity, wealth and family composition. Delineating social maps is only feasible at smaller scales, or when the number of households in a territory is not too large. In the last decades, a new set of methods emerged that combine participatory learning and action research (PLAR) and geographical information systems (GIS), collectively known as participatory GIS (PGIS) (Corbett and Keller 2005). Mapping methods vary in their complexity and precision, from ephemeral maps drawn on the ground or sketch maps drawn from recollection, to geo-referenced scale mapping, three-dimensional modelling or multimedia information linked to a map. Such methods offer the opportunity to capture valuable spatially explicit information from local knowledge. The information can be stored and processed, and used for inventorying, planning, communication or to support negotiation processes over a territory. One example of tools to support negotiation is participatory three-dimensional modelling (Rambaldi and Callosa-Tarr 2002), which uses a scaled relief model of a territory based on a digital elevation map on which people identify and locate landscape features as points, lines or polygons by means of materials such as pushpins, threads or paints. Once the exercise is ready, a grid is placed on the resulting model in order to digitalise the geographical information displayed by the participants.

4.3.3 *Transects*

Village, topographic or landscape transects are used in a diversity of studies to characterise either the biophysical features of a landscape, soils, vegetation and lands use, or the socio-economic diversity in a community, or all of these simultaneously.

They are normally done through a transect walk (or drive) cutting across the various units of a landscape, aiming to capture its spatial diversity. Sometimes a transect walk through the territory can be used to confirm/correct/triangulate the information displayed in a community map. The transect length, the number of transects to represent a territory, the number of observation points per transect and the distance between them depend on the objectives of the study. In ecology, transects are used to estimate variables such as species abundance or habitat diversity, etc., and hence plenty of literature is available on how to design transect sampling (e.g., Cowling, A. 2006. Line-Transect Sampling. *Encyclopedia of Environmetrics*, 2). In vegetation studies, all objects situated at a distance x from each side of the transect (*Transect width* = $2x$) are recorded for the calculation of these metrics, so that a transect actually covers a certain area (*Area covered* = *Length* × *Width*). The number of transects necessary and their length are calculated on the basis of a chosen transect density. For instance, if the density of sampling is 10 m ha^{-1}, and the total area to be sampled is 500 ha, then the total length of transects is 5000 m. According to the characteristics of the landscape in terms of shape, topography, diversity, the study may require 10 transects of 500 meters, or 5 transects of 1000, or two transects of 2500 m, etc.

In a village transect, in which the aim is to characterise the biophysical and socioeconomic context of the target population of farms or rural households to be categorised, the sampling could be less intensive than in ecological studies. The transect trajectory is decided *a proiri*, on a map (perhaps the community map, or a soil map if available), often following a topographic gradient. One may record all objects or features observed during the transect walk (homesteads, crops, animals, trees, relief, water courses, infrastructure, marketplaces, etc.) or decide on 'sampling stations' or 'observation points'. These are places that are representative of the surrounding landscape or where we notice abrupt change, or where we observe particular features that need to be characterised such as a livestock watering point, or a piece of forest, etc. One method to assess landscape diversity is to take a 360° view of the surrounding landscape at each observation point to record all relevant features for our study. When the landscape is heterogeneous, as is often the case in smallholder farming landscapes, the field of vision may be divided in four quadrants and observations made for each quadrant individually. In the example of Table 4.3

Table 4.3 Example of a 360° observation of the agricultural landscape, dividing the field of vision in four 90°-quadrats

Quadrant	Main land use	Coverage	Infrastructure	Vegetation strata	Tree cover
0–90°	Annual crops	80%	Road, fences	2 (crops, trees)	20%
90°-180°	Grazing	100%	None	1 (grass)	No trees
180°-270°	Home gardens	60%	Homesteads, paths	3 (trees, shrubs, crops)	80%
270°-360°	Annual crops	60%	Fences, homesteads	3 (crops, hedgerows, trees)	20%

annual crops cover 80% of the field of vision in the first 90° quadrant, where there is also a road and field fences, and trees that cover about 20% of the soil surface. The number of observation points and hence their interval may be systematised and decided statistically, by computing sample sizes on the basis of confidence intervals and levels (cf. next section). However, in this kind of study, it is recommendable to decide on the basis of direct *observation* of changes in the biophysical and socio-economic landscape associated with its diversity.

Transects can also be used to characterise topographic sequences (or toposequences) of soil types. Figure 4.3 illustrates two types of transects, one in which the major objective is to capture variations in soils and landscapes (the

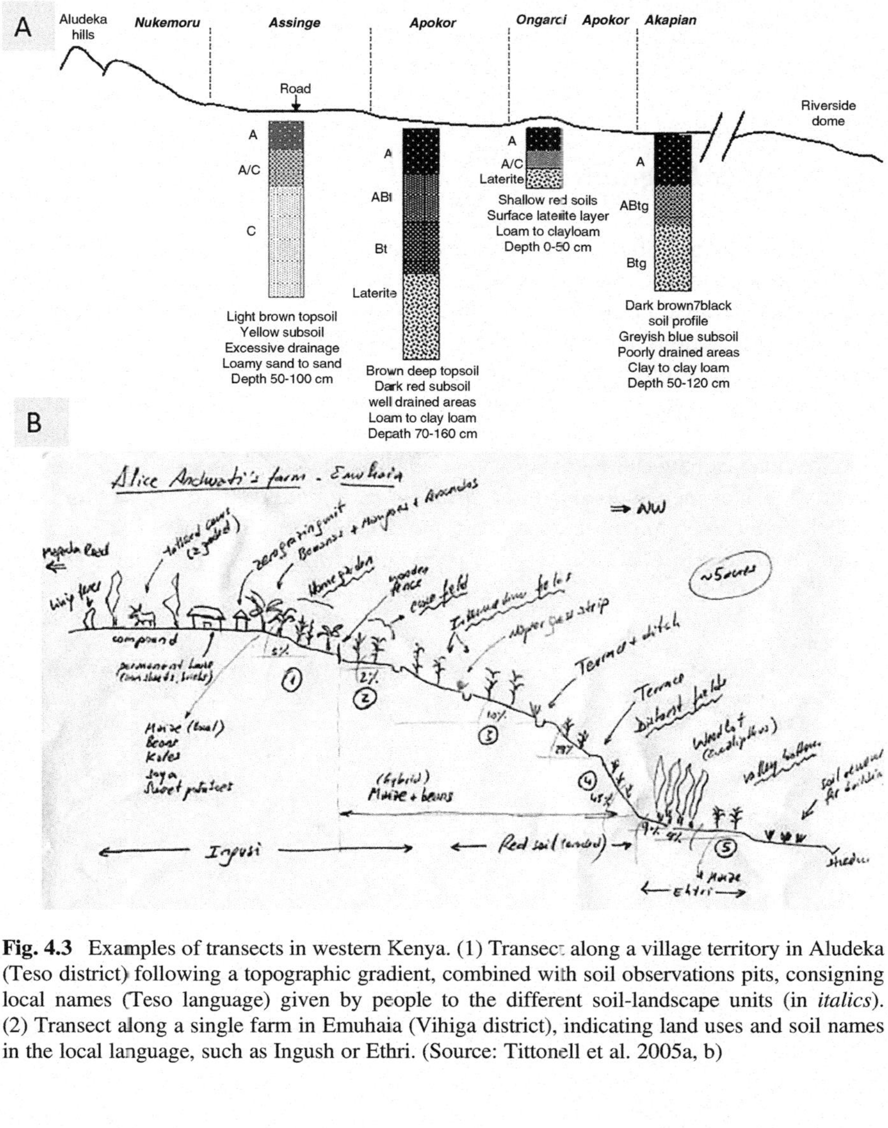

Fig. 4.3 Examples of transects in western Kenya. (1) Transect along a village territory in Aludeka (Teso district) following a topographic gradient, combined with soil observations pits, consigning local names (Teso language) given by people to the different soil-landscape units (in *italics*). (2) Transect along a single farm in Emuhaia (Vihiga district), indicating land uses and soil names in the local language, such as Ingush or Ethri. (Source: Tittonell et al. 2005a, b)

'soilscape' or 'pedolandscape' as termed by Deckers 2002) and a second one in which the focus is on land use. In a transect study of soil landscape units the observation points are located within each representative soil landscape unit. Soil observation pits may be dug in such places. Once the 'modal' soil profile corresponding to each unit has been characterised, further soil monitoring along the transect can be done with an auger, to detect changes in the depth of the soil profile, in texture or in the sequence of soil horizons, their thickness, presence of stoniness or hardpans, or depth of the water table, etc. Substantial changes in these attributes (of more than e.g. 20%) would indicate that we are in the presence of a new soil unit, and a new observation pit may be necessary. When the objective is to characterise land use and socioeconomic aspects of the landscape (land tenure, ownership, communal resources, social diversity of households, etc.) it is generally useful to transect walk the territory together with members of the community and discuss every feature that is recorded in the sketch of the transect.

4.3.4 Venn Diagrams

Venn diagrams are often used as a rapid method to depict the local institutional landscape (e.g. for the study of innovation systems, of social networks, etc.). They are normally drawn during focus group discussions, and can be used to understand the role of institutions and organisations, the existence of potential conflicts between agents, or the possibility of local people to make decisions. Circles or diagrams are drawn to represent each organisation concerned, which may be local or external, and the degree of contact between circles represents the extent of co-action (collaboration, exchange) between them (Fig. 4.4). Ideally, different sub-groups in the community, such as men and women, or different age groups, can draw Venn diagrams of their own perception of the same institutional landscape. This *ad hoc* utilisation of Venn diagrams does not necessarily correspond with the use of Venn or Euler

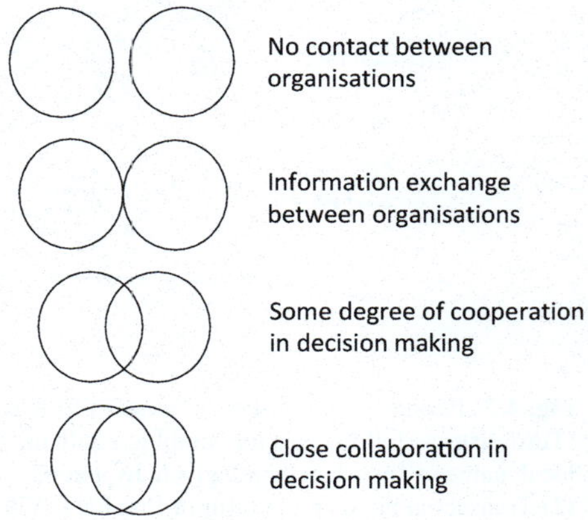

Fig. 4.4 Type of relationship between organisations as represented by Venn diagrams

diagrams in set theory, in which each circle represent a set, and the possible operations between two or more sets include intersection, union, symmetric differences or relative and absolute complementarities (Jech 2003). Set theory will be revisited in the next section, when dealing with farm categorisations.

4.3.5 Timelines and Trendlines

The dynamics of farming systems and their context can be rapidly captured by means of historical timelines (past) and trend lines (future). By recollecting historical events that took place in the territory, as well as the major changes operated in the landscape through time (in vegetation and land use, infrastructure, population density, climate, etc.), timelines offer valuable information on the long-term dynamics of a community (Fig. 4.5). Trend lines, on the other hand, are meant to capture future dynamics as perceived by the local community. They are drawn in the context of focus group discussions, using a Cartesian plane (x = time; y = an indicator variable) and drawing lines or arrows indicating the rate and direction of expected change. For example, trend lines can depict the expected increase in the next decade in the percentage of land cleared for agriculture in a given territory, or in the area available for communal grazing, etc.

The use of timelines and trend lines as visualisation techniques is more difficult when participants are not familiar with methods of abstract representation. In such cases, changes in time or expected changes in the future (scenarios) can be depicted by means of rich pictures, a common soft system methodology in which there is no predetermined syntax or code to be followed. Rich pictures are used to illustrate complex problems or systems by enriching the picture with as much pictorial or textual information as necessary to describe and understand a problem or develop a hypothesis (Fig. 4.6). They can be linked to fuzzy-cognitive mapping in order to model the dynamics of complex systems.

4.3.6 Observation

Observational techniques can be used in isolation or, preferably, complementing all the other techniques described for the characterisation of the local context. There are three types of observational research methods used in marketing and social studies: (i) covert observation, in which the researchers are undercover or observe activities of individuals at a distance; (ii) overt observation, in which the researchers identify themselves as such and explain the purpose of their observation to the community; and (iii) participant observation, in which the researchers participate in the activities being observed. There are advantages and disadvantages with each of these three methods. Many researchers prefer overt or participant observation to avoid deceiving the community, although this may condition the behaviour of those being

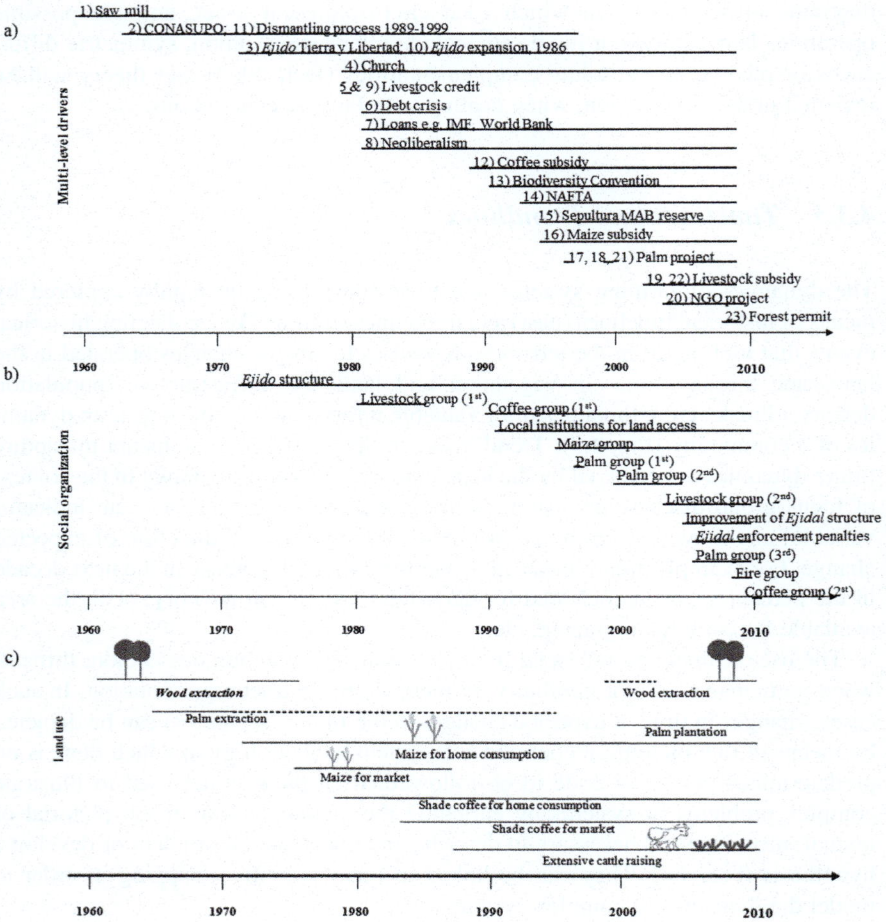

Fig. 4.5 Example of a detailed description of historical events in a territory using timelines. The three timelines describe the recent history at the community Tierra y Libertad, in Chiapas, Mexico in terms of (**a**) multi-level drivers, (**b**) social organization, and (**c**) qualitative land use change, as identified by key stakeholders (1960–2010). Dashed lines refer to illegal land use activities. (Reproduced with permission from: Speelman 2014)

observed. Beyond the peculiarities of this technique, critical observation should be a 'built-in' characteristic of any field researcher aiming to understand diversity and dynamics of rural households. For example, much can be inferred from the observation of homesteads, of infrastructures or of agricultural practices. Interpreting such information requires a solid basis of background information on the historical, cultural, agricultural and/or political aspects of the local context. In any case, it is always good practice not to run into conclusions from simple observations. Observation can, at most, lead to the enunciation of good hypotheses; not to conclusions.

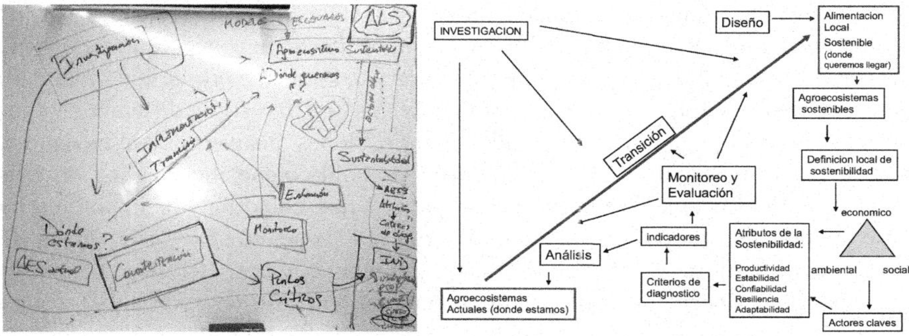

Fig. 4.6 A scheme depicting the possible transition pathway from current agroecosystems (*agroecosistemas or AES actuales*) towards sustainable local food systems (*Alimentacion local sostenible, or ALS*) drawn during a participatory scenario analysis exercise with farmers, researchers, extension agents and members of non-governmental organisations in Vittoria-Gasteiz, Basque Country. Left: scheme drawn on a whiteboard during the group meeting; Right: the same scheme as included in a policy brief for the regional government

Triangulation through different sources of information is always necessary to confirm them. Most importantly, we must keep in mind that when visiting a community, we are only observing the landscape at one point in time, ignoring the short- and long-term dynamics inherent to the social and ecological systems.

4.4 Categorising Farms and Rural Households

Farm and/or rural household diversity can be captured through field surveys. The information thus obtained is processed so that a value for each indicator variable is obtained for each household visited, sometimes through simple calculations. Different variables are used to categorise households. In agroecology, often the number of years in which a farm is in 'transition' or the transition 'level' as described by Gliessman (2007) are used as criteria to categorise farms. According to the discussion about types of variables in the previous section, the level of transition to agroecology is a performance variable, and hence not a good one to categorise households from a systems analysis perspective. We may also be interested in other aspects of the household, beyond their extent of transition to agroecology, such as their motivations and practices (Teixeira et al. 2018). We may want to investigate why a certain household type is more or less advanced in the transition, what are the structural and functional characteristics that explain such performance (El Mujtar et al. 2023).

Examples of variables that are typically collected through household surveys include the land area owned by the household or the number of household members, both of them structural variables; or crop yields and total income, both of them indicators of performance. Yet the total area owned by the household may be poorly informative in certain cases. Related variables, such as the total area cultivated, left as fallow, leased out or hired in, borrowed, lent, occupied, in concession, etc. may be important as well. Sometimes it is also necessary to know the proportion of the farm area under different land uses (e.g., agriculture, pasture, forest), in different geomorphological units (e.g., marshland, valley bottom, slopes, plateau) or soil types (e.g., sandy, clay, gravel). The same applies to household composition, as not only the total number of people living and eating within a household may be important, but also the number of those who work on the farm full time or temporarily, those who work off-farm, those who practice non-agricultural activities, etc. Once the survey information has been collected, it is often pertinent to construct informative secondary variables through simple calculations. These include variables such as land-to-labour ratios, cattle densities, income per capita, food self-sufficiency, etc.

If households are diverse we expect a certain range of variation in the value of the key variables that describe such diversity. This variation will be characterised by a certain distribution, which is not necessarily the same for all the variables concerned. In the simplest of cases, we may assume normal distributions of structural variables, albeit with different degrees of skewedness for each variable (Fig. 4.7). We may then assume the existence of a modal farm, which may be fairly representative of all the farms surveyed. This assumption is often made in policies for the agricultural sector, such as in subsidy schemes, in which the diversity of farms and households is largely disregarded. It must be noticed that while some variables may exhibit more or less normal distributions, other variables measured on the same sample of farms may not be normally distributed. For example, while total land owned may be normally distributed, most of the farms owning agricultural machinery or the highest horsepower asset may be those with the largest areas of arable land (negatively skewed

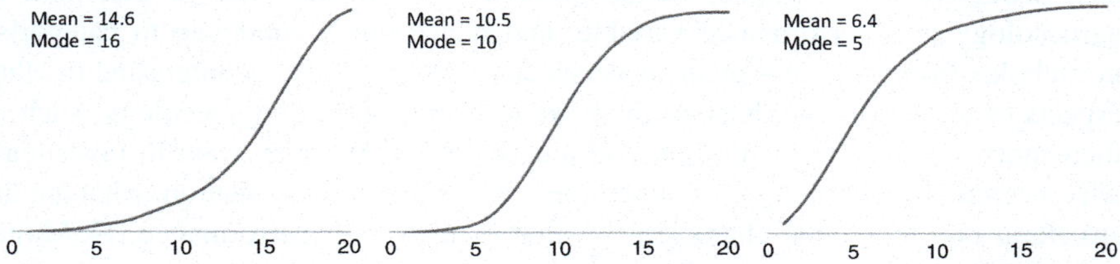

Fig. 4.7 Three types of distribution, illustrating the effect of data skewness on the average and mode values. Above are the frequency distributions and below their corresponding cumulative frequencies. Note that the central distribution is not perfectly normal (it was drawn using actual frequency data). There are statistical routines to test how normal a distribution is

distribution). In smallholder farming situations, livestock ownership often shows a positively skewed distribution, as there are typically many households who own few or no livestock and a few that own many. In such case, a Poisson density function could be more appropriate than a normal one to describe frequency distributions.

Finally, there are cases in which polymodal distributions are observed. Each modal peak may be indicative of a certain farm type, which would make the delineation of a typology rather easy. Yet it is also possible that polymodal distributions are indicative of the co-existence of more than one population of farms in our sample. If farms are too different at simple observation, then delineating a typology may be a redundant exercise. We can often see in the literature examples of typologies that distinguish farms based solely on their production orientation, objectives or main agricultural activity. For example, a typology that would consist of fruit growers, livestock keepers and urban investors in farmland as three major farm types. The main objectives of these three groups are clear (producing fruits, animal products, and real state speculation or hobby farming), as would be their structural characteristics (orchards, grasslands, or leisure infrastructure). There is no point in establishing a detailed typology in this case, as they are clearly three distinct populations of farms. Rather, a typology could be delineated to describe the diversity among fruit growers, for example, in order to categorise their diversity in responses to certain technology or policy measures. The risk of falling in the trap of redundant typologies may be higher when dealing with participatory rankings, in which participants tend to start classifying their peers based on the most obvious characteristics.

4.4.1 Participatory Wealth Categorisation

There are several methodologies for wealth categorisation available in the literature on participatory research. It is not the intention to reproduce them all here. A distinction must be made though between participatory wealth categorisation and participatory wealth rankings. The latter are not recommended, unless strictly necessary. Why would we need to rank the members of a community according to their wealth? This can result in making private, sometimes delicate information public in a local context. When pair ranking in a group exercise or even through semi-secret ranking through key informants, people are generally not willing to reveal their wealth. There is no reason to make this a public exercise. Instead, community participation can be highly valuable in delineating wealth categories. Categories are sometimes hard to define due to their fuzzy boundaries or to the existence of particular cases that are hard to categorise, but these problems are inherent to any form of categorisation, including statistical clustering. Categories could be simply based on structural properties of households, on their assets and resource endowment, such as poor, mid-class and wealthy households, or be more informative of the functional and/or dynamic aspects of households, e.g., livestock, crop and mixed farmers, or original versus recent settlers. The number of categories depends on the objectives of the categorisation as much as on the diversity of households in a community. A rule of thumb that is often used in multivariate analysis to arrive at the number of clusters necessary to explain most of the variance in a sample is the following:

$$k \approx \sqrt{(n/2)} \tag{4.1}$$

Where, *k* is the number of clusters (classes or categories in this case) and *n* the sample size. Note that the variance (or diversity) within the population is not considered in this case. In a participatory categorisation of the households of a certain community, the total number of households in the community may be used as *n*, since there is no real sampling in the proper use of the term. Thus, in a community of 50 households the number of classes or household types should perhaps not exceed five, although this number is simply an indication.

Whether participatory categorisations are better than other forms of categorisation depends, again, on the objectives of the study. Let us keep in mind the four possible objectives outlined in the introductory paragraph to this chapter. Participatory wealth categorisations may be certainly useful for local targeting of development interventions (when not too unbiased). They may be less useful for up-scaling or comparative analysis across locations. The main problems associated with participatory wealth categorisations are their site-specificity and, particularly for systems analytical purposes, the frequent lack of a clear distinction between what are the causes (drivers, structural variables) and the consequences (management and performance variables) of wealth. For example, in several participatory wealth categorisations the wealthiest farmers are identified as 'those who attain high yields'. This is to be expected, in many cases, but attaining high yields says little about the structural and functional characteristics of the farm, which are of prime interest in systems analysis.

In participatory wealth categorisation exercises that took place in several regions of mid- to high agricultural potential across Kenya and Uganda, farmers selected 'wealth' and 'farm management' indicators that were not always consistent across localities (24 of them in total – Tittonell et al. 2010). In four different localities in western Kenya, indicators pertaining to food security, cash crops, livestock, labour and input use and timely crop management (closely associated with labour availability) were selected quite consistently by different groups of farmers, as illustrated in Table 4.4.

Land availability, income sources and commitment to farm work were alternatively selected in three of these four localities, whereas access to information, educational level, family size and the type of housing, among other broadly-used indicators, were less consistently selected (Table 4.4). Therefore, the participatory categorisation of households based on these criteria was different for each of these localities, as shown in Table 4.5. The proportion of households in the wealthier class varied from 5 to 13% across localities. The proportion of those in the poorest class, by contrast, ranged widely from 30 to 80%. If a policy or development intervention to target the poorest households in western Kenya would be designed on the basis of

Table 4.4 Farmers' criteria to classify households in relation to resource endowment and farm management during participatory wealth categorisations in Vihiga (Ebusiloli and Emusutswi) and Siaya (Nyabeda and Nyalugunga) districts, western Kenya (adapted from Tittonell et al. 2010)

Criteria	Categories chosen by farmers
Indicators selected by farmers in the 4 localities	
1. Food security	Months of food self sufficiency (Poor: 0–2 months; Mid-class: 3–5 months; Wealthier: 8–12 months) and having or not food surplus to sell on the market
2. Labour availability	Depending exclusively on family labour, complemented with hired labour or using exclusively hired labour
3. Cash crops	Presence and acreage of tea plantations (> or < 1 acre); presence of tobacco, sugar cane, tomatoes; level of input use and maintenance
4. Livestock	Type and number of livestock heads owned (e.g. 3–5 improved dairy cows in the wealthier class) and management system (stall fed, free grazing)
5. Use of fertilisers	Regular, occasional or no use of organic and/or mineral fertilisers; applied in most fields or only in homegardens; only basal or basal plus topdressing applications
6. Timing of farm operations	Timely planting and weeding, ownership/ capacity to hire oxen for ploughing vs. hand hoeing; labour hired for timely weeding
Indicators selected by farmers in 3 of the 4 localities	
7. Land availability	Farm size (variable acreages across localities); hire-in, use own or hire-out land for cultivation
8. Use of quality seed	Use of certified seeds, maize hybrids; use certified in long rains and local seeds in the short rains
9. Income	Annual income (e.g. KSh 80,000-100,000; 30,000–50,000 or < 10,000 for wealthier to poorer class, respectively, in Nyabeda); main source of income (on-farm vs. non/off-farm); permanent vs. intermittent off-farm income
10. Commitment to work	Hardworking vs. idlers; need to work for other farmers or commit to other occupations
11. Soil conservation	Presence and maintenance of permanent or semi-permanent (grass strips) soil conservation measures
Indicators selected by farmers in 2 of the 4 localities	
12. Access to information	Having regular or sporadic access to agricultural information and knowledge, seeking extension services
13. Planting method	Planting in lines using oxen furrows or ropes vs. broadcasting
14. Weeding frequency	Weeding once or twice in the season or not at all, in all the fields vs. a few of them
Indicators selected by farmers in only 1 of the 4 localities	
15. Type of house	Permanent brick houses vs. huts, tin roofing vs. thatched, maintenance
16. Transport means	Ownership/ hiring wheelbarrow, bicycle, wheel carts
17. Veterinary services	Contracting veterinary services vs. using herbal treatments
18. Household nutrition	Number of meals a day (1, 2 or 3) throughout the year, balanced diets vs. starchy diets, meat consumption
19. Family size	Small families vs. large, polygamous families

(continued)

Table 4.4 (continued)

Criteria	Categories chosen by farmers
20. Education level	Level of education (primary, secondary) completed plus additional training; well educated and informed
21. Postharvest storage	Presence of storage facilities (permanent) or use of drums, pots, sacks; use of chemicals vs. traditional methods for preservation

Table 4.5 Participatory categorisation of households in four localities in western Kenya, in absolute and relative terms per locality, based on indicators selected by farmers (cf. Table 4.4)

Locality in western Kenya	Wealthier households, best farm managers	Moderately endowed, regular farm managers	Poor households, poorest farm managers
Ebusiloli	49 (10%)	277 (60%)	138 (30%)
Emusutswi	19 (5%)	58 (16%)	285 (79%)
Nyabeda	32 (13%)	125 (49%)	97 (38%)
Nyalugunga	29 (9%)	180 (53%)	132 (39%)

this information, what should be the size of the target population, 30 or 80% of the households? This is a large difference. Although some of the criteria that farmers selected in these exercises represent structural variables (e.g. availability of land and labour), others were simply a consequence of these (e.g. timely weeding, use of hybrid seeds or veterinary services) and were highly correlated with each other. Nevertheless, farmers' criteria and particularly the site-specific value of thresholds proposed for different indicators are highly valuable information to complement and refine other means of categorising households, such as formal statistical clustering.

4.4.2 Statistical Methods

Farm or rural household categories can be identified with the aid of statistical analysis. The analysis may vary in complexity, from calculation of descriptive statistic for groups of households formed on the basis of observational criteria to statistical grouping by means of different clustering techniques. Any form of categorisation should preferably start from a hypothetical typology that needs to be tested (as in any other form of statistical inquiry). Such hypothesis is derived through observation, during data collection, surveys, field visits or expert knowledge, and responds strongly to the objectives of the study. It is of course possible to establish a typology 'blindly', that is, from the mere statistical analysis of the surveyed data, without having seen a single farm in reality. This is however not recommendable, and should be avoided as much as possible, as no method for statistical clustering is completely objective. Subjective decisions are often made during the analysis, as we will see, and they can be best informed by expert knowledge. The hypothesis to be tested could be enunciated, for example, as follows:

H0: All farms are similar, of the same type
H1: There are three types of farms based on resource endowment, rich, medium and poor

This is a simple example. The alternative hypothesis H1 could be made as complex as necessary. Before embarking in testing this hypothesis, it is necessary to understand the structure of the dataset, which depends on the nature of the data itself, the type of variables and their distribution, their degree of co-variance, and the extent to which they represent the actual population. Since these factors determine to a large extent the type of statistical analysis to be chosen, as well as the interpretation of the results of such analysis, let us first consider some relevant aspects that relate to sampling and to the nature of variables.

4.4.2.1 Influence of Household Sampling Methods

Sampling theory and methods, and particularly those for household sampling, can be found in any manual on statistics or research methods for social surveys. A rather complete overview is provided in the practical guidelines developed by the Department of Economic and Social Affaires of the United Nations (http://unstats.un.org/unsd/demographic/sources/surveys). The sampling of households should be designed in such a way that the most relevant factors that determine household diversity in a certain context are considered, that all possible household types are surveyed, and that a good balance is found between costs (time) and precision. A sample must be stratified within geographical sub-units or population sub-groups, so that no particular sub-unit or sub-group is overrepresented with respect to others. Cluster sampling of neighbouring households may be desirable in certain geographical situations, especially when human and material resources are scarce. However, cluster sampling may introduce high levels of co-variance when neighbouring households are too related or similar in terms of their structure and function. Sometimes logistical issues may influence sampling procedures, as when e.g. poor accessibility makes sampling too time consuming and thus only households located close to roads or markets are surveyed. When access to households in a community is only possible through 'introduction' of the survey team by a community member or local extension agent, often lists of households to be surveyed are developed by a local key informant. The survey can be done only on the households defined in such initial list, or extended by asking each surveyed household to introduce the team to their neighbours for surveying, which results in a sort of snowball sampling scheme. In some other cases, only those households designated by a village chief can be surveyed, which makes the sampling far from being probabilistic.

The above-described aspects inherent to (smallholder) farming landscapes must be considered when designing procedures for rural household sampling, or when analysing an interpreting survey data that has already been collected. In addition, the spatial heterogeneity of the landscape and the distribution of farm households must also be considered. In heavily dissected topographies, farms may be dispersed on a spatially heterogeneous landscape, which may often correspond to spatially variable soil types. In such landscapes, smaller farms may be located within one single landscape unit, while large farms may comprise several landscape units within their boundaries. For these reasons, sampling is not always done following the precepts of probability sampling. In probability sampling, each element must have a known mathematical chance (greater than zero and numerically calculable) of being sampled. However, the chance of each element of being sampled does not necessarily have to be equal. According to the objectives of the study, we may want to favour certain (more frequent) cases over others to avoid overrepresentation of the latter in the dataset. For example, knowing a priori the approximate distribution of wealth classes in a community – from national census for instance – one may decide to execute a form of quota sampling. If it is known, for example, than the wealthiest households in the community represents only 10% of the entire population, random sampling may lead to an overestimation of this wealth class. Overestimation of largest farms occurs also when sampling is based on spatial randomisation, as the largest farms occupy more of the space sampled and have therefore greater chances to be selected (Fig. 4.8).

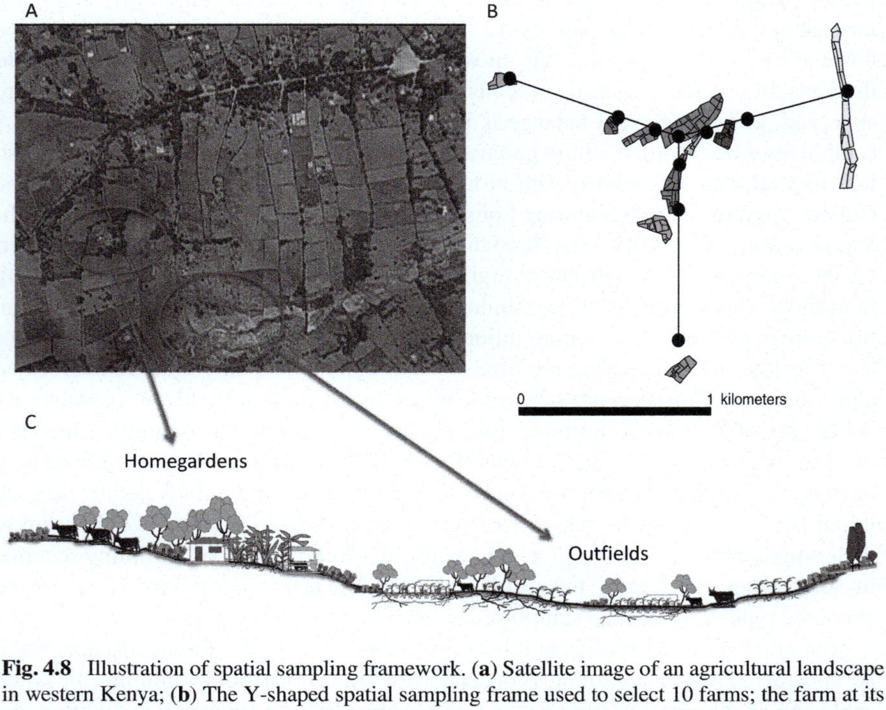

Fig. 4.8 Illustration of spatial sampling framework. (**a**) Satellite image of an agricultural landscape in western Kenya; (**b**) The Y-shaped spatial sampling frame used to select 10 farms; the farm at its centroid was selected randomly, and all other 9 farms were those intersected by three 1200-m long transects departing 120° from one another (each set of contiguous polygons represents an individual farm); (**c**) A transect drawing of the area showing main land use patterns. Randomization based on space gives more chances to the largest farms to be selected in the sample. (Sources: Henry et al. 2009; Tittonell et al. 2010)

Probability sampling allows for statistically inference to be made on the entire population based on the sample, and sampling errors to be calculated from the sample data. None of this is possible with non-probability sampling. The principles of probability sampling must be used in all stages of sample selection, from the selection of sites and villages or geographical units to the households and persons to be interviewed. Failure to use probability methods in selecting households means that the sampling will be biased. In the simplest of cases, sample size can be calculated on the basis of confidence levels and intervals. The confidence interval (δ) is the maximum allowable error of the estimate (1/2 tolerance), or the maximum difference between the mean of the population and that of the sample:

$$\delta = Z_{\alpha/2} \times \sigma / \sqrt{n} \tag{4.2}$$

Where, $Z_{\alpha/2}$ is the critical Z-score corresponding to the confidence levels (or Type I error) in a standard normal distribution (of mean zero and variance one), σ is the standard deviation of the population, and n is sample size. From this equation, knowing σ and assuming a value for δ, sample size can be calculated as:

$$n = \left(Z_{\alpha/2} \times \sigma / \delta\right)^2 \tag{4.3}$$

Commonly used confident levels are 90, 95 or 99%, resulting in $\alpha = 0.1$, 0.05 or 0.01, respectively. For example, if we are interested in estimating the average farm size in a region in terms of area (ha), with an error no greater than ±0.5 ha, with a confidence level of 95% ($\alpha = 0.05$), and knowing from previous surveys that the standard deviation is around ±2.5 ha, then the necessary sample size can be calculated as:

$$n = (1.96 \times 2.5/0.5)^2 = 96\,\text{farms} \tag{4.4}$$

In general, it is unlikely that the standard deviation would be known if the mean of the population is not known. Yet, this formula could still be used as a general indication to orient decisions on sample sizes. Figure 4.9 shows the variation in recommended sample sizes for the same example above as a function of the standard deviation (σ) and the tolerance level (δ). Sample sizes are less sensitive to σ when the tolerable error is relaxed. When σ is not known from previous studies, a pilot quick survey can be done to estimate it.

4.4.2.2 Exploring the Data

Prior to any form of statistical grouping, it is advisable to examine the structure of the variance in the survey data. Data exploration is done by means of descriptive or explorative techniques, such as correlation, principal component, factor or correspondence analysis, etc. These are not hypothesis-testing techniques, in the sense

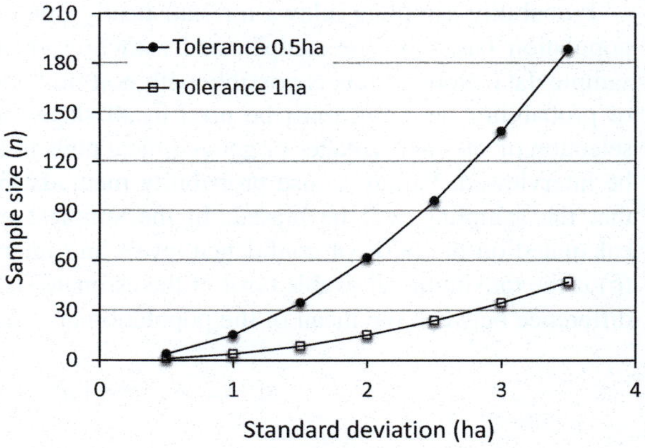

Fig. 4.9 Example of sample sizes calculated to estimate average farm area with tolerable errors of 0.5 and 1 ha, with a confidence level (1 – α) of 0.95, as a function the standard deviation in farm sizes of the entire population

that there is no assumption about causation, as is the case when fitting regression models in which dependent and independent variables are defined. The choice of data exploration technique depends on the nature of the data, in terms of number of observations, number of variables, type of variables, expected frequency of outliers or skewed distributions, etc. The number of variables in a dataset determines its dimension; for example, a dataset containing information only on family size and area owned by a household is a two-dimensional (2-D) dataset. A dataset with 10 variables is a 10-D dataset, and so on. Datasets consisting of continuous variables can be explored by means of correlation matrices or principal component analysis. Datasets combining different types of variables, continuous, discrete and/or categorical variables, can be explored by means of correspondence analysis.

The information provided by these techniques is essentially the same. They provide a measure of the co-variance between the variables in the dataset. Unlike correlation, principal component and correspondence analysis reveal the underlying structure of the variance in the data and are also dimension-reduction techniques which help us to retain only those dimensions in the dataset that contain information (variation). As we will see, principal components can be used in cluster analysis replacing the original survey data, thereby reducing the number of dimensions to be dealt with. Principal component and correspondence analysis are typically used when there are no or poorly specified hypotheses concerning the relationship between variables in the dataset.

Household survey data often comprise a diversity of variable types, from continuous variables (e.g. farm area in ha) or ratios (e.g., crop yield), to nominal (e.g., land use types: forest, pasture, cropland) or categorical (e.g. high, medium, low). In some cases, binary variables (0 or 1) are defined to indicate presence/absence attributes. Very often scores are used to provide a semi-quantitative measure to visual assessments (e.g., animal status: 1 = very poor, . . ., 5 = very good) and rankings to reveal the relative importance of different activities (e.g., a ranking of the five most important cash crops for the household). Scores result in interval variables, while rankings result in ordinal variables. In farming system characterisation, categorical variables or scores are generally used to replace variables that are possible but difficult to measure with some degree of precision.

For example, a 'resource use intensity' (RUI) score was used in the study published as Tittonell et al. (2008) to summarise the intensity of use of nutrient inputs of different sources by smallholder farmers. Sources ranged from mineral fertilisers to animal manure, fresh or mature, compost, biomass transfers, kitchen waste, etc. As it was difficult to quantify the exact rates of application of these various resources just from asking farmers to recall their past actions, a scoring system from 0 to 3 was developed such that a value of 0 indicated 'no use' of any nutrient input and 3 indicated annual applications of different sources of nutrients equivalent to or higher than the recommended rates. When working with such type of variables, it is important to pay special attention to their distribution. Variables that depend on natural physical processes tend to exhibit a normal distribution, as does the experimental error. Variables that describe inherently asymmetrical attributes such as those closely associated with wealth classes (e.g., land area farmed, the number of livestock owned or input use) tend to exhibit positively skewed distributions (cf. Fig. 4.7).

4.4.2.3 Assessing the Structure of Variation

A simple correlation analysis measures the degree of association between two variables in a coordinate system (x, y). The sign and magnitude of the correlation coefficient (or Pearson's correlation coefficient), which ranges from -1 to $+1$, indicate the direction and the strength of the relationship between the two variables. A correlation matrix consisting of all the variables in the dataset provides a good overview of the degree of association between all pairs of variables. High degrees of association between variables may denote redundancy, expected patterns, or totally unexpected trends. Some associations may be obvious, such as a positive relationship between the area cropped and the total area of the farm. A positive correlation between age of the household head and farm-scale plant species richness, for example, might be pointing to a more interesting trend or hypothesis on household diversity and its relationship with ecosystem services that would deserve to be further explored. Conversely, low degrees of association between variables indicate that they vary independently from each other. In general, a simple correlation analysis can be used to reveal patterns in the dataset when its dimension is relatively small. Large dimension datasets require more sophisticated techniques to reveal the underlying structure of their variance.

A correlation analysis between two variables reveals their degree of co-variation only in the coordinate system in which these variables are defined (e.g., number of household members and years of schooling). The same two variables may have a stronger correlation when plotted in a different coordinate system. Canonical correlation is used to identify such new coordinate systems in which pairs of variables correlate most. As in the case of principal component analysis, the new axes are

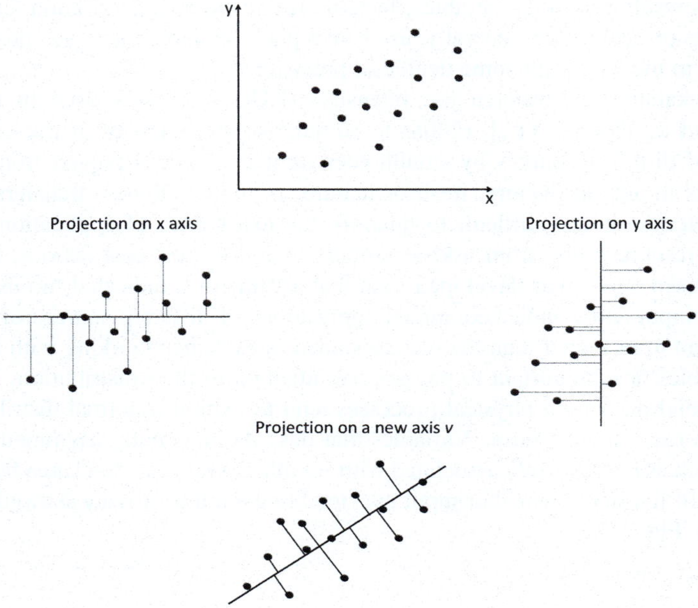

Fig. 4.10 An illustration of how an eigenvector is delineated to increase the amount of variance in the *x, y* relation that can be described. Neither the projection of the cloud data points on *x* or on *y* captures the maximum breadth of the variation. Vector *v*, or eigenvector, is introduced in a direction such that it captures the maximum variability in the data when these are projected on it. Vector *v* is a linear combination of *x* and *y*

found through the calculation of eigenvectors and eigenvalues, through matrix algebra. The mathematical underpinning of these calculations can be found in any book on multivariate statistics. Here, graphic representations are used to illustrate how the eigenvectors are obtained.

Consider the data plotted on a *x, y* coordinate system in Fig. 4.10. Neither the projection of the data points on *x* nor on *y* captures the full breadth of the dispersion of the data (variation). To capture the maximum range of variation in the data, it is necessary to introduce a new vector *v*, which is a linear combination of *x* and *y* known as eigenvector. The projection of data points on this new axis *v* yields the widest spread, and a measure of such spread is indicated by the eigenvalue. Mathematically, this requires some knowledge on matrix operations. Let us simply say that if we multiply a square correlation matrix[1] *A* by a vector *v* then the resulting vector *Av* is a number λ times the original vector *v*, so that:

[1]The Eigen Decomposition Theorem applies to square ($k \times k$) matrices, and consist in diagonalizing the matrix by multiplying it by a vector (eigenvector)

$$A \times v = \lambda \times v \tag{4.5}$$

The number λ is an eigenvalue of A (for those interested in the maths, note that this operation implies a post-multiplication) and each eigenvalue is paired with an eigenvector. Back to the example in Fig. 4.10, the eigenvalue indicates the exact direction of vector v necessary to capture most of the variation in the data. In other words, the eigenvalue is a measure of the amount of variance explained by an eigenvector. Eigenvalues greater than one indicate that the vector accounts for more variance than the original variable[2] in standardised data. The eigenvector with the highest eigenvalue is the principal component.

Any dataset can be decomposed into eigenvectors and eigenvalues; the number of possible eigenvectors is equal to the dimension of the dataset. Although in Fig. 4.10 we found a vector v that explains most of the variation in the data in one direction, there is still some unexplained variation represented by the distance of the data points to vector v. The data in this example consists of two variables, x and y, and thus it is possible to decompose it into two eigenvectors. A second vector u can be then introduced, which is orthogonal[3] to v, to capture the remaining variation. The eigenvalue associated with this new vector is smaller than the first one ($\lambda_v > \lambda_u$), as only the remaining variation is being captured. Graphically, this looks like the left hand side scheme in Fig. 4.11. The vector u is the second principal component and the fact that it is orthogonal to v implies that both vectors are uncorrelated or independent from each other. The data can now be rotated so that it is represented in a new coordinate system, defined by vectors v and u, as in the right hand side of Fig. 4.11. The coordinates of each data point can be then defined in terms of v and u units, as indicated graphically for the data point θ in Fig. 4.11. These are known as scores. The data cloud in Figs. 4.10 and 4.11 has approximately the shape of an oval, or ellipse, indicating that the data has a distribution that is approximately normal. Note that this technique is highly sensitive to outliers and extreme values. As in the case of computing Pearson correlation coefficients, principal component analysis assumes normality in the distribution of the data. This is not a strict requisite, as principal components or Pearson coefficients can be calculated even if the data are not normally distributed, particularly for data visualisation purposes. This is not a smart thing to do in the case of statistical inference, however, since the lack of normality will hinder the interpretation of the analysis.

Principal components are thus orthogonal directions that capture most the variation in the data. The data are not changed in their nature; they are only looked at from a different perspective. Data corresponding to a three dimensional dataset (e.g., number of household members, years of schooling, months of food self-sufficiency) can be graphically represented in a three dimensional space x, y, z and decomposed

---

[2]This is often used as a cut-off threshold to decide on how many eigenvectors to retain in a data analysis

[3]Adding a second vector at 90° from the first one is the most effective way to span the entire x-y plane

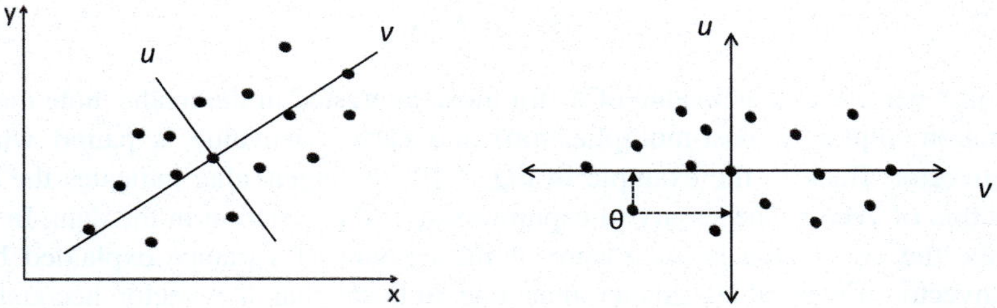

Fig. 4.11 Illustration of how a second eigenvector, orthogonal to the first one, is introduced to capture the remaining unexplained variance. The right hand side figure shows the same data but rotated so that they are now projected on vectors *v* and *u*, or principal components, instead of the original coordinates *x* and *y*. Any data point, such as *θ*, can be now expressed in *u* and *v* units (or eigenvalues)

Table 4.6 Loadings of the five first principal components (PCs) with respect to seven farm survey variables in a sample of 250 smallholder households in East Africa

Variable	PC1	PC2	PC3	PC4	PC5
Off/non-farm income (%)	−0.998	0,053	0,006	−0,005	−0,027
Age of the household head	0.054	0,996	−0,003	−0,041	−0,049
Total area farmed (ha)	0.006	−0,022	0,498	−0,169	−0,130
Family size	−0.004	0,046	0,360	0,885	0,129
Food sufficiency (months)	0.023	−0,036	0,316	−0,024	−0,867
Total number of livestock	−0.003	0,013	0,558	−0,294	0,346
Number of local cattle	−0.007	0,013	0,365	−0,219	0,277

From: Tittonell et al. (2010)

into three eigenvectors *u*, *v*, *w* (or principal components). The data are thus projected on different axes, which are linear combinations of the original ones. The next questions are, what do these principal components represent? To what extent do they relate to the original variables? How many principal components are necessary?

The first two questions can be answered by computing the loadings of each principal component with respect to the original variables. The loadings or component coefficients (or factor coefficients in factor analysis) are the correlation coefficients between the variables and the principal components. As in the case of Pearson's coefficients, the squared loading is the percentage of variance in the variable explained by the principal component. In the example presented in Table 4.6 the first two principal components (PCs) have very high loadings with respect to the variables off/non-farm income (99.6% of the variance explained) and age of the household head (99.2%), respectively, and very low for the rest of the variables. In this example, it may be stated that PC1 represents access to off/non-farm income and PC2 the age of the household head. The third principal component has relatively high loadings with respect to total farm area (24.8%) and total number of livestock (31.1%), two classical variables associated with wealth in smallholder systems.

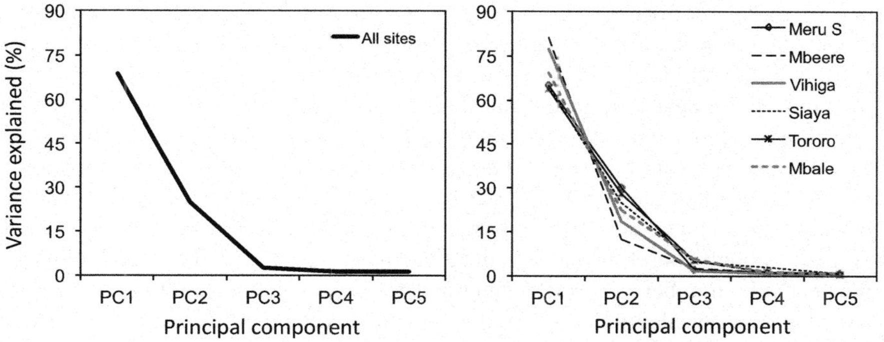

Fig. 4.12 Scree plots showing the proportion of the total variance in the data explained by each principal component in Table 4.6

At this point, it is time to explore the third question posed above: how many principal components are necessary? This is decided on the basis of the eigenvalues of each eigenvector, or in other terms, on the basis of the amount of variance accounted for by each principal component. The left hand side pane in Fig. 4.12 shows a scree plot from the same case study presented in Table 4.6, in which PC1 explains 67.7% of the variance in the data, while PC2 explains 25.7% and PC3 only 2.7%. The graph describes a typical 'elbow' at PC3, which is often used as cut-off point to decide how many PCs must be retained.[4] In this case, the first two principal components account for 93.4% of all the variance in the data. It would make little sense to retain more PCs. However, since in this example age of the household head may be associated with wealth in different ways and thus masking valuable information, it is worth exploring the data again by removing this variable. When removing age of the household head, the first PC still has a high loading with respect to off/non-farm income, but the second one has now high loadings with total farm area (0.547), total number of livestock (0.589), months of food self-sufficiency (0.429) and family size (0.338). The analysis in this example shows, in principle, that access to off/non-farm income is an overruling variable when it comes to describing household diversity in this region. To further explore if this trend is consistent across different sites in the region, or only true for one or a few of the sites (outlier), the analysis can be repeated site by site, as shown in the right hand pane of Fig. 4.12. In all sites in this example the first principal component had high loadings with respect to off–/non-farm income, and it explained between 60 to 80% of the variance in the data.

As expressed earlier, principal component analysis is a means of exploring the latent structure of the variance in the data, and it is also a dimension reduction technique. In the example above the information contained in seven variables was 'condensed' in two principal components. Each observation (each household in this

[4]Often the cut-off threshold is defined for eigenvalues smaller than one for the standardised data

case) will have a certain score with respect to the first two principal components. These PC scores can be now used to cluster households, instead of using the original seven variables. This makes clustering easier, irrespective of the method used. Both functions of principal component analysis, revealing the structure of the variance and reducing the dimension of a dataset, are used very often in agroecosystems analysis (not only to delineate household typologies). There is no strict recommendation in the literature as to the minimum number of observations necessary to conduct a principal component analysis. Certain authors recommend a minimum of 300 data points, although household surveys in agroecosystems analysis are often limited to a much smaller number of cases. When the dataset is small, the structure the variance can be easily revealed through simpler statistical methods of through direct observation. When the number of observations in the dataset is limited, the number of variables to be included in the analysis should not be large. A rule of thumb would be to keep a ratio of minimally 3–5 to 1 between the number of observations and the number of variables to be included.

As indicated before, principal components are guaranteed to be independent only if the data set is jointly normally distributed, and the analysis is sensitive to the relative scaling of the original variables (e.g., when including variables such as soil organic matter, expressed in %, and household income, which is some currencies corresponds to thousands or millions). Transforming the data prior to the analysis can be of help. Principal component analysis is not suited to deal with qualitative (nominal, categorical) variables, for which multiple factor analysis or multiple correspondence analysis can be used. Correspondence analysis is a special case of principal component analysis in which the data are tabulated (contingency tables). For details on canonical correlation, factor and correspondence analysis the reader is referred to the book of Sharma (2001).[5] When dealing with large datasets consisting of continuous, nominal and categorical variables and when it is known that most variables have thresholds, extreme values and skewed distributions, the data can be analysed (and grouped) by means of classification and regression trees (e.g. Tittonell et al. 2008). The groups or classes obtained through this type of analysis can match the hypothesis formulated on household diversity, or typology.

4.4.2.4 Identifying Groups or Clusters

Before embarking on the description of statistical methods to group similar households, let us go back to our hypothesis about farm diversity. A farm typology hypothesis is first formulated based on our observations in the field, on previous knowledge, on expert advice or through participatory categorisations. This initial categorisation can be merely structural or also functional. The groups of similar farms that will be identified through statistical clustering will be mostly based on

[5] A lot has been written about multivariate analysis in the past 20 years. I however still recommend the old text book from where I learnt these techniques, as I found it extremely didactic.

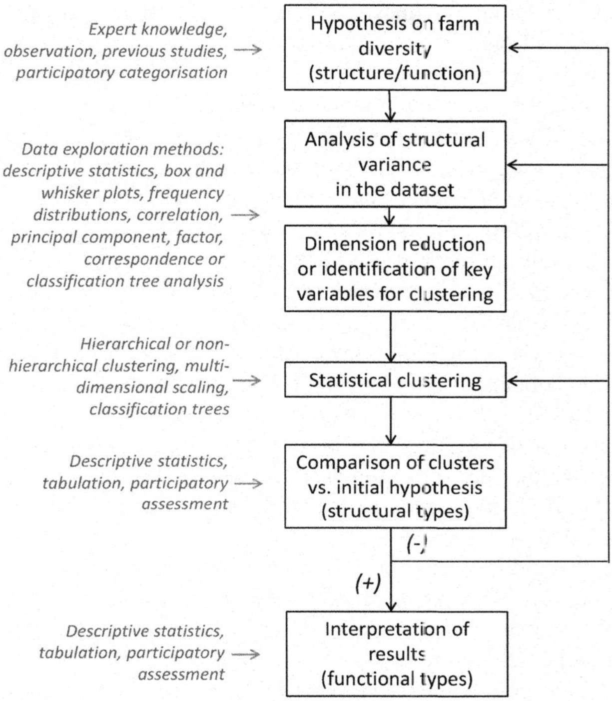

Fig. 4.13 Methodological steps to arrive at a functional farm and rural household typology, based on the exercise published by Tittonell et al. (2010) for East Africa and subsequently used in a diversity of contexts and applications, e.g. Tittonell et al. (2013), Cortez-Arriola et al. (2015), Bhattarai et al. (2017), Alvarez et al. (2018)

structural characteristics, and may match the predefined farm types or not. Already the exploration of the structure of the variance in the dataset through correlation or principal component analysis may have provided signals that confirm or reject our initial hypothesis. This is equivalent to say that the delineation of farm or rural household typologies, even when it is done by means of statistical methods, is an iterative procedure.

The scheme in Fig. 4.13 shows the various steps necessary to arrive at a functional typology, based on the approaches described so far in this chapter. When the clusters obtained through statistical methods do not match the groups defined in our hypothesis, it may be necessary to revise either the clustering operations (including accounting for misclassifications), the latent variance in the dataset and/or the choice of key variables, or even our initial hypothesis. A typology is rarely obtained in just one go! Here, the phases of data exploration and statistical grouping are kept separately for didactic purposes. In reality, both steps are done almost simultaneously by most statistical software packages of today. Classification trees, for example, are both a way to explore the data and at the same time form homogeneous groups. When the resulting clusters match the structural typology that best represents our initial hypothesis, the analysis can progress towards the interpretation of such types from a functional point of view. Yet, further feedback loops may still be necessary from this point.

Clustering consists of grouping the observations in such a way that those within a group are as similar as possible in terms of a specified set of attributes, and at the same time as dissimilar as possible from observations contained in other groups or clusters. Clustering itself is not a single algorithm. It is a notion, it is a task, and it can be accomplished through various algorithms that differ in the way they define a cluster as well as in the method used to identify it. Clustering is not an automatic method, but rather an iterative process of learning and discovery through iterations following the principle of trial and error. Often after an initial cluster analysis, it will be necessary to revise the assumptions of the model or the variables used until a desirable result is obtained. Clusters can be formed on the basis of distances between observations, of density areas in the dataset or of statistical distributions. The proper clustering algorithm as well as the number of clusters depend on the objective and/or on the hypothesis that drive the analysis. In the case of household typologies, the number of clusters should ideally match the hypothetical types that were predefined (cf. Figure 4.13).

Clustering methods are classified as hierarchical and non-hierarchical. Hierarchical clustering produces classifications in which small clusters of highly similar entities are nested within large clusters of less similar ones. Hierarchical clustering can be agglomerative (bottom-up), in which entities are progressively merged to form larger groups, or divisive (top-down), in which the initial cluster that represents the entire dataset is progressively subdivided. Non-hierarchical clustering partitions a dataset into non-overlapping groups that have no hierarchical relationships among them. The various techniques differ in the heuristics used to evaluate the partitioning, and the main categories of methods include single-pass, relocation and nearest neighbour. Relocation methods (such as *k*-means clustering) assign data points to predefined clusters, evaluate the outcome, and iteratively reassign them to find the most efficient clustering.

As in the case of data exploration methods, the reader is referred to the vast body of specialised literature on cluster analysis (cf. Sharma 2001). Here, only a few techniques commonly used in agroecosystems analysis will be described and illustrated with examples. Hierarchical or connectivity-based clustering relies on the principle that the distance that separates objects from each other defines the degree of similarity between them. Algorithms are used to connect objects in order to form clusters, and the number of clusters to be formed depends on the maximum distance necessary to connect objects within a cluster. The shorter the distance, the greater the number of clusters that will be formed. This is often represented in a dendrogram as the one presented in Fig. 4.14, in which the y axis indicates the distance at which the objects merge to form a cluster, expressed in this case as the level of similarity among the members of a group or cluster.

The within-group similarity threshold of 50% adopted in the example of Fig. 4.4 leads to the formation of five clusters. A similarity threshold of 40% will result in three clusters in which the objects they contain are less similar among themselves, whereas a similarity threshold of 60% will yield 11 clusters (*NB*: here is

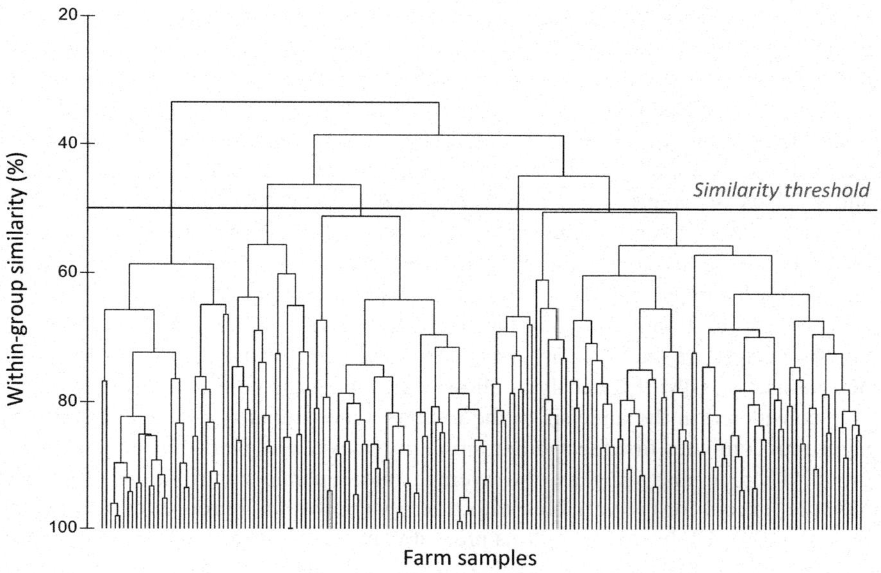

Fig. 4.14 Cluster dendrogram grouping a sample of smallholder farms from the Mid Zambezi Valley of Zimbabwe. (From Pacini et al. 2014)

where a solid initial hypothesis on household diversity can save us plenty of time and guide our decisions). A common measure of similarity is the squared Euclidean distance. The squared distance between two observations, *i* and *j*, is calculated as:

$$D^2{}_{ij} = \sum_{k=1}^{p} \left(x_{ik} - x_{jk}\right)^2 \tag{4.6}$$

Where, x_{ik} is the value of the *k*th variable for the *i*th observation, and the total number of variables is represented by *p*. In a two dimensional database defined for example by the variables number of household members and months of food self-sufficiency, the squared Euclidean distance between an observation (household) with four members and six months of self sufficiency, and another one with seven members and five months is calculated as:

$$D^2 = (4-7)^2 + (6-5)^2 = (-3)^2 + (-1)^2 = 10 \tag{4.7}$$

The various methods available for hierarchical clustering differ in the way in which 'distances' are calculated.[6] They include the centroid method, the nearest-neighbour

[6]Other measures of similarity use association coefficients and correlation coefficients instead of statistical distances.

or single linkage method, the farthest-neighbour or complete linkage method, the average linkage-method and Ward's method (which does no compute distances but a measure of homogeneity within clusters). Bray-Curtis distances, which are commonly calculated in ecology, are non-metric coefficients calculated from multidimensional, quantitative and qualitative, standardized variables (Bray and Curtis 1957). The Jaccard similarity coefficient is used for comparing the similarity and diversity of sample sets of binary variables (e.g. presence/absence) based on the size of the intersection and the union of the sample sets (Jaccard 1901). These two methods are particularly mentioned here because they are the ones used in the analysis of Pacini et al. (2014)[7] from which Fig. 4.14 was derived.

The next question is, if two or more observations are found to be similar, respect to what exactly are they similar, to a single variable or to all of them? The similarity percentage (or, SIMPER) analysis allows calculating the contribution of each variable considered in the analysis to the total average similarity of each cluster. This algorithm computes first the average similarity between all pairs of samples within a group. Then, it disaggregates the average into separate contributions from each variable (except for the variables with a zero value within a group). In the example from the Mid Zambezi valley farms presented in Fig. 4.14 the similarity percentage for each of the five clusters identified was calculated on the basis of the average similarities computed with the Bray-Curtis coefficient. The average similarity within clusters ranged from 60 to 70% and the contribution of the different variables to each cluster similarity can be seen in Table 4.7. A number of variables contributed to the average similarity within *Cluster A*, but the number of adults working off farm had the greatest contribution (63.5%). This variable did not contribute to the average similarity within the other four clusters. We may therefore infer that *Cluster A* groups all or most of the households in which off-farm income is an important livelihood strategy.

The non-hierarchical, centroid-based clustering technique known as *k*-means partitions the data into k groups. The number of clusters k must be known a priori. The analysis starts by defining the initial cluster centroids or seeds and assigning the closest observations to each of them, then reassign methods based on a set of rules. As indicated earlier, the various methods of non-hierarchical cluster differ in the method used to identify the seeds as well as in the algorithm used to reassign observations. It is not the intention here to dive any further into non-hierarchical cluster analysis, but just to warn the reader that the non-hierarchical clustering algorithms are highly sensitive to the initial definition of centroids. A clustering algorithm that minimises some form of statistical criterion is basically an optimisation algorithm, for which an objective function is defined. Since different starting points can be used, the final solution may result in local optimisation of this function, which would be undesirable. One way of identifying the initial centroids is to first

[7] For those interested in further reading on alternative methods, this paper provides a good example of how satistical and participatory clasifications can be used in combination to cluster households using multi-dimensional scaling techniques.

Table 4.7 Results of similarity percentage analysis for a sample of 176 farms from the Mid Zambezi Valley,

Variable	Contribution to group similarity (%)	Cumulative Contribution (%)
Cluster A (68.9% similarity)		
Nr of adults working off-farm	63.5	63.5
Nr of adults working on-farm	8.1	71.6
Nr of household members	7.6	79.2
Total area cultivated (ha)	6.0	85.2
Area under cotton (ha)	4.2	89.4
Nr of ruminants	3.4	92.7
Land preparation method	2.4	95.1
Nr of casual workers	1.8	97.0
Nr of fertiliser applications	1.6	98.6
Nr of draft cattle and donkeys	1.4	100.0
Cluster B (62.9% similarity)		
Nr of adults working on-farm	31.6	31.6
Nr of household members	30.8	62.4
Total area cultivated (ha)	25.4	87.8
Area under cotton (ha)	8.2	96.0
Nr of ruminants	4.0	100.0
Cluster C (62.8% similarity)		
Nr of casual workers	41.9	41.9
Nr of adults working on-farm	20.3	62.2
Nr of household members	20.1	82.4
Total area cultivated (ha)	10.7	93.1
Nr of ruminants	3.4	96.5
Area under cotton (ha)	3.3	99.8
Land preparation method	0.2	100.0
Nr of draft cattle and donkeys	0.0	100.0
Cluster D (70.3% similarity)		
Nr of draft cattle and donkeys	24.8	24.8
Nr of fertiliser applications	18.9	43.7
Area under cotton (ha)	13.3	56.9
Total area cultivated (ha)	10.9	67.9
Nr of ruminants	10.7	78.5
Land preparation method	7.7	86.2
Nr of household members	7.0	93.2
Nr of adults working on-farm	6.8	100.0
Cluster E (60.3% similarity)		
Land preparation method	23.4	23.4
Total area cultivated (ha)	14.0	37.4
Area under cotton (ha)	13.8	51.3
Nr of household members	12.9	64.1
Nr of adults working on-farm	12.4	76.6

(continued)

Table 4.7 (continued)

Variable	Contribution to group similarity (%)	Cumulative Contribution (%)
Nr of ruminants	7.5	84.1
Nr of draft cattle and donkeys	6.8	90.9
Nr of fertiliser applications	4.9	95.8
Nr of casual workers	4.2	100.0

Zimbabwe, adapted from Pacini et al. 2014 (original data from Baudron et al. 2012)

run a hierarchical cluster analysis, and use the resulting clusters as seeds for the non-hierarchical one. This combination is recommended for household clustering through non-hierarchical algorithms when dealing with large datasets or when the person analysing the data has not been in the field or is not familiar with the diversity of farming systems in the particular case under study. The number of classes can also be determined on the basis of set theory. A first rough approximation can be obtained with Eq. 4.1.

4.4.2.5 Interpretation of Clusters

Once the clusters are obtained through statistical means, and once all necessary iterations took place to arrive at a satisfactory grouping of households, it is time to interpret the results in the light of our initial hypothesis and of the new insights from the statistical analysis. Figure 4.15 shows an example of clusters formed in a dataset consisting of ca. 40 smallholder households from Siaya, in Nyanza Province of Kenya. The clusters were obtained through the hierarchical method, using principal component scores to replace the original variables (note that this is a sub-sample of the data from East Africa used in the examples of Table 4.6 and Fig. 4.12 scree plots). Five clusters were obtained, which differ in terms of the average level of off-farm income of the households (note that the first principal component has a 98% loading with respect to this variable). The numbering of the clusters in Fig. 4.15 is not chosen by chance. In fact, they correspond approximately to the five farm types described in Table 4.2 at the beginning of this chapter. Table 4.8 presents average values for some key socio-economic variables used in the analysis. Farms of type 1 and 5 (clusters 1 and 5, respectively) are those that show a greater reliance on off-farm income than the rest of the farm types. They are placed at the positive end of the first principal component in Fig. 4.15. The opposite is true for farms of type 2, in which on-farm activities and especially cash crops represent their greatest income.

In another example, Cortez-Arriola et al. (2015) used principal component analysis in combination with agglomerative hierarchical cluster analysis, based on the Ward's agglomeration method mentioned earlier, to group smallholder dairy farmers in a district of Michoacán, Mexico. The initial farm typology hypothesis in this case was the national typology of dairy farms used by the Mexican government, namely family-based, semi-specialised, specialised and dual-purpose dairy farms.

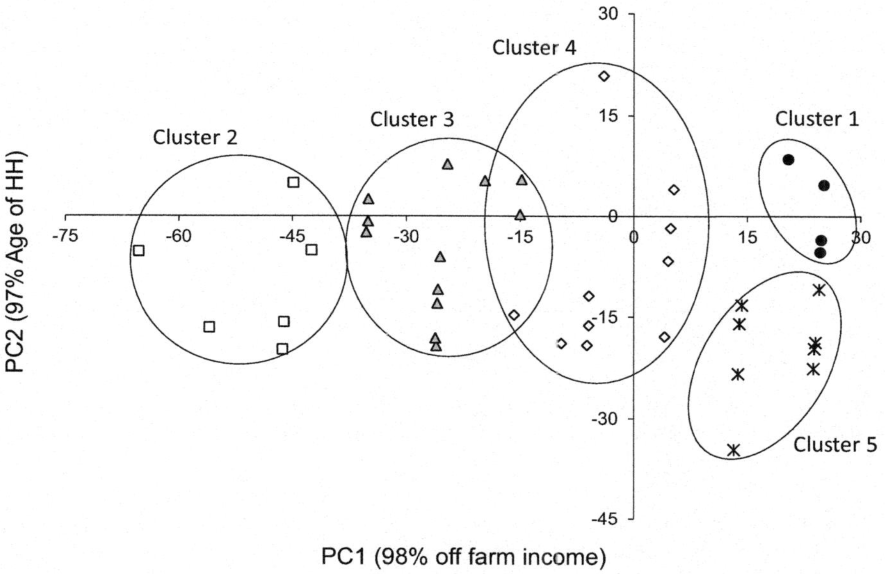

Fig. 4.15 Clusters of smallholder farm households (n = 40) in Siaya, Nyanza Province of Kenya, presented in a principal component (PC) coordinate system. The percentage between brackets represents the loading of each PC with respect to the variables indicated

Table 4.8 Average value of key socio-economic variables per cluster (Farm Types) presented in Fig. 4.15

Cluster	Household distribution	Total area owned	Owned cattle	Land available (ha) per family*		Off -farm income	Food self-sufficiency
	(%)	(ha)	(TLU)	member	labour	(%)	(months)
1	10	1.6	2.5	0.44	1.52	35	7.3
2	13	3.2	7.2	0.59	1.10	12	8.6
3	28	1.4	2.5	0.34	0.71	16	8.7
4	30	1.0	1.0	0.20	0.41	26	7.2
5	20	0.7	0.1	0.16	0.32	31	5.3

[a]Calculated as land cropped over the total number of family members or the number of those working on the farm, respectively

The dendrogram from the cluster analysis is presented in Fig. 4.16, where it is shown that a between-group dissimilarity level of 50% yields four clusters. The average value of key socioeconomic variables for each of the four clusters is presented in Table 4.9. In this case, the comparison of the groups of farms obtained through statistical clustering against the reference typology used by the government revealed that not all the types defined in this typology occurred among the farms surveyed. In fact, only the types family-based and semi-specialised had been surveyed, as the other two types did not exist in the study area.

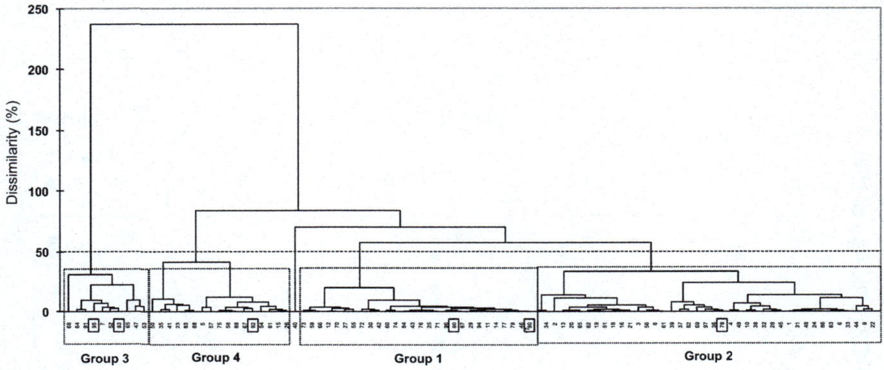

Fig. 4.16 Dendrogram from the agglomerative hierarchical cluster analysis of a survey of 97 smallholder dairy farms in Marcos Castellanos, Michoacán, Mexico. (From Cortez-Arriola et al. 2015)

Table 4.9 Average value of key socio-economic variables for each of the clusters obtained from the sample of 97 dairy farms in Michoacan, Mexico in the study of Cortez-Arriola et al. (2015)

Group	Land owned (ha)	Land rented (ha)	Grazing land (ha)	Livestock (LU)	Off-farm income (%)	Hired labor (Labor-day y^{-1})	Family labor (Labor-day y^{-1})
1	12	15	21	35	0	56	772
2	15	15	24	35	60	166	1001
3	139	42	167	144	25	905	477
4	26	78	72	70	18	281	974

Further, the authors identified wide variability in groups 1 and 3 (not shown in Table 4.9), and decided that a new classification criterion was necessary to refine the original typology. This new criterion was the level of intensification of the dairy farms, defined in three categories as Extensive, Medium and Intensive, and determined by the average cattle stocking rates (animals per unit area) and milk production levels. The typology resulted in six groups, namely family-based extensive (FBE), medium (FBM) and intensive (FBI), and semi-specialised extensive (SSE), medium (SSM) and intensive (SSI). Groups 2 and 4 were identified as intensive systems, with stocking rates greater than 1.25 heads per hectare, and milk production per cow greater than 4300 L per year. They were labelled as FBI and SSI, respectively. Groups 1 and 3 were subdivided with respect to the level of intensification of the farms, resulting in FBE and FBM from group 1 and SSE and SSM from group 3. Although the ramifications in the agglomerative hierarchical cluster dendrogram were not followed anymore, this subdivision done 'by hand' using expert knowledge improved the classification of households considerably with respect to the initial clustering. The resulting groups were more homogeneous with respect to key socioeconomic variables (Table 4.10).

Table 4.10 A typology of dairy farms from Michoacán, Mexico (Cortez-Arriola et al. 2015) and average values for several variables used in the analysis; family-based extensive (FBE), medium (FBM) and intensive (FBI), and semi-specialised extensive (SSE), medium (SSM) and intensive (SSI)

Group	Type	Land owned (ha)	Land rented (ha)	Grazing land (ha)	Livestock units (LU)	Off-farm income (Index)	Hired labor (Labor-day y^{-1})	Family labor (Labor-day y^{-1})
1	FBE	0	57.9	55	18.9	0	0	548
	FBM	11	35.5	45	35.2	0	185	365
2	FBI	24	0	18	31.2	0.5	387	148
3	SSE	95	0	76	59.2	0.5	1110	114
	SSM	101	31.8	109	141.7	0	1208	365
4	SSI	12	11.3	19	37.3	0	1000	532

4.5 Functional Farm Typologies as Archetypes

As explained earlier in this Chapter, functional typologies are those that aim to capture decision-making by farmers, as well as their behaviour in the face of changing contexts or socio-economic or ecological situations (Mettrick 1993). Delineating functional typologies may serve to identify patterns in the response of households to different drivers, such as natural disasters, ecological stressors (e.g., droughts, soil erosion, etc.), price shocks, or the effects of new policies, rules or institutional settings. The type of information to be collected through household surveying for functional typologies differs from the common variables used in structural typologies. Data collection is normally done through a conversation with the household members or other informants rather than through a structural questionnaire. The information collected is often ´narrative´ data. For example, to a question such as ´*what would you do with your herd if there is a long-lasting drought this summer?* ´, household members may answer: ´I will move it somewhere else ´, ´I will sell part of it´, or ´I will purchase concentrate feeds´, and so on. Collecting this type of information tends to be time-consuming, which limits the number of households that can be included in a typology. In other words, building functional typologies typically implies dealing with limited number data points (observations) and with a combination of quantitative and qualitative data. Such qualitative data may be expressed as ordinal, scores, ranks, nominal or binary variables, or simply as ´strings´ of information.

Multivariate analysis in combination with clustering techniques have been the most commonly used statistical approach to identify farm types in the last decades, as illustrated earlier in this Chapter. This includes principal component, multiple correspondence or factor analysis (e.g. Bhattarai et al. 2017), multi-dimensional scaling (e.g., Pacini et al. 2014) or use of Bayesian systems (e.g. Tiffin 2006). These methods tend to group farm households around an observation (centroid) that represents ‘a type’ and around which the rest of the observations are grouped or clustered. The scores of the principal component (or similar dimension-reduction technique) are used to create the clusters, such that the variance within the clusters is minimised while that between clusters is maximised. The larger the number of observations, the stronger the power of the method to create distinct clusters. One way to approach the identification of functional types has been to use functional variables instead of structural ones in the multivariate analysis and clustering exercise (e.g., Tittonell et al. 2010; Alvarez et al. 2018).

However, one limitation of the multivariate/clustering methods is that the central concepts used to form the clusters, or centroids, tend to exhibit close-to-average values in most dimensions. The extremal cases in the sample of households tend to be allocated within one of the selected clusters and 'lost' from sight. Yet, such salient cases are often those that reflect distinct behaviour or new strategies. These are typically the most innovative households in a community, the early adopters, the influencers, etc. In functional responses to external drivers, and their diversity, extremal cases or apparent outliers may be as informative or more than ´typical´ ones. The use of archetypal analysis to form clusters, instead of classical clustering techniques, has been proposed as a way of capturing the diversity or extremal cases (cf. Tittonell et al. 2020). Archetypal analysis has been used in recent years in combination with GIS techniques to identify and categorise recurrent processes at regional scale, such as archetypes of vulnerability to climate change (e.g. Sietz et al. 2017), or archetypes of land use change (e.g., Oberlack et al. 2019), etc. In the study of Tittonell et al. (2020), archetype analysis was used for the first time to categorise household responses to environmental hazards. But before describing this example, let us first examine how archetypal analysis works conceptually.

Figure 4.17a, b reproduces what has been already illustrated in Fig. 4.11 of this Chapter, the way in which principal components (*u* and *v*) are calculated as linear combinations of the original variables (x, y). Figure 4.17c illustrates how archetypes are identified from the same set of data. In its classic form (e.g., Cutler and Breiman

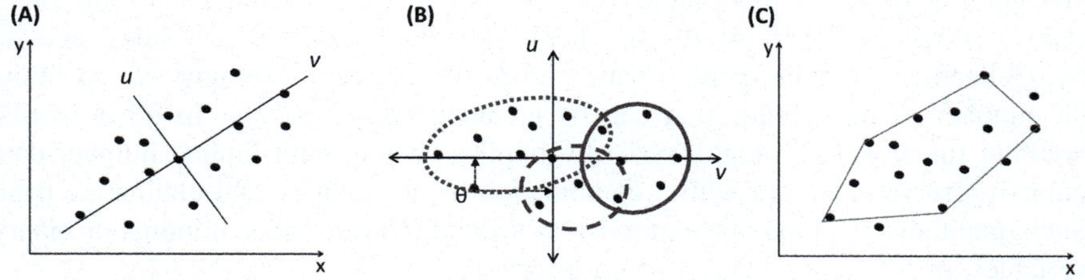

Fig. 4.17 (**a**) Observation set plotted in a plane defined by two variables, x and y; (**b**) The same observation set plotted with respect to two orthogonal vectors (e.g., principal components), *u* and *v*, that represent linear combinations of x and y, and grouped into three possible clusters; (**c**) The same observation set plotted against the original variables x and y and surrounded by a convex hull defined by six extremal observations or archetypes. These archetypes are used as starting points to build the clusters

1994), archetypal analysis is an unsupervised learning method that seeks extremal points in the multidimensional data and creates convex combinations of observations.[8] Convex combinations are linear combinations of extremal points, as illustrated by the convex hull approximated with six archetypes in Fig. 4.17c. The archetypes represent ideal functional types, yet an archetype (a vertex in Fig. 4.17c) does not necessarily represent the strategy or response of a ′real′ farmer or household. An archetype depicts the main features in the responses found in the sample of households. These features emerge from individual strategies, which are associated with an archetype only in a probabilistic manner. Archetypes represent thus ′ideal′ types that symbolize diverse responses in a community or group of farmers in a study area.

Together with colleagues from the Bariloche branch of the Instituto Nacional de Tecnología Agropecuaria (INTA), we used archetypal analysis to categorise rural household responses to stress generated by several years of persistent droughts and a volcanic ash fall event (2011) in northern Patagonia, Argentina. In other words, a functional typology of how households responded to natural hazards, of the strategies they deployed to cope with or adapt to them. We interviewed 23 households, to find out about their perception of droughts and climatic hazards, their decision-making processes and adaptation strategies (cf. Solano-Hernandez et al. 2020). To identify archetypes, we first transformed the narrative information into categorical variables, as shown in Table 4.11. The number of classes varies from one variable to the other, from binary to eight classes, and they are not ordinal. For example, if we take the variable 'Income diversification', the numbering proposed for each class may change (e.g., Permanent off arm income becomes #8 while Handicraft work becomes #2), yet the result of the analysis would be exactly the same. In this case, we used 10 variables to analyse 23 households. This is also an advantage over multivariate analysis, which normally requires a minimum of 3 to 5 observations per variable included in the analysis.

The new variables shown in Table 4.11 were used as inputs in the archetype analysis. Once the analysis is run,[9] a decision has to be made concerning the number of archetypes to select to best categorise the diversity of responses. In this case, we used the Akaike Information Criterion (AIC), starting from two archetypes (the minimum possible) and increased stepwise until the value of the AIC began to increase. The output with the lowest AICc criterion was selected, which in our case resulted in three archetypes (A, B and C). Each household had different loadings with respect to the three archetypes, as shown in Fig. 4.18. For example, households #13 and #16 had high loadings with archetype A, #4 and #19 with archetype B, and #2, #8, #9, #14, #20 and #22 with archetype C. Some households had similar loadings with more than one archetype (e.g. #17, #21, #23). Likewise,

[8] This is but one among several different methods used nowadays for archetype analysis (cf. Oberlack et al. 2019)

[9] In this case, using the *py_pcha* module for Archetypal Analysis in python (Jensen and Schinnerl, 2017), which implements the algorithm from Mørup and Hansen (2012)

Table 4.11 Example of how to transform drought-related narrative information into scores for the archetype analysis

Variable	Classes and description
Livestock production losses due to drought	1. No loss 2. < 20% 3. 20–40% 4. 40–60% 5. 60–80% 6. > 80%
Supplementary livestock feeding	1. No 2. Yes, a few times or punctual 3. Yes, a fraction of the heard or certain species 4.Yes, all herd and/or frequently
Herd size reduction (elimination of less productive animals)	1. No 2. Yes
Grassland resting period	1. No 2. Yes, certain areas or some years 3. Yes, as a permanent practice, or management of summer/winter grasslands
Practice of transhumance	1. No 2. Yes, punctually, or in earlier times 3.Yes, transhumance as a management strategy
No change in management over the last 15 years	1. No, there were changes 2. Few, almost no change 3. Yes, no change at all
Building of new infrastructure	1. No 2. Yes, hay barn 3. Yes, shed or shelter for animals
Found new water sources	1. No 2.Yes, but failed or still in construction 3.Yes, functioning water well
Improved water use efficiency	1. No 2. Water pump and7or surface irrigation 3. Reservoir, tank 4. Rainwater harvest
Income diversification	1. No 2. Permanent off-farm income (salary) 3. Transitory of-farm income 4. Pension or retirement 5.Family member with a job (children) 6.Combinations of 1, 2, 3 or 4 7. Farm income diversification 8. Handicraft or hand-made products

From Tittonell et al. (2020)

each variable had a different degree of participation in defining each archetype; for example, income diversification was more frequent in archetype A than in the other two. The reader is referred to the original paper by Tittonell et al. (2020) for further details on the analysis and results from this exercise.

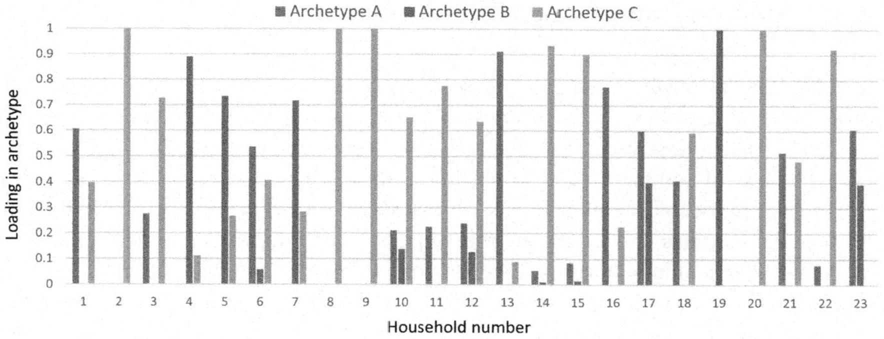

Fig. 4.18 Loading of the 23 households with respect to the three archetypes identified in the analysis of household responses to natural hazards in north Patagonia. (cf. Tittonell et al. 2020)

What is important to highlight from this example is the interpretation of the archetypes identified. Archetype A corresponded to households that exhibited a limited impact from the frequent droughts and the 2011 volcanic ash fall that affected the region, with average productivity losses below 40%. This archetype corresponded with households that implemented simple technological innovations, such as supplementary animal feeding (concentrates, hay, fodder), infrastructure for animal housing and sheltering (from predators), irrigation capacity in part of their land, new water points for livestock, etc. A diversified portfolio in terms of production activities and sources of income was also associated with this archetype. Archetype B represented farm households that underwent intermediate impact in terms of productivity losses (40 to 60%), that invested heavily in agriculture and livestock intensification, and that depended strongly on off-farm income. Archetype C corresponded to the group of more severely affected farm households, with productivity losses >60%, limited capacity to respond and adapt to droughts and to the consequences of the ash fall, implementing traditional extensive livestock ranching with low levels of income and production diversification.

As said earlier, archetypes highlight extremal cases that represent salient behaviour, which can be categorised into functional responses groups using the analogy of adaptive strategies in ecology: tolerance, resistance, avoidance, diversification and transformation. Ultimately, social-ecological systems such as rural households respond to large-scale, persistent disturbances through a diversity of mechanisms, dependent on their characteristics in terms of vulnerability, resilience and adaptive capacity (structures, functions and states), and on the type and magnitude of the disturbance(s) considered (e.g., Gallopin 2006). Functional typologies obtained through archetypal analysis may contribute to shedding additional light on such responses.

4.6 Summary and Concluding Remarks

Categorising farm diversity requires some basic knowledge on statistical tools in order to be able to decide on which methods to use and to be able to interpret the outcomes of the analyses. Yet statistical methods are only used to assist in the classification of farms or rural households, they are not in themselves an object of study in agroecosystems analysis. We should avoid 'placing the cart before the horse'. First comes the question, the objective, and then the method (although in some cases the question can be of a methodological nature). In this Chapter, we examined different methods to build typologies, from participatory self-categorisation by farmers and communities to statistical methods. The latter included both multivariate analysis in combination with clustering techniques, on the one hand, and archetype analysis on the other. Yet, the different examples shown in this Chapter indicate that common sense and knowledge on the actual farm systems under study are as important as any statistical method. The interpretation of the resulting clusters or archetypes is crucial to understanding farm diversity. The aim is to be able to categorise households that are structurally and functionally different in order to fulfil one or more of the possible objectives described at the beginning of the Chapter: targeting and upscaling. Ultimately, understanding household diversity can more easily orient efforts to target agroecology principles and development initiatives.

References

Adelhart Torop R, Gosselink K (2013) Analysis of positive deviants among organic dairy farmers in The Netherlands. MSc Thesis, Wageningen University, The Netherlands, 53 p

Alvarez S, Timler CJ, Michalscheck M, Paas W, Descheemaeker K, Tittonell P et al (2018) Capturing farm diversity with hypothesis-based typologies: an innovative methodological framework for farming system typology development. PLoS One 13(5):e0194757. https://doi.org/10.1371/journal.pone.0194757

Baudron F, Tittonell P, Corbeels M, Letourmy P, Giller K (2012) Comparative performance of conservation agriculture and current smallholder farming practices in semi-arid Zimbabwe. Field Crop Res 132:117–128

Bhattarai S, Alvarez S, Gary C, Rossing WAH, Tittonell P, Rapidel B (2017) Combining farm typology and yield gap analysis to identify major variables limiting yields in the highland coffee systems of llano Bonito, Costa Rica. Agric Ecosyst Environ 243:132–142

Bray JR, Curtis JT (1957) An ordination of the upland Forest communities of Southern Wisconsin. Ecol Monogr 27:325–349

Cabrera VE, Hildebrand PE, Jones JW (2005) Modelling the effect of household composition on the welfare of limited-resource farmers in Coastal Cañete. Peru Agic Syst 86:207–222

Corbett J, Keller C (2005) An analytical framework to examine empowerment associated with participatory geographic information systems (PGIS). Cartographica 40:91–102. https://doi.org/10.3138/J590-6354-P38V-4269

Cortez-Arriola J, Rossing WAH, Massiotti RDA, Groot JCJ, Tittonell P (2015) Leverages for on-farm innovation from farm typologies? An illustration for family-based dairy farms in north-west Michoacán, Mexico. Agric Syst 135:66–76

Crowley EL, Soule MJ, Carter SE (1996) Off-farm income and farming in western Kenya. Report to USAID/Kenya, Nairobi

Cutler A, Breiman L (1994) Archetypal analysis. Technometrics 36(4):338–347

Deckers J (2002) A systems approach to target balanced nutrient management in soilscapes of sub-Saharan Africa. In: Vanlauwe B, Diels J, Sanginga N, Merckx R (eds) Integrated nutrient management in sub-Saharan Africa: from concept to practice. CAB International, Wallingford, pp 47–61

Diarisso T, Corbeels M, Andrieu N, Djamen P, Douzet J-M, Tittonell P (2015) Soil variability and crop yield gaps in two village landscapes of Burkina Faso. Nutr Cycl Agroecosyst 105(3): 199–216. https://doi.org/10.1007/s10705-015-9705-6

El Mujtar VA, Zamor R, Salmerón F, Guerrero AS, Laborda L, Tittonell P, Hogan R (2023) Lexical analysis improves the identification of contextual drivers and farm typologies in the assessment of transitions to agroecology through TAPE – a case study from rural Nicaragua. Agric Syst 209:103686

Gallopin GC (2006) Linkages between vulnerability, resilience, and adaptive capacity. Glob Environ Chang 16:293–303. https://doi.org/10.1016/j.gloenvcha.2006.02.004

Gliessman SR (2007) Agroecology: the ecology of sustainable food systems. CRC Press/Taylor & Francis, New York, 384 p

Henry M, Tittonell P, Manlay R, Bernoux M, Albrecht A, Vanlauwe B (2009) Biodiversity, C stocks and sequestration potential in aboveground biomass in smallholder farming systems of western Kenya. Agric Ecosyst Environ 129:238–252

Jaccard P (1901) étude comparative de la distribution florale dans une portion des Alpes et du Jura. Bulletin de la Société Vaudoise des Sciences Naturelles 37:547–579

Jech T (2003) Set theory, Springer monographs in Mathematics (third millennium ed.). Springer-Verlag, Berlin/New York, ISBN 978-3-540-44085-7

Jensen UA, Schinnerl C (2017) Fast Python implementation of Archetypal Analysis using Principle Convex Hull Analysis (PCHA). https://github.com/ulfaslak/py_pcha

Landais E (1998) Modelling farm diversity: new approaches to typology building in France. Agric Syst 58:505–527. https://doi.org/10.1016/S0308-521X(98)00065-1

Mettrick H (1993) Development oriented research in agriculture: an ICRA textbook. ICRA

Mørup M, Hansen LK (2012) Archetypal analysis for machine learning and data mining. Neurocomputing 80:54–63

Oberlack C, Sietz D, Bürgi Bonanomi E, De Bremond A, Dell'Angelo J, Eisenack K, Ellis EC, Epstein G, Giger M, Heinimann A, Kimmich C, Kok MTJ, Manuel-Navarrete D, Messerli P, Meyfroidt P, Václavík T, Villamayor-Tomas S (2019) Archetype analysis in sustainability research: meanings, motivations, and evidence-based policy making. Ecol Soc 24(2):26. https://doi.org/10.5751/ES-10747-240226

Pacini GC, Colucci D, Baudron F, Righi E, Corbeels M, Tittonell P, Stefanini FM (2014) Combining multi-dimensional scaling and cluster analysis to describe the diversity of rural households. Exp Agric 50:376–397

Rambaldi G, Callosa-Tarr J (2002) Participatory 3-dimensional modelling: guiding principles and applications. ASEAN Regional Center for Biodiversity Conservation (ARCBC), Los Baños. ISBN 971-8986-47-2

Ruiz Meza L (2014) Adaptive capacity of small-scale coffee farmers to climate change impacts in the Soconusco region of Chiapas, Mexico. Clim Dev 7:1–10. https://doi.org/10.1080/17565529.2014.900472

Sharma S (2001) Applied multivariate techniques. John Wiley and Sons Inc., New York, 512 p

Sietz D, Ordoñez JC, Kok MTJ, Janssen P, Hilderink HBM, Tittonell P, Van Dijk H (2017) Nested archetypes of vulnerability in African drylands: where lies potential for sustainable agricultural intensification? Environ Res Lett 12(9):095006. https://doi.org/10.1088/1748-9326/aa768b

Solano-Hernandez O, Bruzzone J, Groot L, Laborda A, Martínez P, Tittonell MHE (2020) Convergence between satellite information and farmers' perception of drought in rangelands of North-West Patagonia. Argent Land Use Policy 97:104726. https://doi.org/10.1016/j.landusepol.2020.104726

Speelman (2014) Gaming and simulation to explore resilience of contested agricultural landscapes. PhD Thesis, Wageningen University, The Netherlands, 221 p.

Speelman EN, Groot JCJ, García-Barrios LE, Kok K, van Keulen H, Tittonell PA (2014) From coping to adaptation to economic and institutional change – trajectories of change in land-use management and social organization in a Biosphere Reserve community, Mexico. Land Use Policy 41:31–44

Teixeira HM, van den Berg L, Cardoso IM, Vermue AJ, Bianchi FJJA, Pena-Claros M, Tittonell P (2018) Understanding farm diversity to promote agroecological transitions. Sustainability 10(12)., 20 p.:4337

Tiffin R (2006) Bayesian clustering of farm types using the mixtures model. J Agric Econ 57: 547–562. https://doi.org/10.1111/j.1477-9552.2006.00064.x

Tittonell P (2003) Soil fertility gradients in smallholder farms of western Kenya. Their origin, magnitude and importance. Quantitative Approaches in Systems Analysis No. 25, ISBN 90-6754-713-1, Wageningen, The Netherlands, pp. 233

Tittonell P, Vanlauwe B, Leffelaar PA, Rowe E, Giller KE (2005a) Exploring diversity in soil fertility management of smallholder farms in western Kenya. I. Heterogeneity at region and farm scale. Agriculture. Ecosyst Environ 110:149–165

Tittonell P, Vanlauwe B, Leffelaar PA, Shepherd KD, Giller KE (2005b) Exploring diversity in soil fertility management of smallholder farms in western Kenya. II. Within-farm variability in resource allocation, nutrient flows and soil fertility status. Agriculture. Ecosyst Environ 110: 166–184

Tittonell P, Shepherd KD, Vanlauwe B, Giller KE (2008) Unravelling the effects of soil and crop management on maize productivity in smallholder agricultural systems of western Kenya – an application of classification and regression tree analysis. Agric Ecosyst Environ 123:137–150

Tittonell P, Muriuki AW, Shepherd KD, Mugendi D, Kaizzi KC, Okeyo J, Verchot L, Coe R, Vanlauwe B (2010) The diversity of rural livelihoods and their influence on soil fertility in agricultural systems of East Africa – a typology of smallholder farms. Agric Syst 103:83–97

Tittonell P, Muriuki A, Klapwijk CJ, Shepherd KD, Coe R, Vanlauwe B (2013) Soil heterogeneity and soil fertility gradients in smallholder agricultural systems of the East African highlands. Soil Sci Soc Amer J 77:525–538

Tittonell P, Bruzzone O, Solano-Hernández A, López-Ridaura S, Esdale MH (2020) Functional farm household typologies through archetypal responses to disturbances. Agric Syst 178, 102714. https://doi.org/10.1016/j.agsy.2019.102714

van der Ploeg JD (2012) The genesis and further unfolding of farming styles research. Hist Anthropol 20(3):427–439. https://doi.org/10.7788/ha.2012.20.3.427

Chapter 5
Production Functions and Factors in Agroecosystems

Abstract Agroecological production systems are often described as being knowledge-intensive or management-intensive. However, what does the term 'intensity' mean in agroecosystems analysis? It can be simply defined as the degree to which a production factor is used in a production process, relative to other production factors and output productivity. Land, labour, and capital are the classical production factors in agriculture, and the way in which they are combined to realize a certain production determines different levels and forms of agricultural intensification and productivity. Nowadays, knowledge is considered as a fourth production factor. The use of agricultural knowledge—traditional, scientific, popular, etc.—may replace or reduce the need for other production factors to achieve similar levels of productivity. A common misconception related to intensification is the assumption that family or smallholder agriculture is less intensive than industrial agriculture. This chapter will provide concepts and tools to analyse that. Another concept closely related to production factors is that of resources. In agricultural economics, resources are classified as fixed or long-term (such as land, machinery, infrastructure, irrigation systems, etc.) and operational or short-term resources, which are consumed completely during one production process (e.g., seeds, fertilizers). This view is strongly rooted in the industrial approach to agriculture that emerged during the green revolution. In agroecology, we tend to see most resources as being reproducible (e.g., land can be restored, seeds can be produced locally, nutrients can be recycled, etc.), and we distinguish between internally sourced and externally sourced resources, as well as between biotic resources (e.g., genetic resources, biodiversity, pollen) and abiotic ones (water, nutrients). The differences between these views have several implications for the way in which resource use efficiencies are calculated, especially because a single resource can be used in several processes in the agroecosystem, within a single season or over time. However, to understand how this can be done in agroecology, it is first necessary to know how factor allocation, resource use, and productivity are analysed in classical agronomy. This chapter provides system analytical concepts and methods to assess agricultural intensification, factor productivity, and resource allocation. However, resource allocation analyses are restricted here to land and labour. Financial resources are left out of this chapter, as they are often analysed in light of economic theory, which represents a completely different paradigm compared to systems analysis. Patterns concerning

P. Tittonell, *A Systems Approach to Agroecology*,
https://doi.org/10.1007/978-3-031-42939-2_5

the spatial allocation of biomass and nutrients are discussed in Chap. 7, while trade-offs around the allocation of financial versus other resources are addressed in Chap. 9.

5.1 Agricultural Intensification and Production Factors

The term "intensive agriculture" has been used to describe production systems with a low fallow-to-crop ratio or no fallow at all. These systems involve the intensive use of labour and capital inputs, as well as the application of large quantities of synthetic fertilizers and pesticides per unit area or in relatively small land areas. Intensive agriculture is often associated with factory farming practices, such as intensive pork, poultry, or dairy systems where animals are raised in confinement. In economics, intensification is defined in relation to the production factors land, labour, and capital, involving the *replacement* of one factor with another. For example, this can occur through the substitution of land with capital through investing in glass-houses. Terms like "input-intensification" or "labor- and capital-intensive systems" are commonly used to describe these intensification processes.

Throughout history, intensification has been a response to either land scarcity, aiming to increase productivity per unit area, or labour scarcity, aiming to increase productivity per worker. These two processes often occur sequentially, as we will explore further later on. Intensification in response to land scarcity typically involves increased labour or capital investments per unit area. On the other hand, intensification in response to labour scarcity occurs when mechanization or other labour-saving technologies (capital) are employed to reduce labour requirements. However, it is worth noting that sometimes the term "intensification" is erroneously used to simply refer to high levels of output per unit area, such as cereal or milk yields, regardless of the methods used to achieve those outputs. Now, let us delve into some classical perspectives on agricultural intensification that are still valuable for analysing intensity in certain contexts.

5.1.1 A Classical View on Intensification Trajectories

The study of long-term trajectories of farm intensification has been a key focus for various schools of thought, including the French school on agrarian systems dynamics (e.g., Mazoyer and Roudart 2006). Figure 5.1 illustrates an intensification trajectory of an agroecosystem by plotting land-to-labour ratios against agricultural income per capita. The level of income per capita determines the viability and attractiveness of the agroecosystem when choices are available. Viability is defined by the income level that ensures the survival of the household, serving as the lower limit, while attractiveness is determined by the income per capita that can be achieved through off-farm employment, often found in towns. At point A on the

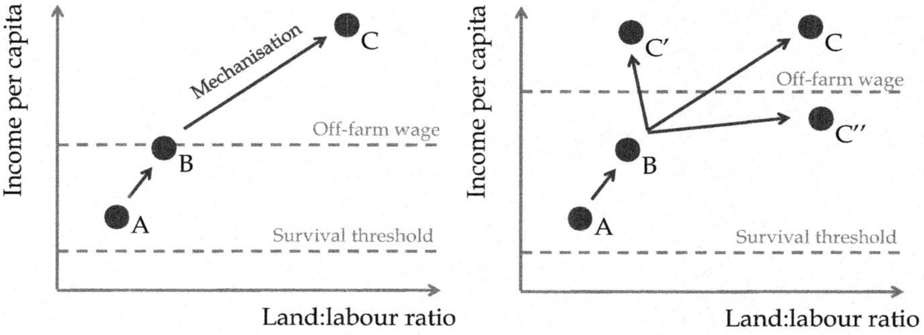

Fig. 5.1 A classical view on the long-term trajectories of agroecosystems in terms of land:labour ratios and incomes when subject to labour-saving technologies, under two off-farm wage scenarios. (Adapted from Mazoyer and Roudart 2006)

graph, the agroecosystem is viable but less lucrative compared to off-farm employment. In such cases, household members may choose to partially or completely exit agriculture, resulting in an increase in the amount of land available per remaining unit of labour in agriculture. This surplus land may be utilized by the same household or acquired or utilized by other households. This increase in the land-to-labour ratio can, in itself, lead to higher per capita income, up to point B, with the same level of production technology. Alternatively, the introduction of new technologies can enhance labour productivity, generating surpluses that enable the acquisition or access to additional land. The net effect of such technologies may eventually lead to the displacement of people from agriculture as farm labour becomes redundant, less productive, and less remunerative.

Regardless of the trajectory systems have followed from point A to B, a new dilemma arises. At point B, agricultural income per capita has increased and is now comparable to off-farm wages (as shown in Fig. 5.1). However, simply absorbing more people as agricultural laborers is not a viable option at constant prices, as the system is likely to regress back to point A. If off-farm wages decrease, agriculture may become attractive again, although trajectories of stepping out of agriculture are not always reversible (see Chap. 4 for further discussion). In other words, when families or individuals migrate to urban areas, they may not be inclined to return to the countryside, even when agricultural incomes improve. However, it should be noted that there are cases worldwide where back-migration is possible and considered as an option during times of economic crisis or unemployment. It is important to differentiate between countries, regions, and societies that predominantly remain rural, such as several regions in sub-Saharan Africa, and predominantly urban countries, like many European nations. In the latter, rural-urban migration has been occurring since the industrial revolution, and even earlier, from the seventeenth century, driven by better off-farm wages.

Referring back to Fig. 5.1, if off-farm wages increase, more individuals will be inclined to exit agriculture. The resulting increase in land-to-labour ratios may

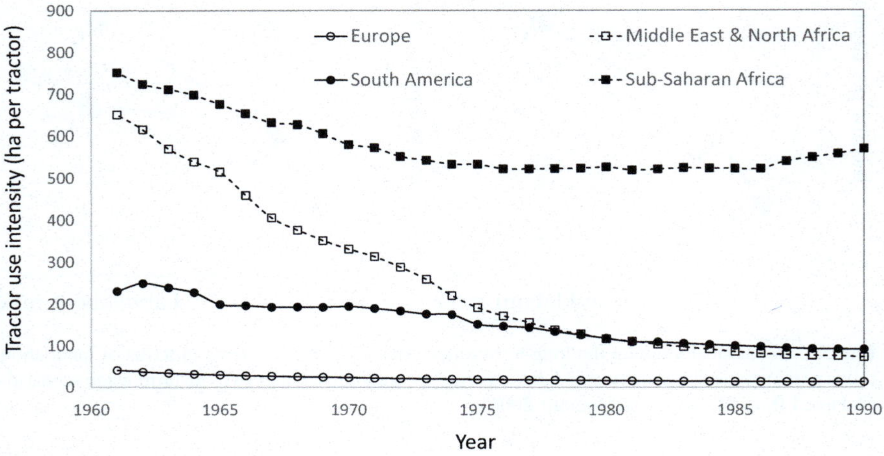

Fig. 5.2 Tractor use intensity in four world regions during a 30-year period corresponding approximately with what is known as the green revolution in agriculture. Tractor use intensity is an index calculated by FAO that relates the number of hectares of agriculture and the number of tractors in a country. The lower the value the greater the intensity of tractor use. (Source: FAOStat)

enable slightly higher income per capita from agriculture. However, the system may soon reach a new limit in production possibilities dictated by the amount of land that can be efficiently managed per worker using similar levels of technology. If technologies remain unchanged, labour productivity will tend to decrease as land-to-labour ratios increase, and income per capita may once again fall below the threshold of off-farm wages. One common response to this phenomenon is the mechanization of agricultural activities (as depicted in Fig. 5.2). Mechanization can lead to significant increases in labour productivity. The area of land that can be cultivated per unit of labour, as well as income per capita, will be much higher, shifting the agroecosystem to point C in Fig. 5.1.

The stability of this new situation is influenced by several factors. As the productivity of labour and land increases due to new technologies, agricultural output prices tend to decline. Consequently, the purchasing power of off-farm wages increases as households need to spend relatively less on food and agricultural products. This leads to a decrease in agricultural incomes per capita. As off-farm wages become more appealing, more individuals are enticed to leave agriculture, resulting in a larger rural population with more land available for cultivation but lower prices received for their produce. To raise labour productivity once again, further technological advancements or mechanization may be required. It is important to note that both the survival threshold and the off-farm wage threshold are not fixed values; they fluctuate based on relative prices of inputs and outputs, policies and regulations, socio-cultural changes, international trade, and the broader economic context. Additionally, the relationship between land-to-labour ratios and income per capita is not constant over time due to the emergence of agricultural innovations or the degradation of the natural resource base, such as soil erosion. Sophisticated and knowledge-intensive technologies can significantly increase productivity and per capita incomes without the need for expanding land-to-labour ratios. In many parts of the world, particularly in developed and urbanized regions, governments have implemented policies to sustain artificially high prices for agricultural products and/or provide subsidies for agricultural technologies and inputs. These measures are implemented to prevent the depopulation of rural areas and maintain their economic viability.

While the dynamics described above serve to illustrate the past evolution of agroecosystems in different parts of the world, notably in rainfed temperate Eurasian regions, current smallholder agroecosystems may exhibit different dynamics. Figure 5.3 shows the relationship between land:labour ratios and per capita income for two smallholder agroecosystems in sub-Saharan Africa, indicating the survival and off-farm wage thresholds expressed in local currencies. In both cases, subsistence and market-oriented households coexist in the same territory, and the differences between survival and off-farm thresholds are too narrow relative to the observed variation in incomes per capita. In the case of the Kanu Plain (Kisumu, Kenya) a number of households appear to be existing – oddly enough – below the 'survival' threshold (Fig. 5.3a). Most of the farmers in the Kanu Plains that grow sugar cane through contract farming with sugar factories, and those who grow vegetables for the market, achieve income levels that are higher than the opportunity cost of labour or off-farm wage. Market-oriented farming, according to this analysis, appears to be a more remunerative activity than the available off-farm jobs. This is even more accentuated in the Benoué basin (Garoua, Cameroon), where virtually all farms growing cotton lay above the off-farm wage threshold (Fig. 5.3b).

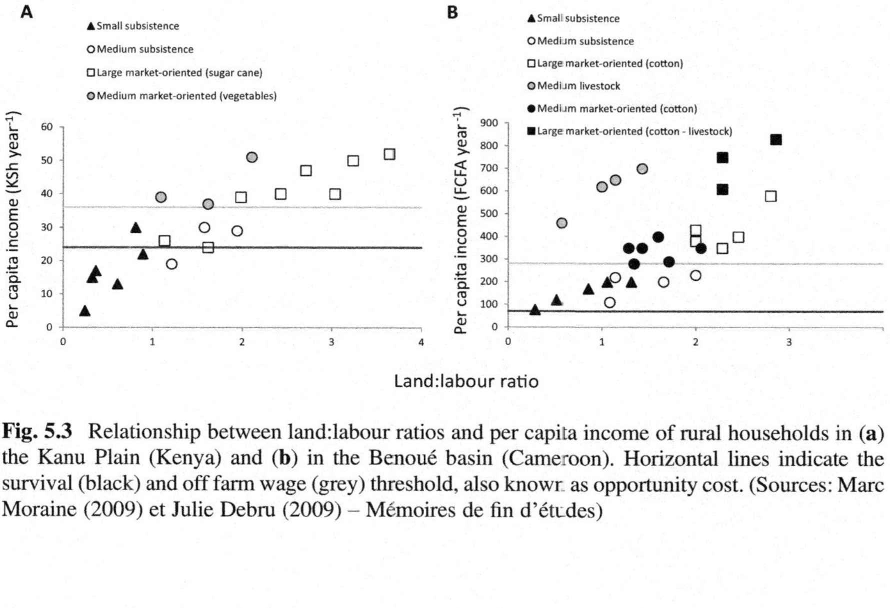

Fig. 5.3 Relationship between land:labour ratios and per capita income of rural households in (**a**) the Kanu Plain (Kenya) and (**b**) in the Benoué basin (Cameroon). Horizontal lines indicate the survival (black) and off farm wage (grey) threshold, also known as opportunity cost. (Sources: Marc Moraine (2009) et Julie Debru (2009) – Mémoires de fin d'études)

In the case of Cameroon, farmers who own livestock, and hence operate within a different production function, may generate higher agricultural incomes compared to cotton farmers, even if they have less land available per unit of labour. This is because the grazing of livestock often occurs in communal lands that cannot be attributed to a single household. When natural resources are shared within a community and decisions are made collectively, the assumptions and principles of classical farm economics may not be readily applicable.

Furthermore, in non- or incipiently industrialized economies, the opportunity cost of labour or the off-farm wage threshold depicted in Fig. 5.1 may be simply hypothetical. This is because viable off-farm labour opportunities may not be available or the wages offered off-farm may be too low. In smallholder contexts, technology adoption and mechanization can be hindered by dysfunctional markets, inadequate service provision, or poorly developed value chains. Often, technologies and mechanization options are not designed to suit the realities of smallholder family agriculture, as they are developed in regions where the farming system and agricultural context differ significantly from the circumstances faced by smallholders in the tropics. Lack of incentives, limited access to information, and insufficient training can also impede technology adoption in specific contexts, particularly when the technologies are knowledge-intensive, such as on-farm production of bio-fertilisers.

It is important to emphasize that the survival and off-farm thresholds proposed by Mazoyer and Roudart (2006), while conceptually useful, are challenging to be defined in the case of current smallholder agriculture. Additionally, in certain cultures, land holds greater significance beyond being merely an agricultural production factor. It may play a role in delineating power structures, serve as a means to ensure community membership, or be associated with family funerals and burials, among other cultural practices. In many African cities, for example, second-generation urban dwellers still refer to their ancestral rural areas as "home."

5.1.2 Traditional, Subsistence, Low Input and Family Agriculture

Traditional, peasant, family or indigenous agriculture are terms often associated with low intensity (extensive?[1]) production systems, in which either productivity per unit area or per unit labour is low. However, when speaking of 'traditional' agriculture – a vague term if any – it is important to differentiate self-subsistence from market-oriented agriculture. The latter may be characterised by the production of surpluses of food that can be sold on the market, or by the production of cash crops or animal products produced exclusively for the market. The level of intensification, in terms

[1] In languages other than English, the term extensive systems is generally used to describe low intensity agriculture, and 'extensification' as a trajectory towards lower intensities in the use of production factors.

of both inputs and outputs per unit area, tends to increase from self-subsistence to market-oriented systems. The production of crops such as coffee, saffron or vanilla is *traditional* to many tropical smallholder agroecosystems. Yet, in spite of being done by smallholders, the production of such crops is always necessarily market (export) oriented. The production of surplus for the market – or to pay tribute or taxes – has also been common across traditional agroecosystems. Without such surpluses, urbanisation, the division of labour, the formation of professional armies or the industrial revolution would not have been possible. The association between traditional, smallholder and subsistence agriculture is thus not necessarily pertinent. Nor is it true that subsistence agriculture is necessarily 'extensive' in terms of output per unit area (Fig. 5.4a). For example, the highly diverse and productive Asian home gardens and rice paddies are good examples of very intensive smallholder, yet traditional, agriculture (Fig. 5.4b).

In many of the less favoured regions of the world, poor farmers often engage in subsistence farming, producing food crops primarily for their own consumption. This may not always be their first choice, but due to resource limitations, inadequate technologies, lack of motivation, or limited knowledge, they may struggle to achieve high levels of productivity. Consequently, the association between traditional, smallholder, subsistence farming, and low-intensity agriculture is commonly generalized. In such contexts, the term "intensification" is often used interchangeably with "high population density", although this association is not necessarily always causal.

Denser rural populations naturally lead to land fragmentation and smaller farm sizes. However, farming practices may not necessarily intensify in terms of factor allocation. While this is not a universal rule, regions with higher ecological potential, characterized by inherently fertile soils and ample rainfall, tend to have denser rural populations that can sustain higher yields. A prime example of this is the placement of the city of Paris in France, situated in a highly fertile valley. In fact, many capital cities around the world were established on fertile agricultural land. However, there are instances where high population densities occur in regions with relatively poor soils and low rainfall due to socio-cultural or historical reasons. For a more in-depth analysis of the relationship between population density, farm sizes, and levels of intensification, we recommend delving into the extensive literature debate that contrasts the views of Malthus (1798, 1826) and Boserup (1965, 1981). They provide differing perspectives on whether intensification in agricultural transitioning societies can be attributed to technology or population growth.

There is a common assumption that high-input systems are more productive and efficient, while systems relying on organic inputs or processes are considered less intensive and less efficient. However, this is a misconception. One of the challenges when discussing intensification and efficiency is that the production factors being analysed need to be specified. For example, an agroecosystem could be highly efficient in terms of labour utilization, with high labour productivity and returns on labour investments. Simultaneously, it may be intensive in terms of mechanization, energy usage, and agrochemical inputs. Such a system could be labour-efficient but not necessarily economically efficient, especially if it relies on external subsidies to function. Efficiency in labour utilization does not guarantee efficiency in the use

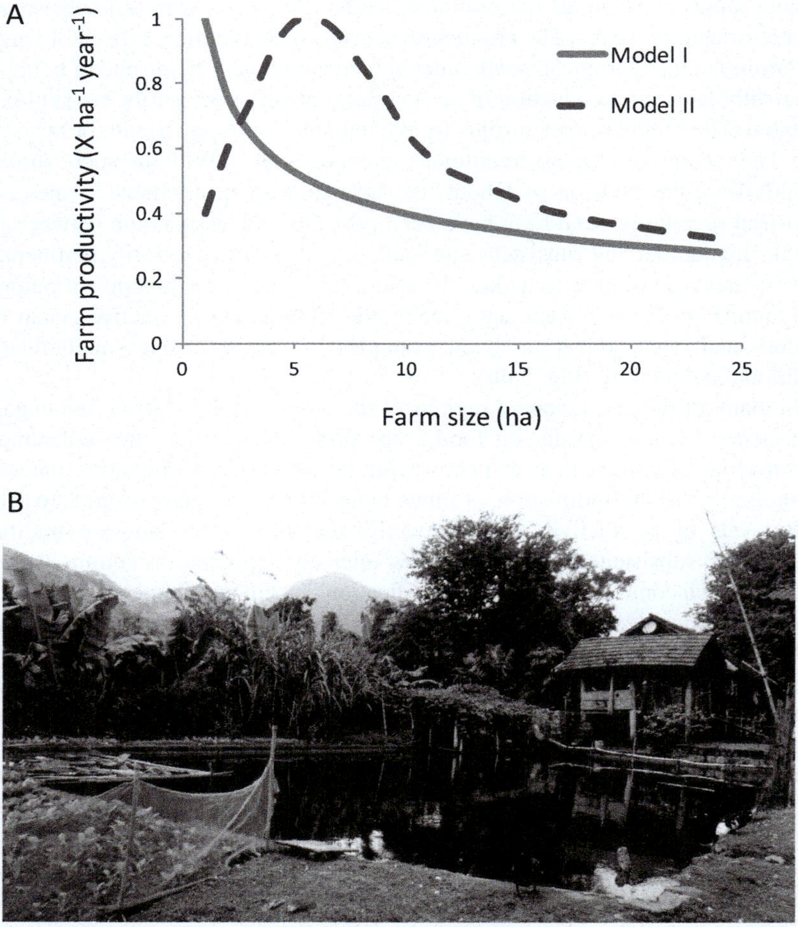

Fig. 5.4 (**a**) Stylised relationship between farm size and farm productivity depicting two conceptual models often found in the specialised literature (e.g. Carter 1984; Cornia 1985). Model I assumes that farm productivity (production of X per ha, per year) decreases with farm size. Model II assumes that a minimum threshold of farm size is necessary to achieve maximum productivity, which then falls again for larger farm sizes. The values in the y and x axes are fictitious. (**b**) An image from a highly intensive smallholder subsistence farm in Northern Vietnam that integrates several plant and animal production systems. (Photo: P. Tittonell)

of other resources, such as water or nitrogen. The relationship between intensity and efficiency is complex and multifaceted.

This is demonstrated in Fig. 5.5a, which distinguishes between input intensity and output intensity. For instance, conventional wheat agroecosystems in OECD countries often receive high levels of inputs in terms of energy, nutrients, and agrochemicals. However, it is not accurate to categorize these systems universally as intensive.

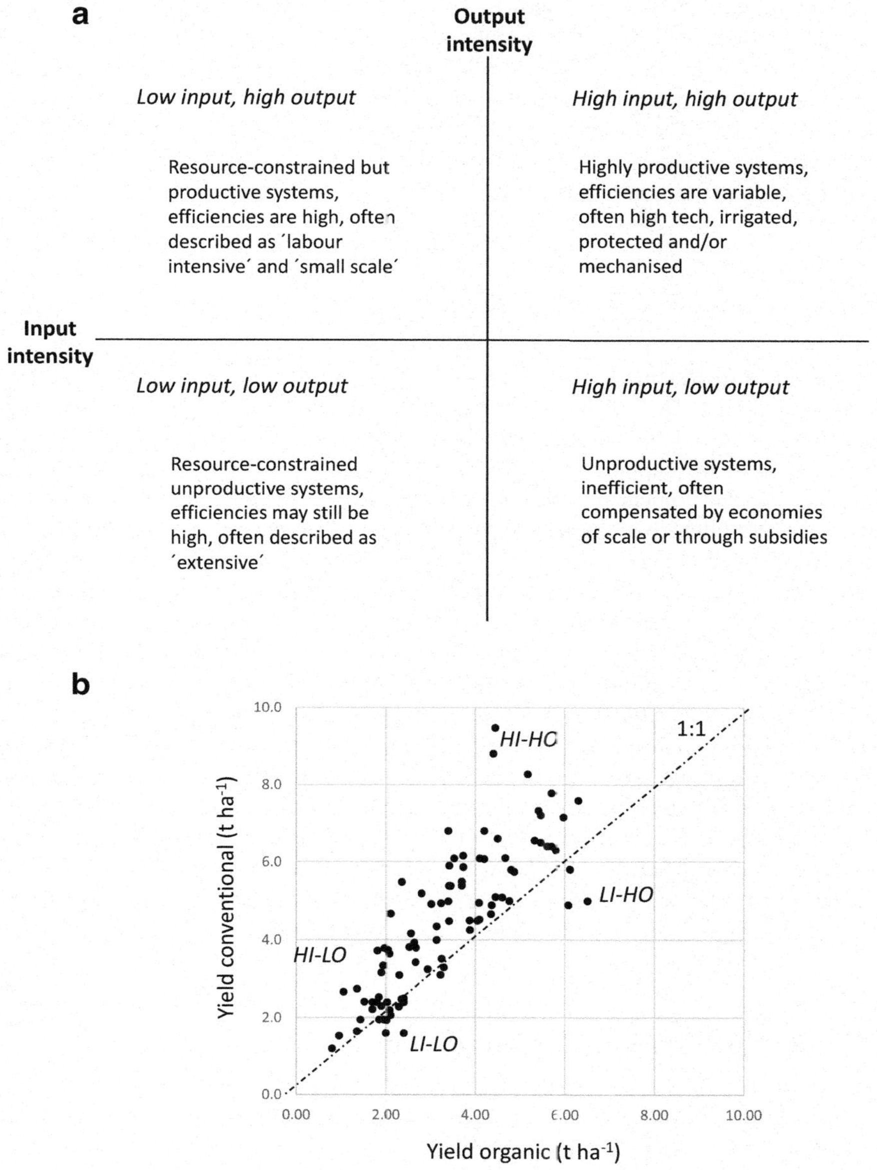

Fig. 5.5 (**a**) Simple classification of agroecosystems based on their input and output intensity; (**b**) Yields of wheat in OECD (developed) countries, under organic versus conventional management (source: de Ponti et al. 2012), indicating areas in the cloud of points were systems exhibit patterns of high input-high output (HI-HO), high input-low output (HI-LO), low input-high output (LI-HO) and low input-low output (LI-LO). The fact that most points are above the 1:1 line indicates that yields under conventional management are equal or higher than under organic management (20% on average). Using the same dataset Ponisio et al. (2015) demonstrated that when yields with equal level of N input are compared then the average difference in favour of conventional was merely 5%

Depending on their yields (as shown in Fig. 5.5b), they can exhibit high-input-high-output or high-input-low-output patterns. Similarly, their efficiencies can vary. Conventional agricultural systems, often labelled as intensive, do not always guarantee the highest productivity or efficiency. The assumption that high-input systems are inherently more productive and efficient while organic or process-oriented systems are less intensive is flawed. The relationship between intensity, efficiency, and productivity is context-dependent.

5.1.3 Ecological Intensification

In the farming systems literature, there are two additional ad hoc uses of the term "intensification." The terms "ecologically intensive systems" or "ecological intensification" are employed to describe farm management systems that rely on deep knowledge of the supporting and regulatory processes offered by the natural ecosystem. Agroecological production systems exemplify ecologically intensive systems and are often characterized as knowledge-intensive, emphasizing the focus on processes rather than inputs. In agroecology, intensification occurs through the substitution or replacement of external inputs (capital) or labour with ecological processes (Tittonell 2014, 2018). In the rest of this chapter, the concept and methods used to analyse resource allocation in agroecosystems are presented, featuring examples from smallholder systems that represent a range of intensification scenarios. The term "intensification" is not used in this book with the negative connotations associated with intensive industrial agriculture or the agricultural practices linked to the green revolution. Instead, a distinction is made between agroecosystems that are more or less resource-intensive in terms of labour and capital, and those that are more or less intensive in terms of knowledge or ecological processes. Agroecology is not about maximizing the efficiency of input use, or the productivity or individual factors, but rather about optimizing resource allocation, harnessing ecological processes, and leveraging knowledge to enhance sustainability and productivity.

5.2 Production Functions and Models

The relationship between physical output from a certain process and physical inputs or production factors used in such process is known in economics as production function. Production is the process of combining inputs and/or factors to create outputs (goods and services) that satisfy human needs, while productivity is the change of output associated with each unit of input used in the process. Figure 5.6a illustrates a possible production function that describes two main phases, defined by increasing or decreasing marginal productivities (the reader is referred to any basic book on economics for further details). Marginal productivity is the first derivative of the production function, and is plotted against the level of input used in Fig. 5.6b.

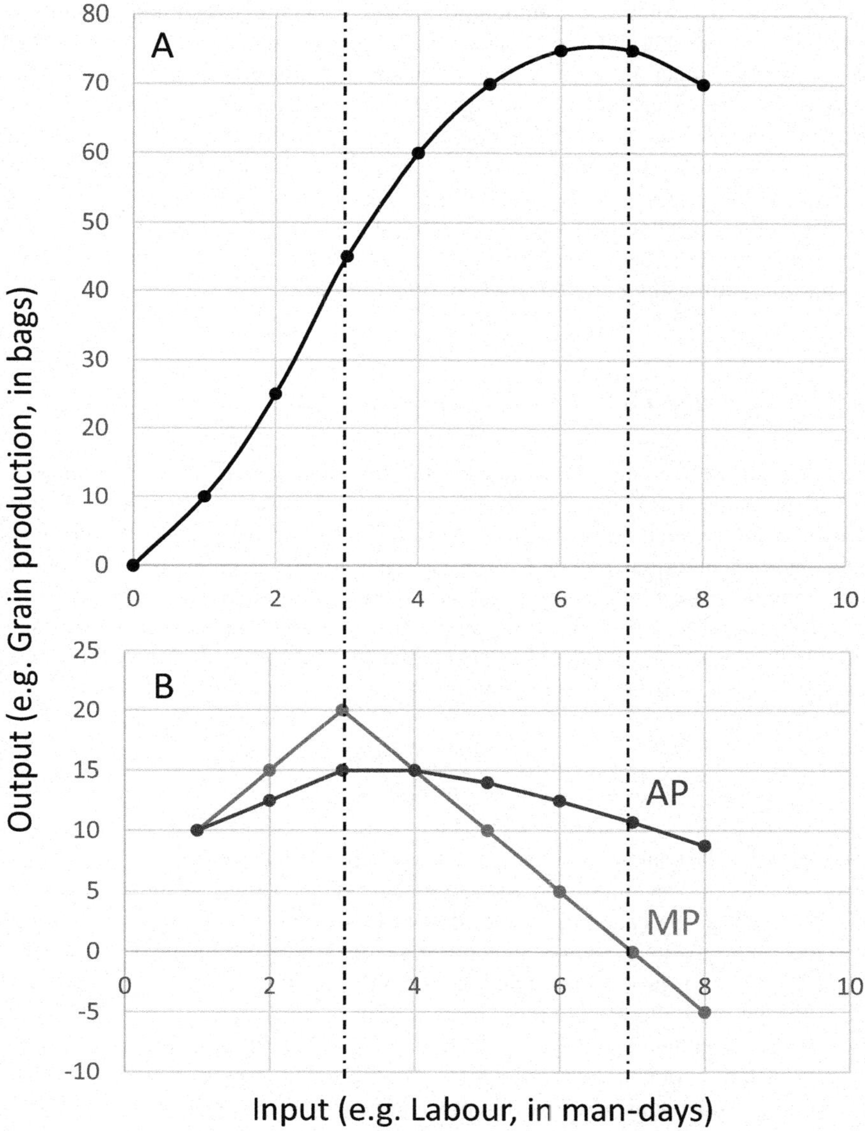

Fig. 5.6 (**a**) Production function describing the relation between inputs (labour) and outputs (grain) in a certain production process; (**b**) The corresponding marginal productivity (MP) and average productivity (AP) at each input level. Vertical dash-dotted lines separate the phases of increasing, diminishing and negative returns to labour

If we assume that Fig. 5.6a represents the relationship between, for example, labour (input) and grain production (output) over a certain period (e.g., an agricultural season), then the marginal productivity in Fig. 5.6b corresponds to the marginal productivity of labour, or the amount of grain produced per each additional unit of labour (e.g., man-day) invested over the entire season. In this example, the shape of the production function curve indicates that small labour investments are associated with high marginal labour productivity, but the total grain production remains low. Beyond a certain threshold of labour investment, production increases almost linearly with every new unit of labour invested, and the average productivity of labour remains constant, until a new level of labour investment is reached beyond which average labour productivity decreases. The phases in which marginal labour productivity increases, decreases and becomes negative correspond with respectively increasing, diminishing and negative marginal returns.

5.2.1 Average Factor Productivity

Another way of assessing productivity, which should not be confused with marginal productivity, is the average factor productivity. This is calculated simply as the relationship between total output (e.g. total grain production) over total input (e.g. labour) invested, and it is also depicted in Fig. 5.6. In our example, the maximum average productivity of labour is obtained at levels of labour investment that correspond to a portion of the production function that describes diminishing marginal returns. The formulae used to calculate marginal productivity (*MP*) and average productivity (*AP*) by relating inputs (*I*) to outputs (*O*) are, respectively:

$$MP = dO/dI \tag{5.1}$$

$$AP = O/I \tag{5.2}$$

Where, dO is the change in output per each unit input dI invested in the process. We just used labour as input and grain production as output, as an example, but productivity can be calculated in relation to other inputs such as land, machinery services or nutrients. Systems that differ in their level of intensification exhibit typically different production functions (or input-output relations, cf. Chap. 2), as do different agricultural activities. Even within a single farm, differences in soil quality between fields may lead to different production functions for the same activity. Both outputs and inputs can also be expressed in monetary units, which is useful for comparisons across different socio-economic contexts or between different agricultural activities. For example, we may be interested in comparing the economic return to labour ($ worker-day^{-1}) that could be obtained through rice production against that of fruit production, or gross margins per unit area ($ ha^{-1}) of dairy versus beef production, or the return to investments ($ $^{-1}$) in open field versus greenhouse vegetable production, etc. Figure 5.7 shows two different production functions derived from real data from smallholder farms in Zimbabwe. In all cases, both outputs and inputs are expressed on a per hectare and per annum or per season basis, as is standard in agricultural systems analysis.

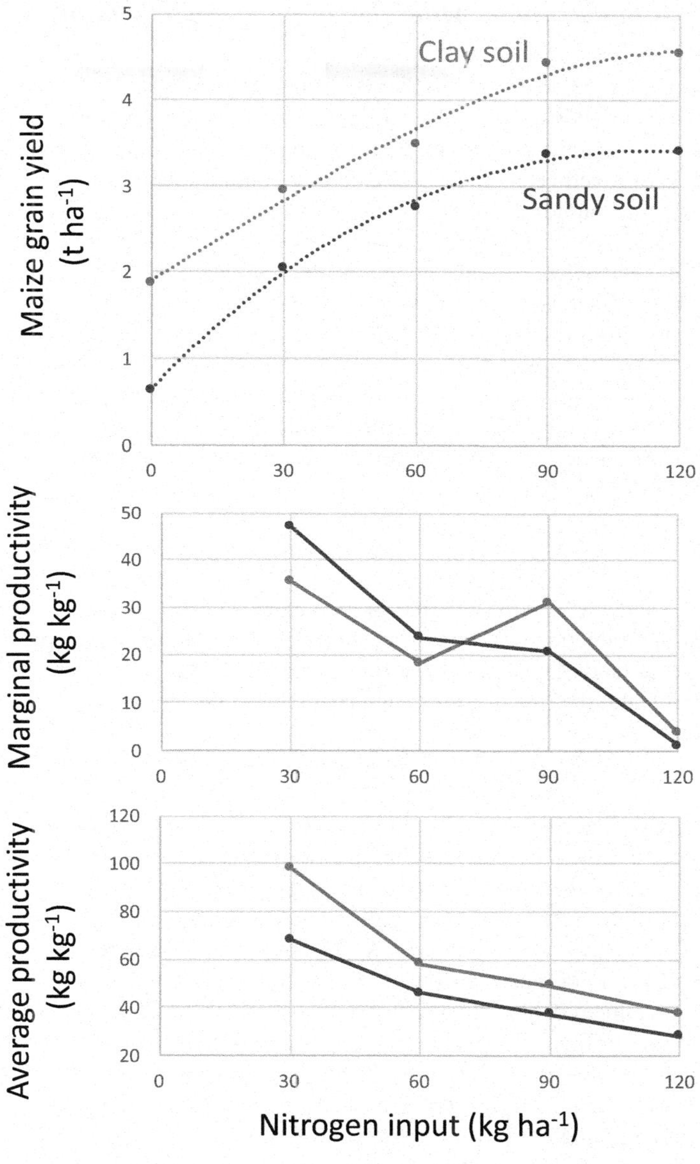

Fig. 5.7 Production functions showing the response of maize to nitrogen inputs in two soils of smallholder farms in North East Zimbabwe (source: Zingore et al. 2007). Both the marginal and the average N productivity differ between these two soil types. At higher N input rates the marginal N productivity is better in the clay soil

5.2.2 Multiple Inputs, Multiple Outputs, and the Long Term

The relationships presented so far, between single output responses to single inputs apply mostly to short-term analyses in which input-output ratios are assumed to remain constant. In the long term, the nature of the relationship between inputs and outputs, i.e. the average and marginal land or labour productivities, may change due to changes in technology, in the climate, in the status of natural resources, etc. For example, labour productivity may decrease with time as soil degradation proceeds, or increase when degraded soils are regenerated, or it can increase with time as crop breeding leads to greater crop yields, or increase as an agroforestry system becomes mature and requires gradually less labour. Changes in agricultural production over time are studied through agricultural growth models; here I will follow the approach proposed by the FAO to analyse such dynamics (https://www.fao.org/3/s9752e/S9752E02.htm). If we define agricultural production (*P*) as the product of area (*A*) times production per unit area or yield (*Y*), then the relative rate of growth in production (*GP*) is defined as:

$$G_P = G_A \times G_Y \tag{5.3}$$

Where G_A and G_Y are the relative growth rates in area and yield, respectively. Agricultural production (*P*) can also be defined as a function of multiple factors and inputs such as land (*A*), labour (*L*), mechanisation level (*M*) and nutrients (*N*), $P = f(A, L, M, N)$, and the rate of growth in production (*GP*) can be expressed as:

$$GP = \theta_A G_A + \theta_L G_L + \theta_M G_M + \theta_N G_N + G_{TFP} \tag{5.4}$$

Where, $\theta_{A\text{-}N}$ are the cost-share weights of each factor or input, and G_{TFP} is a residual term reflecting growth in total factor productivity, which is the growth in production that could be expected in the absence of changes in *A*, *L*, *M* and *N*. In other words, G_{TFP} reflects the extent to which agricultural production growth can increase through a more efficient use of these factors. Agricultural growth has been explained from different perspectives, namely (i) the Malthusian or resource theory, (ii) institutional change, (iii) human capital, (iv) best practices and (v) adaptive invention perspectives. The Malthusian theory is built upon the classical concepts described so far in this chapter: limited land and water resources and diminishing returns to labour as land:labour ratios increase. The second two perspectives emphasise reduction in transaction costs and increases in efficiency as a result of skills, legal systems and rights. The last two emphasise the role of technologies, their adoption and adaptation. Their detailed treatment exceeds the scope of this book; these are well-known concepts in classical agricultural economics (e.g. Ellis 1993).

Short-term, fixed input production functions can also be defined with respect to more than one type of input. Typically, agricultural production is expressed as a function of both labour and land when studying effects of scale on production. Both production functions combined will define a surface instead of a curve, known as production/utility (or Cobb-Douglas) function. If we assume, for example, that production (P) depends on labour (L), capital (K) and machinery (M), then the function looks like this:

$$P = a^* L^{\alpha *} K^{\beta *} M^{\gamma} \tag{5.5}$$

Where a represents total factor productivity (cf. Eq. 5.4) and alfa, beta and gamma are the output elasticity of labour, capital and machinery, respectively. Output elasticity is the percentage change in output with respect to the percentage change in input (e.g., $dP/dL * L/D$, in the case of labour), is determined by the type of technology available, and is an indicator of returns to scale. When the sum of alfa, beta and gamma is smaller, equal or greater than one we are in the presence of respectively decreasing, constant or increasing returns to scale. In the last case, increasing returns to scale may exhibit decreasing, constant or increasing marginal production/utility ratios. Further, when two inputs are considered, they may exhibit a certain degree of substitution. For example, the same level of agricultural production could be achieved on a small and highly productive piece of land than on a larger but less fertile area. When various combinations of both inputs to obtain the same level of output are possible, such that they describe a continuous function, input substitution can be described through isoquant or indifference curves (*NB*: here again the reader is referred to any basic book on economics for further detail on this). This is rarely the case when dealing with biological systems, but it can be the case in practical farming applications. For living entities such as plants or animals, it is impossible to substitute inputs of nitrogen for phosphorus, or dietary protein for fibre, etc. Yet, we can substitute two types of nutrient sources, such as different combinations of organic and mineral fertilisers that could lead to the same level of crop yield in the short term, or different combinations of forage species and concentrates in the diet of animals that may lead to similar weight gains.

5.2.3 Marginal Productivity and Elasticity

In the context of smallholder agroecosystems, we are interested in assessing the impact of farmer (or community) decision-making on agroecosystem functions, including agricultural production. Their room for manoeuvring in terms of increasing the area of suitable land for farming is often very small in highly populated areas, or when farming around protected areas such as nature reserves or state-owned forests, or in areas where the expansion of large-scale commercial farming exerts pressure on land resources. This is why, when analysing smallholder farming systems, we often assume land to be a quasi-fixed factor and pay more attention to other resources. Let us then assume that grain production (G) can be expressed as a function of labour (L) and nutrient (N) inputs, which is more realistic in many smallholder contexts:

$$G = f(L, N) \tag{5.6}$$

$$MP_L = dG/dL \tag{5.7}$$

$$MP_N = dG/dN \tag{5.8}$$

Where, MP_L and MP_N are the marginal productivity or factor elasticity of labour and nutrients, respectively. Unlike in the manufacturing sector, factor or input productivities in agriculture are largely governed by natural phenomena such as solar radiation or rainfall. The marginal productivity of labour or nutrients will vary for the same crop between a dry and a wet year (or between soil types – cf. Figure 5.7). In the case of nutrients, a difference must be made between nutrients added to the soil by the farmer as organic or mineral fertilisers and nutrients supplied by the soil itself (thereby reducing its natural stock). Due to soil supply of nutrients, production functions describing crop responses to added nutrients exhibit a positive intercept (the yield that may be obtained on a certain soil without nutrient inputs), as we see in Fig. 5.7. Within a certain location, the magnitude of this intercept depends on soil quality and crop management practices. Bearing in mind the spatial heterogeneity and temporal variability that characterise agroecosystems (cf. Chaps. 2 and 7), plus the complexity of plant-soil-plant interactions that take place in polycultures such as intercrops or agroforestry, it should not surprise us that the application of production function models requires a large number of assumptions in agroecology and smallholder farming.

5.2.4 Why Is All This Relevant to Agroecology?

First of all, because these concepts portrait the limitation of classical agronomy and farm economics to capture the complexity of actual agroecosystems, and hence the ability to understand how farmers make resource-allocation decisions. Allocation decisions by smallholder farmers tend to be seen as sub-optimal when judged in the light of classical farm economics. Yet the two models presented above, the single input-single output curve (Eqs. 5.1 and 5.2), and the multiple-input single output surface (Eq. 5.5), fall short of describing the actual model that represents agroecological production: multiple (and less) inputs, and multiple outputs (Fig. 5.8). For example, in agroecology, every 'molecule' that enters the system in the form of input (e.g. N fixed by legumes) is meant to be used in different processes, this season and the following ones. This results in less losses (negative externalities) and in a diversity of outputs, which reduces risks and increases opportunities for farming families. In addition, the spatial heterogeneity of smallholder agroecosystems may lead to extra gifts in terms of production that cannot be explained by the additive effect of single inputs (Ruben and Pender 2004). When adequately managed, diverse agroecosystems offer opportunities for emerging processes and synergies, such as biological regulation of pests and diseases, soil fertility restoration, greater animal

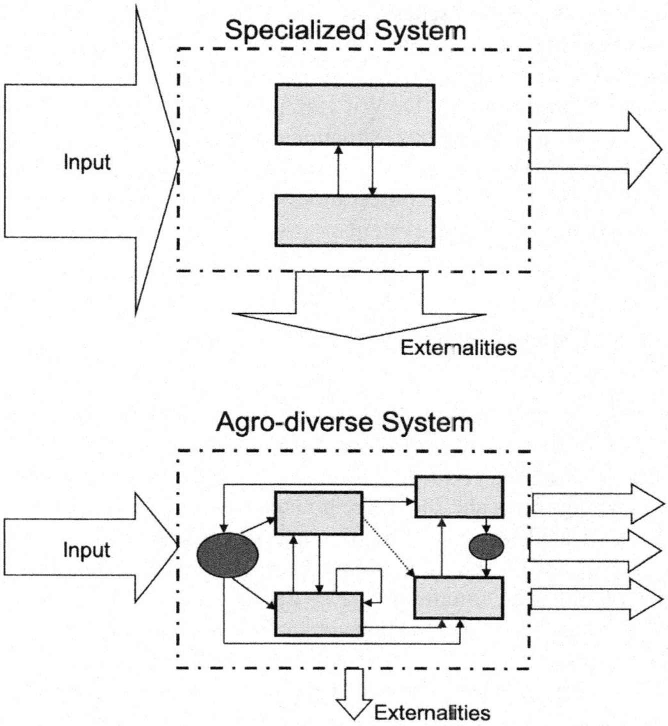

Fig. 5.8 A schematic representation of specialized (single output) and agro-diverse (multi-output) agricultural systems. Externalities refer to negative impacts on the environment, biodiversity, the climate or human health. Grey boxes represent sub-system; blue circles represent emerging processes or synergies. Diversity does not only refer to outputs, but also to diverse components and processes within the system

welfare, etc. Optimising or prioritising one output over others, or one process over others, or short- versus long-term goals, are important decisions to be made in complex agroecosystems. Smallholder decisions are made daily considering such complexity, and they tend to be seen as suboptimal under the perspective of classical economics.

The few basic concepts borrowed from agricultural economics that were presented in the previous paragraphs will be used in this chapter and throughout the rest of the book to analyse how farmer decision-making on resource allocation may affect positively or negatively farm productivity, profitability and efficiency. Further elaboration of concepts and their mathematical derivation can be found in disciplinary textbooks on economics. Some of these basic concepts are good analogies to how resources interact in biological systems. For instance, the notion that marginal productivity of one factor depends on the availability of a second one has been used to represent limitations to crop yield as determined by the interactions between multiple nutrients and water availability in the model FIELD (Tittonell et al.

2009). This stems from Leibscher's law of the optimum (1865), as opposed to Liebig's law of the minimum (1840) used in classical agronomy. We will see too that the model of optimum resource use efficiency that constitutes the backbone of agricultural production ecology (De Wit 1982; van Ittersum and Rabbinge 1997), is less applicable in heterogeneous situations where nutrients or water are in sub-optimal supply and/or under inter-specific plant competition. Let us examine now how decision-making on resource allocation may affect the functioning of the crop, animal and household sub-systems.

5.3 Factor Allocation in Agroecosystems

Allocation of land, labour and capital to the various activities of the farm system constitutes the backbone of farmer decision-making. The allocation of these resources may also be the result of collective decisions at community level, when for example grazing rights and routes are granted to different pastoral households, or when land is allocated to a newly-wed couple in a village. Allocation takes place through conscious design and through management decisions, but it may also be the result of negotiations at community (e.g. Zingore et al. 2011) or at household (Michalscheck et al. 2020) level. Management decisions are categorised according to the time horizon they concern as operational (day to day), tactic (short to mid-term) and strategic (long term). The factor capital may represent financial resources as such, for example when considering cash allocation to different activities, investment priorities, etc. Also, capital allocation concerns the resources that can be purchased with cash or other capital means, such as the allocation of external inputs of nutrients or energy, the allocation of water when it is paid for, or the investment in fixed resources as infrastructures. Land, labour and capital allocation decisions have consequences for the flows of energy, carbon or biomass, and nutrients in the agroecosystem (Fig. 5.9). However, in the following sections, we will explore aspects of decision making concerning the allocation of only two of the key resources in agroecosystems: land and labour. We will discuss examples that illustrate methods that can be used to analyse land and labour allocation patterns. Assessing capital allocation requires mobilising concept from the realm of agricultural economics, which not necessarily compatible with systems analysis, and exceed the scope of the present book.

5.4 Land Allocation

The allocation of land to different activities can be done as part of a strategic decision, as when the area of a farm is divided into cropland and permanent grassland, or at the establishment of perennial crops such as fruit trees, or of irrigation schemes. Land is also allocated to different activities on a year-to-year

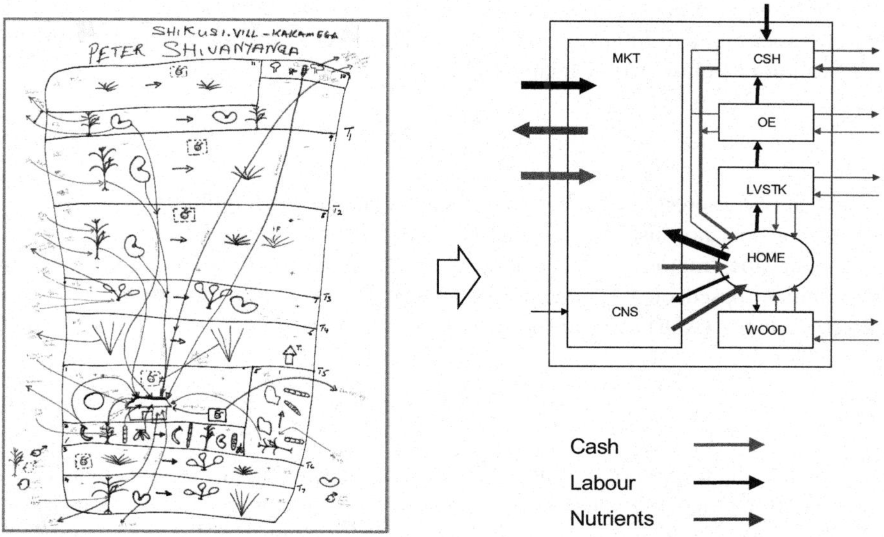

Fig. 5.9 A participatory resource flow map depicting resource allocation patterns in space and time in a smallholder farm of western Kenya (left), and its translation into a semiotic representation of the flows of cash, labour and nutrients between the various components or the farm system and with the exterior as inputs and outputs (right). *CSH* Cash crops, *MKT* market-oriented food crops, *CNS* food crops for self-consumption, *LVSTK* livestock, *WOOD* woodlots, *HOME* Households, *OE* Other enterprises. The size of the boxes and the thickness of the arrows indicate their relative importance in the system. (Modified from: Tittonell 2003)

basis or through a crop rotation plan. Planning becomes particularly challenging when dealing with spatially heterogeneous land and/or with more than one cropping season per year. The intensity and diversity of land utilization has been classically characterised by calculating a series of indexes that differ in complexity and in the type of information they convey (Dalrymple 1971). The simplest one is the land occupation index (LOI), which is simply the number of days that crops occupy a certain area of land divided by 365 days. This is equivalent to the percentage of area cropped (= 100 * area cropped/total area owned) when the entire area of a farm or a territory that is suitable for cultivation is taken as reference. This is equivalent to the multiple cultivation index (MCI) proposed also by Dalrymple (1971):

$$MCI = \sum a_i/A \tag{5.9}$$

Where, a_i is the area of each individual crop and A is the total area of land that is cultivable. Farmers may expand the area of land available for cultivation though renting or borrowing land from other farmers, leading in such cases to a percentage area cropped greater than 100.

The cultivated land utilization index (CLUI) considers the sum total of the areas of the crops (a_i) and their duration (d_i), with respect to the total area available for cultivation (A) times 365 (Chuang 1973):

$$CLUI = \sum a_i * d_i / A^* 365 \quad (5.10)$$

CLUI can be expressed as a fraction or as percentage when multiplied by 100.

To consider not just the areas but also the diversity of crops grown on the farm (or landscape, region), Ray et al. (2005) modified the diversity index used in economics (based on the diversity of enterprises on a farm) to calculate an area diversity index (ADI) as follows:

$$ADI = 1/\sum \left(a_i/\sum a_i\right)^2 \quad (5.11)$$

The ADI is the inverse of the sum of squares of the fraction of the area of each crop relative to the total cultivated area. To illustrate how this works, let us suppose that the same farm can be grown to a variable number of crops as indicated in Table 5.1. In the first case the totality of the area available for cultivation is grown to 13 different crops, and in the second and third land allocation plans the total number of crops is 7 and 3. The ADI decreases accordingly from Plan 1 to Plan 3. In the last case, in Plan 4, the number of crops is also 3, as in Plan 3. However, Plan 4 exhibits a more

Table 5.1 Three different crop allocation plans on the same farm, which has a total of 2.85 ha of cultivable land, and calculation of the area diversity index

	Areas (ha)			
Crops	Plan 1	Plan 2	Plan 3	Plan 4
Banana	0.1	0.2	0.2	0.01
Red beet	0.2			
Taro	0.34	0.48		
Cabbage	0.1			
Beans	0.2	0.7	1	0.01
Carrot	0.1			
Sweet potato	0.4	0.5		
Maize	0.6	0.9	1.65	2.83
Tomato	0.2			
Potato	0.3	0.06		
Pepper	0.1			
Papaya	0.01	0.01		
Onion	0.2			
Total area (ha)	2.85	2.85	2.85	2.85
Number of crops	13	7	3	3
Sum of squared fractions	0.11	0.22	0.46	0.99
ADI	8.77	4.45	2.15	1.01

uneven distribution since almost the totality of the land is allocated to maize and only small patches are allocated to bananas and beans. Such unevenness is also reflected in a lower value of the ADI.

A number of additional indexes of land use and diversity can be calculated when spatially explicit data are available through remote sensing and using geographical information systems. Land cover richness is an indicator used to assess diversity in land cover from spatial data. It is calculated as the average number of land cover types in a 250 m transect. According to Frank et al. (2006), this is an ′entropy′ measure, and reflects the evenness of the distribution of several land-use types within a region.

5.4.1 *Land Equivalent Ratio (LER)*

The various indexes discussed above do not capture the impact of polycultures, or the association of two or more crops in the same field, on the intensity and diversity of land utilization. In principle, however, the area diversity index could be calculated for a single field plot where different crops are grown together. Another index commonly used to assess land use diversity and its impact on land productivity is the land equivalent ratio (LER) which relates the productivity of a polyculture with respect to the productivity of each single crop in monoculture:

$$\mathrm{LER} = \sum (\mathrm{Yp_i}/\mathrm{Ym_i}) \tag{5.12}$$

Where, $\mathrm{Yp_i}$ is the yield of the ith crop in the polyculture and $\mathrm{Ym_i}$ is the yield of the same crop in monoculture. A simple example to illustrate how the LER is calculated is presented in Fig. 5.10a. The sole crops, let us assume they are a legume and a cereal, yield respectively 4 t ha^{-1} and 6 t ha^{-1}. When growing together, they yield respectively 2.5 t ha^{-1} and 3.5 t ha^{-1}. The resulting LER is 2.5/4 + 3.5/6 = 1.2, which means that the intercrop yields 20% more than the sum of both sole crops. In other words, two hectares of the intercrop will yield 5 t of the legume and 7 t of the cereal, which is 20% more than the 4 and 6 t that could be achieved by growing one hectare of each of the sole crops. However, not always the combination of two species in an intercrop results in yield synergies. Two crops growing together may have neutral effects (LER = 1) or even compete with each other (LER $<$ 1), resulting in the latter case in lower yields than when growing both crops separately (Fig. 5.10b). In some cases, one of the two crops may be negatively or positively affected by the intercrop while the other remains neutral. Also, facilitation or competition may occur at different crop proportions, or in different growing environments, soil types, sowing dates, seasons, etc. While crops such as maize have been domesticated and bred over centuries as an intercrop in *Milpas* (together with a.o. beans and squash), current high yielding cultivars of most crops are bred as monocultures and hence they do not always benefit from intercrops.

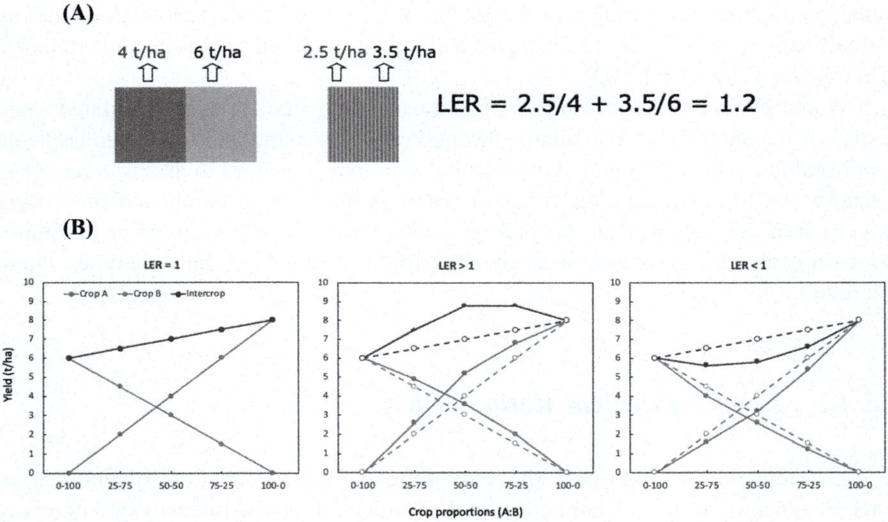

Fig. 5.10 (**a**) A simple example of the calculation of the land equivalent ratio (LER). (**b**) An illustration of how the LER may fluctuate above or below 1 at different crop proportions in the intercrop (generically, crops A and B), depending on whether the crop components facilitate (LER > 1), are neutral (LER = 1) or compete (LER < 1) with each other. Dotted lines in the centre and right panels indicate neutral (LER = 1) yields

Table 5.2 Calculation of the land equivalent ratio in a homegarden by comparing yields of each crop grown in monocultures and their respective yields in polyculture

Crops	Yield in monoculture	Yield in polyculture	Ratio Yp_i/Ym_i
Banana	16.9	5.7	0.34
Maize	4.1	0.6	0.15
Taro	13.6	2.4	0.18
Beans	0.9	0.4	0.44
Sweet potato	12.1	1.4	0.12
Papaya	38.0	4.1	0.11
Kale	12.3	3.8	0.31
Land equivalent ratio			1.64

Next to yields, farmers are sometimes interested in intercrops for other reasons, such as long-term soil fertility maintenance, pest control, production diversification, risk reduction, etc. These are some of the objectives for which farmers grow polycultures. Let us examine the example in Table 5.2, where the mix of crops represents a typical home garden in a tropical agroecosystem. The sum of the relative yields of all the crops results in a LER value of 1.64, which means that the polyculture yields 64% more than would all these crops in monoculture. In other words, if a farmer needs 16.4 ha of land to produce all these crops in separate fields, he/she would need only 10 ha to produce the same total yield of all these crops in polyculture. The design of a polyculture responds to specific objectives. A family or an individual farmer may have a certain target in terms of e.g. maize, beans or kale production that is determined by household consumption needs or by market demands, by the need to produce straw or grain to feed livestock, etc. These decisions will shape the allocation of crops in space and time in a polyculture as well as their individual area share and hence their absolute and relative yields.

In the particular example of Table 5.2, banana in polyculture yields 34% of what it yields when cultivated as a sole crop. Maize in the mix of crops yields only 15% of what it yields as a sole crop, whereas beans yield 44%. Although these values would reflect a different sensitivity of the different crops to the polyculture, which is real and observable in practice, these sensitivities are inherent to this particular home garden and cannot be extrapolated to other situations. Not just because of differences in ecological conditions, such as rainfall or soil types, but also because of design and management effects. On the same soil, and the same year, the relative yields (Yp_i/Ym_i) will vary enormously by changing, for example, the relative proportion of different crops in the mix through their area share, or their allocation in space (e.g., avoiding or promoting shading, belowground effects, etc.), the pruning and canopy architecture of the perennial plants (e.g. Ocimati et al. 2019), planting densities for the annuals, choice of cultivars, fertilisation, etc. Each polyculture is an ´environment´ in itself; it responds to environmental factors at the same time as it contributes to modify them.

5.4.2 *LER in Cereal-Legume Intercrops*

Intercropping of annual cereal and legume crops is a widespread practice around the world. In China, intercropping is at least 1000-year old, and the surface under intercropping estimated about 10 years ago amounted to 28–34 million hectares (Knörzer et al. 2009). Although one of the major advantages of such intercrops is the additional N benefits to the cereal crop provided by the legume, intercropping of cereals with non-legume crops is also practiced. For example, Qian et al. (2018) measured average LERs of 1.31 on farmers' fields cultivated with oat/sunflower intercrops in semiarid regions of Northeast China. The management of N in cereal legume intercrops is crucial to attain LER values >1. If N is amply available, especially in early stages of the intercrop, cereal growth is stimulated in such a way that it suppresses the growth of the legume. The legume plants will produce less grain than in monoculture, will fix less N, and will still be there to compete with the cereal, acting almost as weeds. Studies on LER of pea and durum wheat intercrops in France showed that the LER decreased as the application of nitrogen fertiliser increased in conventional production, and were below 1 when N application rates were in the order of 140–180 kg ha^{-1} (Bedoussac and Justes 2010). An example of how N fertilization may affect both components of a cereal legume intercrops differently is illustrated in Fig. 5.11. In organic farming, however, the average LER of cereal legume intercrops measured in 58 farmers' fields in France and

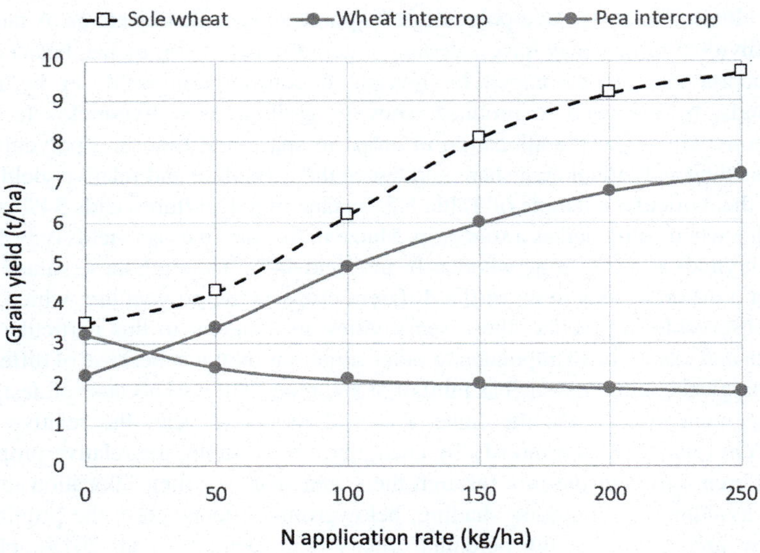

Fig. 5.11 An example of how increasing fertilization with nitrogen favours wheat over pea growth in an intercrop, from facilitation to competition effects between both crops, based on the work of Eric Justes at INRA, France; (https://www.researchgate.net/project/Agroecology-applied-to-arable-crops). Wheat yields in the intercrop are always lower than in sole wheat cropping, with or without fertiliser, while the intercrop provides 2–3 extra tones of pea yield. From an economic perspective, whether the pea yield compensates or the loss of wheat yield will depend on their relative prices and production costs. From an ecological perspective, the intercrop tends to favour soil fertility and biocontrol of pests and diseases. Without fertilisers, or at N application rates below 100 kg/ha, it appears more attractive to the farmer to grow wheat and pea intercrops than sole wheat, whereas the breakeven point seems to be at N application rates of 150 kg ha^{-1}

Denmark was 1.27, suggesting a 27% advantage over sum of the respective monocultures (Jensen et al. 2015).

In Central America, the traditional *Milpa* system is still widely used among smallholder farmers, in which maize is usually intercropped with common bean (*Phaseolus vulgaris* L.), and squash (*Cucurbita* spp.), although the species composition in the intercrop may vary across environments (Lozada-Aranda et al. 2017). Lopez-Ridaura et al. (2021) reported on a survey of 989 smallholder farms in the western highlands of Guatemala that practiced the *Milpa* system in two thirds of their 1257 field plots, with LERs ranging from 1.06 to 1.9. Next to bringing about productivity advantages, the *Milpa* contributes crucially to the food and nutritional security of farming families (Isakson 2009; Ibarrola-Rivas and Galicia 2017). According to Mann (2006), the crops grown in the *Milpa* are nutritionally and environmentally complementary, whereas Leatherman et al. (2020), surveying rural villages in the Yucatan peninsula, found greater deficiencies in zinc, vitamin A and C in those communities where *Milpa* cultivation were less frequent. However, several studies point to a gradual decline in the number of hectares of *Milpa* in Mexico and Central America (e.g. Ibarrola-Rivas and Galicia 2017; Novotny et al. 2021a), with an estimated 6% of households that can fully cover their needs from

Table 5.3 A relative comparison of labour productivity and caloric carrying capacity between Milpa (Maize, beans, squash) against sole crops of common beans and maize in two contrasting environments in Oaxaca, Mexico

	Low yielding environment		High yielding environment	
Cropping system	Labour productivity (kg h^{-1} ha^{-1})	Caloric capacity (persons fed ha^{-1})	Labour productivity (kg h^{-1} ha^{-1})	Caloric capacity (persons fed ha^{-1})
Common bean	0.6	0.6	0.8	0.7
Maize	2.0	1.4	5.0	3.5
Milpa	1.8	1.9	3.3	4.2
Relative advantage[a]	38%	90%	14%	100%

Adapted from Novotny et al. (2021b)

[a]Relative labour and caloric advantage of growing *Milpa*, calculated as: (i) investing 2 h of work on the *Milpa* versus investing 1 h of work on the bean plus 1 h on the maize; (ii) growing 2 ha of *Milpa* versus growing 1 ha of bean plus 1 ha of maize

their own production (Putnam et al. 2014). One of the main reasons advocated for the gradual abandonment of *Milpa* is the imports of cheap maize from the US that followed the signature of the NAFTA trade agreement in the 1990s (Speelman et al. 2014). Another reason, often put forward by farmers, is greater labour needs to grow *Milpa*, as compared against maize monocultures, in which weeds can be controlled through mechanisation or use of selective herbicides.

As we will see in the next section, weeding is one of the most labour demanding activities in cropping systems. Yet, beyond looking at labour needs, agroecosystems analysis requires looking at labour productivity or labour use efficiency, or the amount of production that can be obtained per hour of labour. Novotny et al. (2021b) calculated labour productivity and the caloric carrying capacity of *Milpa* versus sole maize cropping in two locations of Oaxaca district, Mexico, which differed in crop productivity levels (Table 5.3). In accordance to what farmers indicate, the data in Table 5.3 shows that labour productivity is greater in a sole maize crop as compared with the *Milpa*, in both low and high yielding environments. The caloric carrying capacity, or the number of people that can be fed per ha, is far greater for *Milpa* than for sole maize or sole bean combined. There seems to be a trade-off between labour productivity and calorie sufficiency when comparing *Milpa* against both sole crops. Yet, when comparing the relative labour advantage of growing *Milpa*, i.e. the advantage of investing two hours of labour in tending the *Milpa* versus one hour on sole maize plus one hour on sole beans, it is clear that *Milpa* is the most sensible strategy, especially in low yielding environments.

The allocation of land to different activities on the farm is not done in isolation from the allocation of labour and capital resources. In non-mechanised agriculture, and when land is not limiting, manpower or animal draft power for soil preparation determine the area of land that can be cultivated per season. In climatically risky environments, farmers often plant large areas of land using few inputs, in the hope that micro-variability in soils and climate will result in heterogeneous crop failure and therefore allow harvesting a variable portion of the cultivated land. Farmers cultivating inherently poor or degraded soils often decide to cultivate as much land as they can fertilise with organic or mineral nutrient sources. When land is a limiting factor, however, its allocation to different activities becomes crucial to farm productivity.

5.5 Labour Allocation

Intensive land use in smallholder agriculture results in high labour demands, to the extent that labour tends to be a limiting resource in these agroecosystems. Labour allocation patterns among different farm and non-farm activities often reflect farmers' priorities over time. In smallholder agriculture, family labour represents the main work force on the farm. Hiring of external labour is also done by smallholder farmers, those who can afford it, temporarily or permanently. Poorer farming families tend to rent out their own work force to other families, creating a shortage of labour on their own farms. The availability of labour becomes a determining factor for the extent and type of production systems that farmers engage in. As different activities tend to overlap in time, creating peaks of labour demand on the farm, such

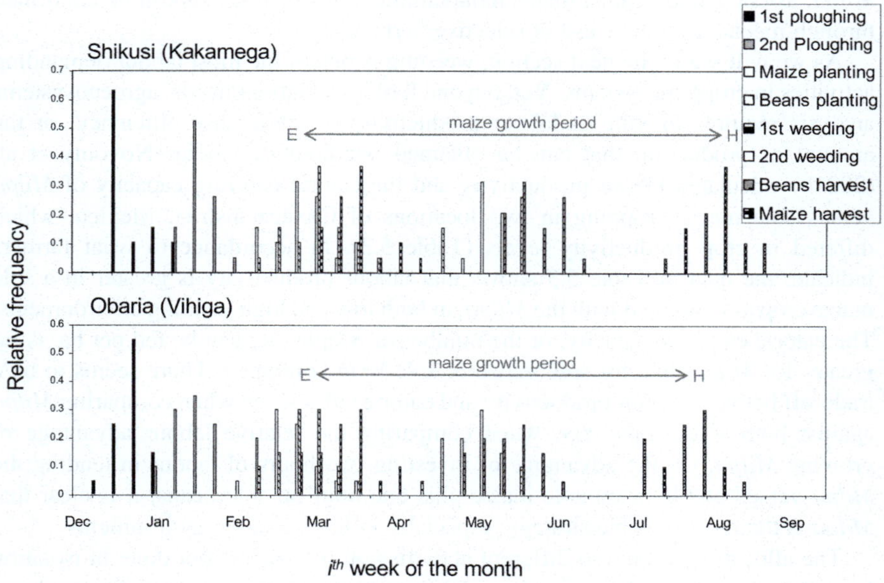

Fig. 5.12 Labour frequency distribution calendars in two villages in western Kenya (from: Tittonell et al. 2007). Frequencies were obtained by interviewing 40 farmers about the timing of their activities for a typical maize growth season. Maize is grown together with beans in this region

peaks are often more critical at determining the choice of farm activities and their success than the total amount of labour available over the year. Labour peaks can be captured through delineating labour calendars, which depict activities along a yearly time line. More informative than a labour calendar is the frequency distribution of different farm activities throughout the season. Figure 5.12 shows an example of the frequency distribution of the activities necessary to grow maize and beans intercrops during the long rains season in the highlands of western Kenya. Such labour frequency distributions, which are built by interviewing a large number of farmers about the timing of their activities (or recording them oneself through participant observation), provide a good measure of labour demands within a certain location. The price paid locally per worker-day of labour often fluctuates over the year, correlating positively with the observed peaks in demand. Management or agroecological design recommendations made to farmers should consider such peak labour demands. Any practice or technology that requires extra labour during peak periods is likely to be poorly adopted by farmers. Conversely, practices or technologies that reduce labour during such peak periods may be more easily adopted, provided that such practices are also effective in other ways.

Calendars can be delineated for the various activities of the farm. Cropping calendars, livestock feeding calendars, manure handling and storage calendars, marketing calendars, etc., may be all very insightful in revealing temporal patterns of labour allocation, as well as other temporal aspects of the agroecosystem. Even

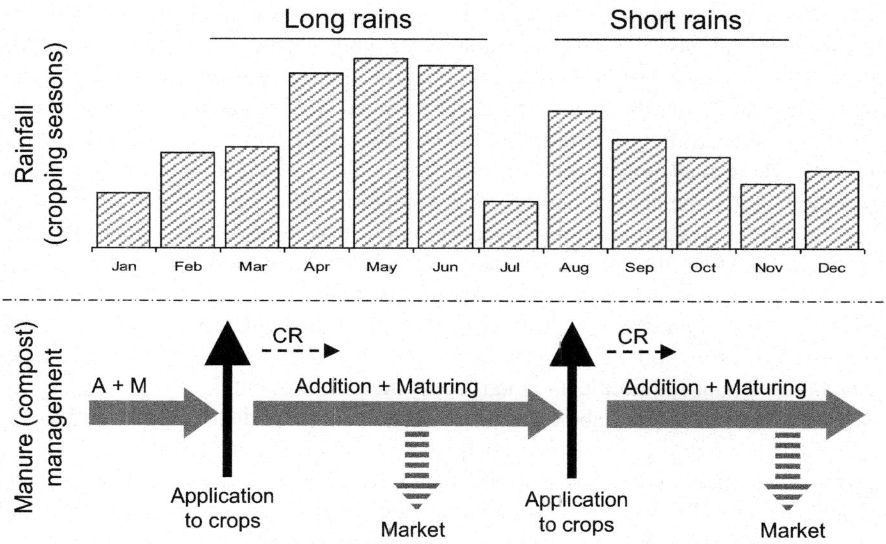

Fig. 5.13 Schematic representation of an operations calendar for manure management by smallholder farmers in western Kenya, depicting the rainfall pattern on top (double cropping season) and the periods during which manure is stored below; the dotted lines indicate the period shortly after harvest during which crop residues (CR) are added to manure

dietary calendars for the household, or calendars recoding the sources of energy used by households, can be of great value. The example of Fig. 5.13 shows how a compost heap is built through manure and crop residue additions throughout the year. This diagram indicates, on the one hand, that manure handling requires labour throughout the year, and on the other hand, that fresh carbon and nutrient sources are being added to the compost heap continuously. The latter has implications for the process of composting and compost maturity. Indeed, continuous additions mean that the compost is never stabilised before application to soil. (This is beyond the objective of this chapter but it is worth noticing in this example that such a way of managing organic resources, which is common amongst smallholder farmers who own a few animals, is suboptimal from the perspective of nutrient management and may lead to poor efficiencies of nutrient cycling on the farm).

The example of Fig. 5.13 also illustrates that the distribution of labour needs overtime is highly determined by the growing season, which depends in turn on the temporal pattern of variation in temperature and rainfall in rainfed agriculture. In dry regions, or at high altitudes, cropping seasons may be short and hence cropping activities tend to be concentrated and peak at certain times of the year, while easing out during the dry or cold seasons. In the dryer farming regions of Southern Africa, for example, male members of the household tend to migrate to towns or even abroad during the long dry season in search for work, only to return at the onset of the rainy season to prepare the land for cultivation. In regions of heavy clayey soils, labour peaks may result from land workability patterns; i.e., soils with expansive clay types should not be too dry nor too wet for proper ploughing, they have a very narrow moisture content, generally near field capacity (0.3 bar), at which they can be tilled. In green house horticulture, labour needs may also be more evenly distributed over time. However, in horticulture in general, both in greenhouse or open-field, labour demands over the year tend to be influenced also by market variability, when e.g. early harvesting of certain products may grant higher prices. In irrigated systems, there may not be water available to cultivate year-round, and hence the season may also be concentrated resulting in strong labour peaks for land preparation, planting, harvesting, etc. Off-season labour needs may also be important, for example for crops that require pruning, such as grapevines or fruit trees. Certain crops require also post-harvest processing, such as coffee, rice or tobacco, or cleaning, conditioning, packaging, storage, transport (including transport from the field to the farm house), etc. These activities should also be considered when evaluating labour requirements in agroecosystem assessments.

In livestock systems, labour tends to be more equally distributed over time, as compared to cropping systems, especially when livestock are fed through cut and carry systems throughout the year (cf. Chap. 3), or in the case of poultry or pork production in stalls, or in dairy systems in which milking cows, tending to calves or cleaning the stalls extend over the year. Animals require care and attention throughout the year, even during weekends and holydays. But livestock systems may also exhibit peaks of labour, when for example sheep are gathered and sheared at a certain moment of the year, or when livestock require seasonal herding. In many regions, specially the dryland and highlands of the world, livestock systems may still include transhumance or be semi-nomadic (e.g., Tittonell et al. 2021). Then, members of the pastoral family pass entire months away from the household herding the animals at very distant places, normally highland meadows or wetlands (e.g. in the Andes), or following the rains over vast regions (e.g. in the Sahel).

5.5.1 Quantitative Assessment of Labour Allocation

In some cases, the distribution of labour activities over time (calendars) or the concentration of activities during peak periods (frequencies) do not provide the necessary information for quantitative agroecosystems analysis. For example, calculating returns to labour, or labour needs and costs, or quantitative trade-offs considering the opportunity cost of labour on the farm, all require a quantitative assessment of labour allocation. Sometimes it may be even necessary to understand labour allocation patterns during a typical day at different times of the year. In such cases, the daily labour clock can be instrumental, as illustrated in Fig. 5.14. This is a tool that can help structuring discussions about labour allocation with farmers. A circle is drawn on a large piece of paper, on sand or on the soil surface. The circle

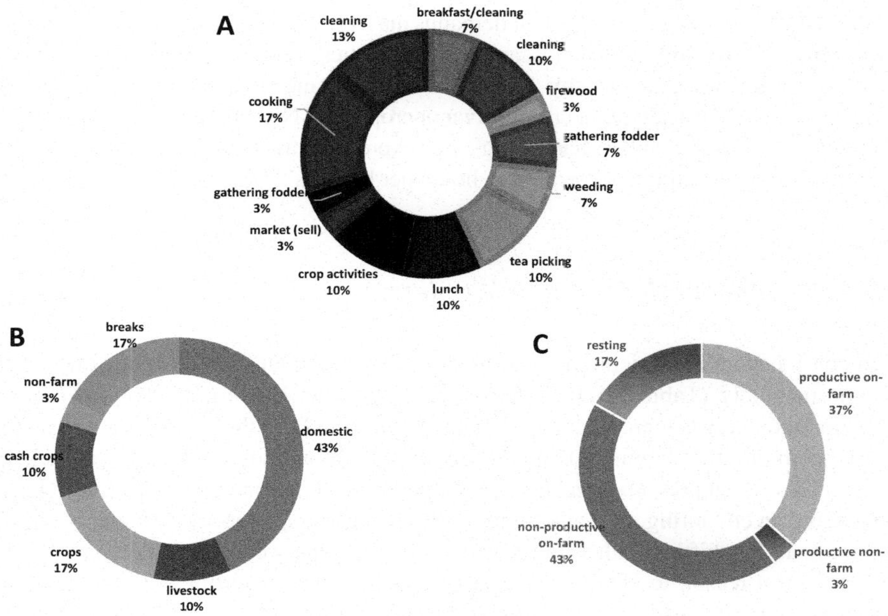

Fig. 5.14 Daily labour allocation of a female farmer in Central Kenya, as gathered through a participatory methodology known as the daily clock. Total time is 15 h, and the percentages are calculated with respect to that. (**a**) All activities as indicated by the farmer. (**b**) Grouping by type of activity. (**c**) Grouping by large categories. (Source: Suijkerbuik 2005)

represents a 24-h clock. Through discussion with the farmer and/or with the entire family, all activities performed during a day are depicted in the graph, from morning to bed time. Then, the total number of hours and the hours spent per activity are computed. In the example of Fig. 5.14, the data corresponds to a female farmer in Chogoria, Central Kenya. The total number of activity hours is 15 (9 h are dedicated to sleep). The day starts off preparing breakfast and cleaning the house, which takes about 1 h (7% of the total time awake). Preparing lunch and dinner take respectively 10 and 17% of her time. Cleaning takes another 23% (10 + 13) of her time during the day, with the end result that on-farm work that is 'non-productive' (i.e., in the narrow sense of not being allocated to crop and livestock production activities) takes up to 43% of this farmer's total daily time (Fig. 5.14b). The time allocated to production activities is 37% of the day time, or about five and a half hours.

As explained earlier, growing different crops requires more or less labour depending on the need for specific crop husbandry activities such as land preparation, seeding, transplanting, pruning, harvesting, etc. Crops that require deep soil tillage and/or hills or furrows tend to be more time-consuming. When activities are done manually, some crops require substantial labour investments also for harvesting, such as cotton, berries, tea or fruit trees, and also crops that present their harvesting produce underground (potatoes, groundnuts, cassava). Mechanisation, even with oxen-drawn tools, may reduce labour needs enormously. Table 5.4 compiles classic data on labour requirements for different crops and activities in family agriculture. These values are presented only as an orientation. Actual labour demands may vary widely across agroecosystems, as affected by soil and landscape characteristics (e.g. hillsides vs. flatlands, heavy vs. light soils), intensity of weed infestation and weed types present in the field, planting density, type of crop cultivar or crop yield. Note in Table 5.4E that labour requirements for manual harvesting are expressed as hours per tonne, while mechanised harvesting is expressed in hours per hectare.

5.5.2 *Labour Allocation to Weeding*

In non-mechanised agriculture, an important sink of labour is weed control which is done manually (Table 5.4D). Weeding of crop fields is not only done to reduce competition between crop and weed plants but also to keep the fields 'tidy'. In many rural communities, having a non-weeded field is synonymous of laziness or lack of care. This social pressure sometimes prompts farmers – especially women – to keep weeding even during dry agricultural seasons in which no crop yield is to be expected. In shifting cultivation systems, increasing weed pressure may be as important a reason to clear up new land as declining soil fertility. In the Amazon region or Brazil, for example, family farmers relying on manual labour tend to abandon fields when perennial 'weeds' (actually, pioneering ligneous plants) start recolonising the fields (Abrell et al. 2022). Due to the importance of weeding as a sink of labour and as a means to avoid yield losses, there is great interest in methods and models to assess quantitatively the effects of timing of weeding operations and

Table 5.4 (A) Examples of total labour demands (operational hours per ha) to grow different arable crops in smallholder agriculture using manual labour or oxen-drawn mechanisation; International Agricultural Research (IAR, 1974). (B) Average labour requirements (hours per ha) for land preparation activities. (C) Average labour requirements (hours per ha) for crop establishment through sowing, planting or transplanting. (D) Average labour requirements (hours per ha) for weeding. (E) Average labour requirements (hours per tonne, hours per ha) for crop harvesting (Van Heemst 1986)

(A) Total

Crop	Manual	Oxen
Maize	285	42
Cotton	565	46
Cowpea	210	77
Millet	203	44
Groundnut	1200	32

(B) Land preparation

Activity	Manual	Draught animal	Light equipment	Heavy equipment
Ploughing	–	28	17	6
Hilling	85	9	9	7
Harrowing	–	24	10	3
Levelling	–	34	4	3
Digging by hoe	300	–	–	–
Spading topsoil	500	–	–	–

(C) Establishment

Activity	Manual	Draught animal	Power equipment
Sowing maize, sorghum, millet	80	15	10
Sowing groundnuts, soya, mung bean, common bean	85	15	10
Sowing cotton	55	15	10
Planting cassava, sweet potato, potato, yam	70–75	–	5
Seeding rice	95	15	10
Transplanting rice	280	–	40
Planting sugar cane	230	–	5
Transplanting tobacco	240		

(D) Weeding

Activity	Manual	Draught animal	Power equipment
First weeding	145	–	–
Second weeding	120	–	–
Third weeding	65	–	–
Mechanical weeder	–	7	2–16
Sprayer	25	2	1–4

(E) Harvesting

Crop	Manual (h t^{-1})	Draught animal (h ha^{-1})	Power equipment (h ha^{-1})
Rice	95	40	13
Maize	110	–	5
Cotton	620	–	10
Groundnut	195	35	22
Cassava	12	100	30

of presence of weeds (biomass) on crop yields. Let us consider an empirical approach to this.

Research done at the Lake Zone Agricultural Research and Development Institute (LZARI) in Kibera district, Tanzania yielded strong quantitative understanding of the relationships between weed pressure, labour demands for weeding, and crop yields. Figure 5.15 shows some mathematical models fitted to the data generated by these experiments, which were run to quantify the amount of labour necessary to weed maize fields that were kept weeded for up to 50 days after planting. The yield of maize decreased as weeding was delayed, and the decrease was relatively well explained through the following linear function (r^2 0.65):

$$\text{Maize grain yield}\left(\text{t ha}^{-1}\right) = 3.2 - 0.03^{*}\text{Days of delay in weeding}$$

The amount of aboveground maize biomass measured in each field decreased with the increase in weed biomass (Fig. 5.15a), which increased in turn with the delay in weeding (Fig. 5.15b). Weed biomass is not always the best indicator of weed competition, as functional traits and plant architecture of both weeds and crops can play a major role in determining their respective competitive ability in a certain environment. These experiments, however, were conducted on farmer fields in which the initial seed banks and the history of use were necessarily different. In controlled experimental settings such functions are built through additive series (i.e., artificially increasing the density of a weed). In the case of Tanzania, the

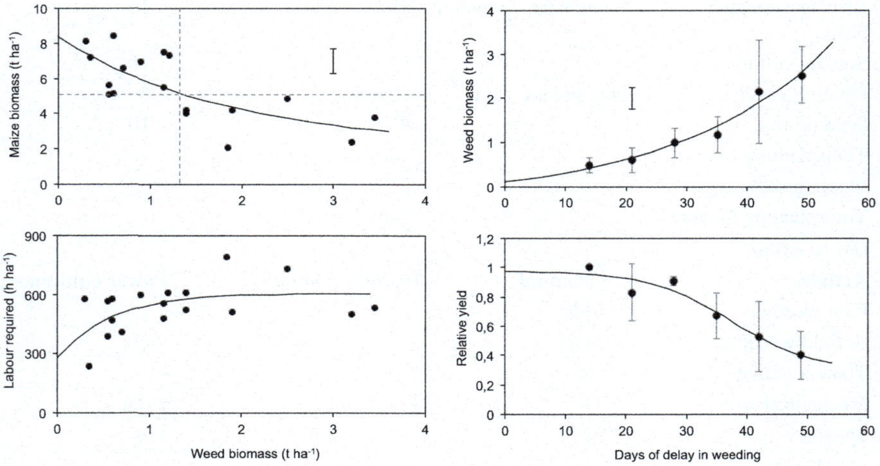

Fig. 5.15 Relationships between (**a**) maize biomass and the biomass of weeds, (**b**) biomass of weeds as a function of delay in weeding, (**c**) amount of labour required for weeding as a function of weed biomass and (**d**) a reduction factor for maize yields as a function of delay in weeding. (Data were obtained from the MSc thesis report of Emilio Righi 2006, Firenze University)

weed biomass that was measured corresponds to the mix of naturally occurring weed species in each field. The results indicate that there seemed to be a threshold around 1.2 t ha^{-1} of bulk weed biomass beyond which maize biomass was always less than 5 t ha^{-1}, and that such a threshold would be reached about one month after planting. The amount of labour necessary to weed these fields clean fluctuated between ca. 300 and 800 h per hectare (far greater than the averages presented in Table 5.4!), with an apparent plateau at 600 h when weed biomass was greater than 1.2 t ha^{-1} (Fig. 5.15c). Estimating labour requirements on such small plots is subject to large methodological error. On the other hand, the type of weed, its density, growth habit and development stage determine how easy it is to remove weed biomass irrespective of the total amount present; e.g., removing broad leaf weeds tends to be easier than *Graminae* or *Cyperaceae*.

The amount of yield reduction due to weed competition is often assessed by calculating the weed-free relative yield, as the yield of a crop affected by weeds divided by the yield of a crop kept weed-free. A simple linear model was fitted to the same data to relate weed biomass to the relative yield of maize and forcing the intercept through one; i.e., relative yield is one when weed biomass is zero ($r^2 = 0.34$):

$$\text{Relative yield} = 1 - 0.18^{*}\text{Weed biomass}\left(\text{t ha}^{-1}\right)$$

This model, which indicates that maize yields decrease by 18% per tonne of weed biomass present in the field, is a rough approximation that is only applicable under the conditions of these experiments. It serves to illustrate how such relations can be derived from empirical data. The weakness of such an empirical relationship resides in the fact that the effect of weed biomass on crop growth is not independent from the development stage of the crop. That is, one tonne of weed biomass has a much greater negative effect at the initial vegetative stages of a crop than at flowering. For this reason, when attempting to simulate the effect of weeding on crop yields through simple empirical models it is preferable to link relative yields to the delay in weeding, as done in Fig. 5.15d:

$$\text{Relative yield} = 0.29 + 0.68/\left(1 + e^{0.147*(\text{Days of delay} - 37.5)}\right)$$

Functions such as this one can be used in models of the agroecosystem in which both biophysical processes (e.g. plant growth) and management decisions (e.g. labour allocation) are considered simultaneously. We will come back to discussing such models in Chap. 10.

5.5.3 *Labour Needs in Alternative Agricultural Systems*

In the debate around models of agriculture intensification, future food security and soil regeneration, there is often the argument – in favour of industrial agriculture – that agroecology-based practices lead to increasing labour demands. Such alternative systems, which are in many cases based on traditional knowledge and practices, include a.o. agroforestry or conservation tillage, relatively new developments such as the system of rice intensification (SRI), or more traditional soil restoration strategies based on indigenous knowledge such as the *zai* planting holes system in the Sahel (Tittonell et al. 2012). These practices are meant to propend to an ecological intensification of agriculture, which is intensive in terms of knowledge and management, in order to improve productivity, reduce the need for fossil fuels and external agrochemical inputs, restore soil fertility and biodiversity, diversify diets and incomes, reduce risks, improve climate change adaptation and mitigation and generate local jobs in rural areas. Yet, it is true, they may require more labour than conventional agricultural practices, especially when the latter rely on mechanisation and herbicides. Important aspects to consider when evaluating alternative agricultural practices are not just the total amount of labour requirements but their distribution overtime, the productivity or labour in terms of physical production and economic returns, the true opportunity costs of the farm labour (i.e., are there real off-farm job opportunities in the region?), and most importantly, the degree of drudgery of different farm activities. Let us examine some of the data available on labour needs in agroecology-based management practices and production systems, such as agroforestry.

Labour needs in agroforestry systems vary enormously depending on the type of annual and perennial crops present. For example, coffee or cocoa agroforestry require substantial amounts of labour, especially because these crops require labour for harvesting and processing as well. Agroforestry systems that consist of an annual crop such as maize sole or intercropped in between rows of trees require labour for tending the annual crops, plus labour for e.g. harvesting and/or pruning the trees, resulting in higher labour demands. Agroforestry systems may also undergo different phases overtime, from establishment to full tree maturity, which require different amounts of labour. Old, mature agroforestry systems may require less labour, barely for maintenance, harvesting, etc. This seems to be the case in the survey conducted by De Giusti et al. (2019) in Kenya, where they compared labour use and productivity between agroforestry systems and arable crop fields (Table 5.5). Their

Table 5.5 A comparison of labour use and productivity between agroforestry and non-agroforestry arable farming on smallholder farms in Kericho, Kenya

Variable	Agroforestry fields[a] (n = 134)	Arable crop fields[b] (n = 544)
Average labour use (h year^{-1})	7	264
Revenue (KSh year^{-1})	3572	33,177
Labour productivity (KSh h^{-1})	705	172
Maximum labour productivity (KSh h^{-1})	30,120	5906

Adapted from De Giusti et al. (2019)
[a]Including also 'trees on farm' (i.e. high density of trees in agricultural plots without a regular agroforestry design)
[b]Under traditional local practises, including maize, beans, sorghum, sugar cane, sweet potato, groundnuts, etc

definition of agroforestry includes 'trees on farm' and home gardens, which are traditional systems that do not present regular agroforestry designs such as alley cropping or parklands. Revenues from arable crop fields were much higher, but so where the labour needs, resulting in lower average labour productivity in this region. While arable crop fields may still be interesting to the household as sources of food or cash, the combination of agroforestry and non-agroforestry fields in a single farm appear to be a sensible strategy in these agroecosystems.

In Putumayo, Colombia, the home gardens were amongst the most labour-demanding agroforestry systems according to a survey done by Palacios Bucheli et al. (2021). Agroforestry systems that included maize and beans were also labour demanding over the year, although the number of observations were few in this survey. While home gardens, silvo-pastoral systems and extractive activities from native forests were mostly done using family labour, agroforestry or annual and perennial crops engaged almost half of the labour used by hiring external workers. The opportunity cost of labour in this study was estimated to be 6.32 Euro per worker-day, which represents the amount a person (normally a male adult) will get by working off farm. The opportunity cost of labour is an assumption that only really works when there are job opportunities outside the farm, locally or at distant places, such as commercial plantations, factories, mining activities or urban jobs. In a context of generalised unemployment, however, the opportunity cost of labour would likely be lower than the one estimated in this study (Table 5.6).

In Zambia, Ajayi et al. (2009) estimated the labour requirement by maize crops without or with fertilisers versus maize in three different legume-tree agroforestry systems, over a period of five years. In their study, applying fertilisers led to a slight increase in labour needs in the sole crop due to the fertiliser application itself plus additional time to harvest the higher grain yield and managing the greater crop residue biomass (Fig. 5.16). Labour needs decline in the second year in agroforestry systems, when maize is not grown, but increase later on, when trees require more labour for pruning and placing their branches as mulching, as they grow. *Tephrosia-*

Table 5.6 Estimates of labour use and cost for different types of agroforestry systems in Putumayo, Colombia

System type	Households	Area (ha)	Annual labour use (days)	Annual labour cost[a] (Euro)	% use of family labour
Home gardens	59	0.6	114	720	81
Silvo-pastoral	21	4.6	19	120	84
Agroforestry with Lulo[b]	15	0.8	32	202	56
Agroforestry with maize-beans	4	0.6	117	739	53
Native forest[c]	67	13.6	2	13	100

Adapted from Palacios Bucheli et al. (2021)

[a]Opportunity cost of labour day estimated at Euro 6.32

[b]*Solanum quitoense*

[c]Extraction of non-timber forest products

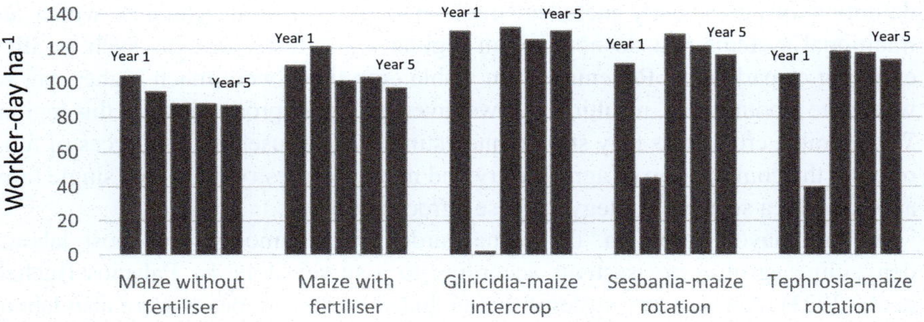

Fig. 5.16 Labour needs for growing maize without or with fertilisers versus maize in three different agroforestry systems, over a period of five years in an experiment in Zambia. (Adapted from Ajayi et al. 2009)

Table 5.7 Labour requirements and economic indicators for sole maize versus different legume-tree agroforestry systems in a five-year experiment in Zambia

System	Total labour over 5 years (worker day ha^{-1})	Years of maize/fallow	Gross margin[a] (NPV: us$ ha^{-1})	Benefit cost ratio
Maize without fertiliser	462	5	130	2.01
Maize with fertiliser	532	5	349–499[b]	1.77–2.65[b]
Gliricidia-maize intercrop	519	3:2	327	3.11
Sesbania-maize rotation	521	3:2	309	3.13
Tephrosia-maize rotation	493	3:2	233	2.77

Adapted from Ajayi et al. (2009)
[a]Net present value catering for inflation rates over the five years of the experiment
[b]Considering respectively fertilisers at market price or subsidised at 50%

maize rotation was the agroforestry system with the lowest labour demand. Although maize was harvested only three of the five years of the experiment from agroforestry systems, these resulted in greater benefit-cost ratios over time as compared with sole maize, even when fertilisers were applied to it (Table 5.7). Maize gross margins with fertilisers were largely dependent on the relative prices of both maize and fertilisers. The maize yields obtained in the different systems can be found in the original publication (Ajayi et al. 2009). They are not presented here as a way to illustrate that, in agroecology, we should look beyond crop yields when analysing and comparing cropping systems.

5.5.4 Family Labour Energy Inputs

When it comes to family labour, it is important to realise that not all the activities on the farm are performed by all family members. Young children may help out with light activities, such as hand weeding, feeding the smaller livestock (chicken, fish, Guinee pigs, etc.), fetching water, or small wood and tweeds to start a fire, taking care of younger siblings, etc. When calculating family labour availability at household level gender and age are factored in through coefficients (fractions of an adult male). Depending on the activity concerned, female and child labour coefficients fluctuate around, respectively, 0.8 and 0.6 (i.e., 80 and 60% of the equivalent of an adult male), which means that an activity that can be performed by an adult male in one full working day will require 1.25 or 1,67 working days if they are performed by a woman or a child, respectively. Such values are however only indicative, and highly theoretical. There are certain activities which are prohibitive for children, due to their level of drudgery (e.g. lifting and carrying heavy loads) or their health risks (e.g. pesticide applications). Also, there are certain activities that are culturally associated with men and women, irrespective of how 'heavy' the tasks may be.

To assess family labour energy inputs to different activities, that is, the amount of energy invested by different household members in performing different activities, Duriaux Chavarría (2014) conducted focus groups discussions with smallholder farming families in southern Ethiopia. With the estimations provided by farmers, and using tabulated values from the literature, he calculated the amount of energy (MJ hour^{-1}) that men, women and children are able to invest in different farm activities (Table 5.8A). Such calculations are useful when estimating energy flows and efficiencies in agroecosystems (cf. Funes-Monzote et al. 2009), but also when calculating factor productivities at farm scale. Table 5.8B illustrates this with data from three localities around lake Hawassa in southern Ethiopia, as calculated by Duriaux Chavarría (2014). To calculate land and labour productivity, this author used not only the female and child coefficients but also information on family composition, farm sizes, production activities and their allocation in space and time, all of which varied across the three localities. The example illustrates the variability in labour productivity that may be expected even across localities that are no more than 50 Km away from each other. However, measuring labour productivity exclusively in terms of energy output (MJ hour^{-1}) may be misleading in agroecosystems analysis. The apparently wide differences in energy efficiency across the three localities may not be indicative of household economic performance or well-being. Indeed, the families in *Wondo Genet* prioritise the cultivation of *Khat*, a stimulant crop that has a low specific energy content but fetches a very high price on the market.

Although gender and age coefficients may be useful to arrive at more realistic labour calculations, such a purely quantitative approach to labour may overlook the more qualitative aspects of intra-household labour sub-division, that comes along the distribution of power, responsibilities, etc. (Michalscheck et al. 2018). For example, activities involving livestock may be almost exclusively performed by men,

Table 5.8 (A) Energy input (MJ hour^{-1}) that men, women and children are able to deliver when performing different farm activities, according to farmers' own estimates through focus group discussions (n = 12) in southern Ethiopia. (B) Calculation of productivity per capita, per area unit and per hour of labour invested on the farm using the above coefficients in three localities around Lake Hawassa in Ethiopia

(A)

Activity	**Men**	**Women**	**Children**
Tillage	1,48	1,19	0,89
Weeding	1,41	1,13	0,85
Harvesting	1,25	0,99	0,75
Fuel collection	0,84	0,68	0,51
Water collection	0,70	0,56	0,42
Fodder collection	0,87	0,69	0,52
Shelling	1,18	0,94	0,71
Threshing	0,94	0,75	0,56
Chopping fodder	0,82	0,66	0,49
Carrying fodder	0,94	0,75	0,56

(B)

Locality	**Productivity per capita** (MJ worker^{-1})	**Area productivity** (MJ ha^{-1})	**Labour productivity** (MJ hour^{-1})
Hawassa Zuria	4908	36,640	4.9
Tulla	2679	35,535	2.2
Wondo genet	1382	28,353	1.3

Adapted from: Duriaux Chavarría (2014)

particularly in traditional pastoralist societies, irrespective of their level of drudgery. The production of food crops for family consumption is often associated with women and their diverse home gardens. Yet, times are changing, a younger generation of farmers is emerging that challenges such cultural barriers, and is highly connected with other rural and urban youth through the internet and social media, challenging the traditional view of 'traditional local' cultures that many still have. Likewise, labour may not always be perceived as a 'cost' on the farm but as a rural employment opportunity, when the conditions are adequate and the tasks motivating. This is an important element to be considered when designing agroecological production systems.

5.6 Summary and Concluding Remarks

A few key concepts from farm economics were introduced in this chapter to illustrate how productivity, efficiency or intensity are assessed in the realm of classical agronomy. Many of these concepts are also useful in agroecology. But most importantly, understanding these classical concepts is necessary to better inform our debates and discussions with different types of stakeholders, from farmers to policy makers, consumers or academics. In particular, critical views on agroecology and other alternative farming systems question their greater labour needs and hence their supposedly lower labour productivity. The examples presented throughout this chapter illustrate that there is not necessarily a direct relationship between economic scale or resource intensity of an agroecosystem and its overall factor productivity (land and labour productivity). Small farms may be more or less productive in terms of land and labour depending on the context, the type production system, the inherent productivity of the environment, the status of the natural resource base, etc., but also on the method used to calculate efficiency (i.e., the system boundaries, assumptions, units). The model of optimum resource use efficiency that constitutes the backbone of agricultural production ecology (De Wit 1982), is less applicable in heterogeneous situations where nutrients or water are in sub-optimal supply and/or under inter-specific plant competition. Besides, smallholder family farmers may prioritise objectives beyond single output productivity. The example of the traditional *Milpa* system originated in Central America indicates the crucial role of polycultures in satisfying several household needs, including productivity, income, stability, nutritional diversity, resilience, resource efficiency, etc. Assessing such complex systems requires different types of indicators, a good example of which is the land equivalent ratio. Although smallholder agriculture is often assumed to be limited by their scale (land, natural resources), a key limiting resource in family farming is labour, either family labour or labour that is hired to work on the farm. Understanding labour allocation decisions is key to the design of agroecological practices that have a chance to be adopted by smallholder farmers, who are indeed mostly labour-limited. This chapter presented several methods and examples on how to assess labour allocation in space and time, quali- and quantitatively. Quantitative assessment of labour allocation may be useful when calculating labour productivity, labour opportunity costs, or the economic return to labour. However, such calculations are subject to several assumptions and are highly affected by farm sizes and productivity, which do not necessarily correlate with labour. Also, quantitative assessments may assume that all labour units available to a household are interchangeable, that they can be allocated to different activities following efficiency considerations. This is not the case in family agriculture, where there may be a strict sub-division of labour activities between the men, women and children of the household.

References

Abrell T, Naudin K, Bianchi F, Aragao D, Tittonell P, Corbeels M (2022) Cassava root yield variability in shifting cultivation systems in the eastern Amazon region of Brazil. Exp Agric 58: E38. https://doi.org/10.1017/S0014479722000333

Ajayi OO, Akinnifesi F, Sileshi G, Kanjipite W, (2009) Labour inputs and financial profitability of conventional and agroforestry-based soil fertility management practices in Zambia. Agrekon 48. https://doi.org/10.1080/03031853.2009.9523827

Bedoussac, Laurent and Justes, Eric, 2010. The efficiency of a durum wheat-winter pea intercrop to improve yield and wheat grain protein concentration depends on N availability during early growth. Plant Soil, vol. 330, n° 1–2, pp. 19–35

Boserup E (1965) The conditions of agricultural growth. Aldine, Chicago, 124 pp

Boserup E (1981) Population and technological change: a study of long-term trends. University of Chicago Press, Chicago

Carter MR (1984) Identification of the inverse relationship between farm size and productivity: an empirical analysis of peasant agricultural production. Oxf Econ Pap 36:131–145

Chuang FT (1973) An analysis of change of Taiwan's cultivated land utilization for recent years. Rural Econ Div, JCRR Rep. 21, Taipei, Taiwan

Cornia GA (1985) Farm size, land yields and the agricultural production function: an analysis for fifteen developing countries. World Dev 13(4):513–534

Dalrymple DF (1971) Survey of multiple cropping in less developed nations. United States Department of Agriculture. Washington, DC, FEDS, 108 p

De Giusti G, Kristjanson P, Rufino M (2019) Agroforestry as a climate change mitigation practice in smallholder farming: evidence from Kenya. Clim Chang 153. https://doi.org/10.1007/s10584-019-02390-0

de Ponti T, Rijk B, van Ittersum MK (2012) The crop yield gap between organic and conventional agriculture. Agric Syst 108:1–9

De Wit CT (1982) Simulation of living systems. Simulation of plant growth and crop production. Pudoc, Laboratorium voor Theoretische Productie Ecologie en Agronomie, Wageningen University, Wageningen

Debru J (2009) L'abandon de la culture du cotonnier est-il momentané ou définitif? Diagnostic agro-économique d'une petite région agricole dans le bassin de la Bénoué, au Nord Cameroun. Mémoire de Master, AgroParisTech, Paris, France, pp 84

Duriaux Chavarría JY (2014) Energy flows in the farming systems of Southern Ethiopia: implications for sustainable intensification. M.Sc. thesis, Wageningen University, Wageningen, Netherlands

Ellis F (1993) Peasant economics. Cambridge Books/Cambridge University Press, number 9780521457118

Frank L, Sallis J, Conway T, Chapman J, Saelens B, Bachman W (2006) Multiple pathways from land use to health: walkability associations with active transportation, body mass index, and air quality. J Am Plan Assoc 71(1):75–87

Funes-Monzote FR, Monzote M, Lantinga EA, Keulen H (2009) Conversion of specialised dairy farming systems into sustainable mixed farming systems in Cuba. Environ Dev Sustain 11(4): 765–783. https://doi.org/10.1007/s10668-008-9142-7

Ibarrola-Rivas MJ, Galicia L (2017) Rethinking food security in Mexico: discussing the need for sustainable transversal policies linking food production and food consumption. Investig Geograficas 2017:106–121

Isakson SR (2009) No hay ganancia en la milpa: the agrarian question, food sovereignty, and the on-farm conservation of agrobiodiversity in the Guatemalan highlands. J Peasant Stud 36: 725–759. https://doi.org/10.1080/03066150903353876

Jensen E, Bedoussac L, Carlsson G, Journet EP, Justes E, Hauggaard-Nielsen H (2015) Enhancing yields in organic crop production by eco-functional intensification. Sustain Agri Res 4. https://doi.org/10.5539/sar.v4n3p42

Knörzer S Graeff-Hönninger B Guo P Wang W Claupein (2009) The rediscovery of intercropping in China: a traditional cropping system for future Chinese agriculture – a review

Leatherman T, Goodman AH, Stillman JT (2020) A critical biocultural perspective on tourism and the nutrition transition in the Yucatan. In: Azcorra H, Dickinson F (eds) Culture, environment and health in the Yucatan Peninsula. Springer, Cham, pp 97–120. https://doi.org/10.1007/978-3-030-27001-8_6

Lopez-Ridaura S, Barba-Escoto L, Reyna-Ramirez CA, Sum C, Palacios-Rojas N, Gerard B (2021) Maize intercropping in the milpa system. Diversity, extent and importance for nutritional security in the Western Highlands of Guatemala. Sci Rep 11(1):3696. https://doi.org/10.1038/s41598-021-82784-2

Lozada-Aranda M, Rojas I, Mastretta A, Ponce-Mendoza A, Burgeff C, Orjuela RA et al (2017) Las milpas en Mexico. Oikos 17:10–12

Malthus T (1798) An essay on the principle of population. In: Elwell F (ed) A commentary on Malthus' 1798 essay on the principle of population as social theory. Mellen Press, Lewiston, pp 127–294, commentary printed 2001

Malthus T (1826) An essay on the principle of population. Cambridge University Press, edited by Patricia James [1989]

Mann CC (2006) 1491: new revelations of the Americas before Columbus, 1st Vintage Books ed. Vintage, New York

Mazoyer M, Roudart L (2006) A history of world agriculture: from the neolithic age to the current crisis. Earthscan Publication Ltd

Michalscheck M, Groot JCJ, Kotu B, Hoeschle-Zeledon I, Kuivanen K, Descheemaeker K, Tittonell P (2018) Model results versus farmer realities. Operationalizing diversity within and among smallholder farm systems for a nuanced impact assessment of technology packages. Agric Syst 162:164–178. https://doi.org/10.1016/j.agsy.2018.01.028

Michalscheck M, Groot JCJ, Fischer G, Tittonell P (2020) Land use decisions: by whom and to whose benefit? A serious game to uncover dynamics in farm land allocation at household level in Northern Ghana. Land Use Policy. https://doi.org/10.1016/j.landusepol.2019.104325

Moraine M (2009) Analyse de l'organisation spatiale à l'échelle de l'exploitation et du territoire : le cas de la région de Kisumu, Kenya. Mémoire de fin d'études Ingénieur de l'enesad en Formation Initiale, Dijon, France, 71 p

Novotny IP, Fuentes-Ponce MH, Lopez-Ridaura S, Tittonell P, Rossing WAH (2021a) Longitudinal analysis of household types and livelihood trajectories in Oaxaca, Mexico. J Rural Stud 81: 170–181. https://doi.org/10.1016/j.jrurstud.2020.10.022

Novotny IP, Tittonell P, Fuentes-Ponce MH, López-Ridaura S, Rossing WAH (2021b) The importance of the traditional milpa in food security and nutritional self-sufficiency in the highlands of Oaxaca, Mexico. PLoS One 16(2):e0246281. https://doi-org.proxy-ub.rug.nl/10.1371/journal.pone.0246281

Ocimati W, Ntamwira J, Groot JCJ, Tittonell P, Ruhigwa B, Blomme G (2019) Banana leaf pruning to facilitate annual legume intercropping as an intensification strategy in the East African highlands. Eur J Agron 110:125923

Palacios Bucheli VJ, Cárcamo Mallen RW, Álvarez Macias A, Coral C, Bokelmann W (2021) Indigenous family labor in agroforestry systems in the context of global transformations: the case of the *Inga* and *Camëntsá* communities in Putumayo, Colombia. Forests 12:1503. https://doi.org/10.3390/f12111503

Ponisio LC, M'Gonigle LK, Mace KC, Palomino J, de Valpine P, Kremen C (2015) Diversification practices reduce organic to conventional yield gap. Proc R Soc B 282:20141396

Putnam H, Godek W, Kissmann S, Pierre JL, Alvarado Dzul SH, Calix de Dios H et al (2014) Coupling agro- ecology and PAR to identify appropriate food security and sovereignty strategies in indigenous communities. Agroecol Sustain Food Syst 38:165–198. https://doi.org/10.1080/21683565.2013.837422

Qian X, Zang H, Xu H, Hu Y, Ren C, Guo L, Wang C, Zeng Z (2018) Relay strip intercropping of oat with maize, sunflower and mung bean in semi-arid regions of Northeast China: yield advantages and economic benefits. Field Crop Res 223:33–40. https://doi.org/10.1016/j.fcr.2018.04.004

Ray S, Sood A, Panigrahy S, Parihar J (2005) Derivation of indices using remote sensing data to evaluate cropping systems. J Indian Soc Remote Sens 33:475–481. https://doi.org/10.1007/BF02990732

Righi E (2006) Optimizing hand-weeding labour productivity in maize (*Zea mais*, L.) production for smallholders in Bukoba District, northwest Tanzania. MSc Thesis Firenze University, Italy, 43 p

Ruben R, Pender J (2004) Rural diversity and heterogeneity in less-favoured areas: the quest for policy targeting. Food Policy 29(4):303–320. https://doi.org/10.1016/j.foodpol.2004.07.004

Speelman EN, Groot JCJ, García-Barrios LE, Kok K, van Keulen H, Tittonell PA (2014) From coping to adaptation to economic and institutional change – trajectories of change in land-use management and social organization in a biosphere reserve community, Mexico. Land Use Policy 41:31–44

Suijkerbuik N (2005) Livelihood strategies; the case of diversification and risk attitudes of smallholders in Central Kenya. MSc Thesis Wageningen Uniersity, The Netherlands, 71 p

Tittonell P (2003) Soil fertility gradients in smallholder farms of western Kenya. Their origin, magnitude and importance. In: Quantitative Approaches in Systems Analysis No. 25, ISBN 90-6754-713-1, Wageningen, The Netherlands, 233 p

Tittonell (2014) Ecological intensification – sustainable by nature. Curr Opin Environ Sustain 8:53–61

Tittonell P (2018) Ecological intensification of agriculture, Ch. 13. In: Serraj R, Pingali P (eds) Agriculture & Food Systems to 2050. Global trends, challenges and opportunities. World Scientific, pp 437–472. https://doi.org/10.1142/11212

Tittonell P, Vanlauwe B, de Ridder N, Giller KE (2007) Heterogeneity of crop productivity and resource use efficiency within smallholder Kenyan farms: soil fertility gradients or management intensity gradients? Agric Syst 94:376–390

Tittonell P, Corbeels M, van Wijk MT, Giller KE (2009) FIELD – a summary simulation model of the soil-crop system to analyse long-term resource interactions and use efficiencies at farm scale. Eur J Agron 32:10–21

Tittonell P, Scopel E, Andrieu N, Posthumus H, Mapfumo P, Corbeels M, van Halsema GE, Lahmar R, Lugandu S, Rakotoarisoa J, Mtambanengwe F, Pound B, Chikowo R, Naudin K, Triomphe B, Mkomwa S (2012) Agroecology-based aggradation-conservation agriculture (ABACO): targeting innovations to combat soil degradation and food insecurity in semi-arid Africa. Field Crop Res 132:1–7

Tittonell P, Hara SM, Álvarez VE, Aramayo VM, Bruzzone OA, Easdale MH, Enriquez AS, Laborda L, Trinco FD, Villagra SE, El Mujtar V (2021) Ecosystem services and disservices associated with pastoral systems from Patagonia, Argentina – a review. Cah Agri 30:43. https://doi.org/10.1051/cagri/2021029

Van Heemst HDJ (1986) Crop calendar, workability and labour requirements. In: Modelling of agricultural production: weather, soils and crops. Wageningen University, pp 251–262, ISBN: 9022008584

van Ittersum MK, Rabbinge R (1997) Concepts in production ecology for analysis and quantification of agricultural input-output combinations. Field Crop Res 52:197–208. https://doi.org/10.1016/S0378-4290(97)00037-3

Zingore S, Murwira HK, Delve RJ, Giller KE (2007) Soil type, historical management and current resource allocation: three dimensions regulating variability of maize yields and nutrient use efficiencies on African smallholder farms. Field Crop Res 101:296–305

Zingore S, Tittonell P, Corbeels M, Wijk MT, Giller KE (2011) Managing soil fertility diversity to enhance resource use efficiencies in smallholder farming systems: a case from Murewa District, Zimbabwe. Nut Cycl Agroecosyst 90:87-103

Chapter 6
Landscape Structure, Functions and Biodiversity

Abstract The concept of landscapes, such as agricultural landscapes, landscape approaches, landscape scale, cultural landscapes, and multifunctional landscapes, has been gaining momentum not only within the realm of agroecology but also in agronomy, nature conservation, and rural studies. It is widely recognized that many of the processes required to sustain the principles of agroecology on a farm or field occur at the landscape level, encompassing ecological and socio-cultural processes. Communities, social collectives, and organizations often operate at this level. Additionally, jurisdictions such as formal territories of counties or municipalities may also be referred to as landscapes. However, it is important to clarify what exactly is meant by "the landscape level" and how it differs from the agroecosystem. Are these two different ways of referring to the same epistemological object? This chapter aims to define landscapes and their spatial properties in an agroecological sense, which means considering their influence and management through human agency. Furthermore, it explores the implications of spatial diversity for species diversity, particularly the functional diversity necessary to sustain ecosystem service provision in agroecosystems and landscapes.

6.1 Landscapes

As discussed in Chap. 2, the agroecosystem offers a more concrete and definable framework in terms of space, time, patterns, and processes compared to the rather ambiguous concept of a "farming system." Diversity within agroecosystems can be examined from various perspectives, including life forms or biodiversity, the distribution of physical resources across space, the influence of land managers leading to diverse farm management or land uses, and the diversity of patches or specific environments often linked to variations in topography, soil type, hydrology, vegetation, and so on. This spatial diversity is typically attributed to a specific portion of the territory known as the landscape, which is roughly delimited by our field of vision. Due to the interplay of social, cultural, and biophysical factors, both agroecosystems and landscapes are not only dynamic but also characterized by spatial heterogeneity.

P. Tittonell, *A Systems Approach to Agroecology*,
https://doi.org/10.1007/978-3-031-42939-2_6

The study of spatial heterogeneity in landscapes is important when it comes to designing management interventions to improve productivity or stability, but also because heterogeneity sustains diversity and may in such a way contribute to overall agroecosystem stability. Heterogeneity can be thus a desirable or undesirable characteristic, and this is largely influenced by the spatial and temporal scales at which heterogeneity is assessed. Heterogeneity is not perceived in the same way by a grazing cow, by a ladybeetle in search for aphids to feed on, or by a farm manager deciding on nutrient application rates to a newly established crop. And the perception of heterogeneity will also differ between a large-scale farmer who manages fields of hundreds of hectares and a smallholder farmer who farms on one acre of land.

The concept of landscape is interpreted differently across various disciplines. However, in most cases, it involves the presence of an observer (subject) and an area of the Earth's surface being observed (object), with a focus on delineating its spatial and visual characteristics. Landscapes serve as the fundamental units of observation in geography. They result from the interactions of multiple factors over space and time, which give rise to visually identifiable patterns. The study of landscapes can involve examining their horizontal structure or profile, or analyzing them as a collection of sub-ecosystems or patches that collectively constitute the landscape (refer to Fig. 6.1). In soil science, landscapes are investigated through toposequences, which are sequences of soil types along variations in topography or geomorphology within the landscape (referred to as a transect or catena). Additionally, landscapes are cultural elements intimately tied to human societies that shape and inhabit them.

Indeed, it can be said that a landscape is a multi-layered entity that emerges from the interactions among physical, ecological, historical, and current socio-cultural layers. These layers contribute to the formation and transformation of the landscape over time. In such sense, the term 'landscape' in English or '*paysage*' in French, bear much more similarity with the Spanish term '*territorio*' than with its literal translation as '*paisaje*' – the latter more often associated with visual arts or outdoor designs. Similarly, what is known in the literature as the 'landscape approach', a set of methods in rural sociology and agrarian studies, is known in Spanish as '*enfoque territorial*'. With the advent and development of geographical information systems (GIS), spatially explicit studies tend to consider a territory or spatial unit that has physical (e.g. watershed) or political (e.g. village, county) boundaries to be a landscape. This way of defining a landscape allows much spatial information to be integrated and analysed, but it departs from the idea that a landscape is a cultural construct, or that it can be delineated by our direct visual range in space.

The concept of spatial heterogeneity lies at the core of landscape ecology, wherein a landscape is recognized as a diverse and heterogeneous entity comprising various elements such as croplands, pastures, woodlands, streams, houses, roads, and more. However, a landscape is not only heterogeneous in space but also exhibits temporal heterogeneity. The combination of spatial and temporal heterogeneity gives rise to a multitude of ecological processes and functions that operate at different scales. It is important to note that heterogeneity depends largely on the resolution at which a landscape is studied. In the realm of commercial farming, the

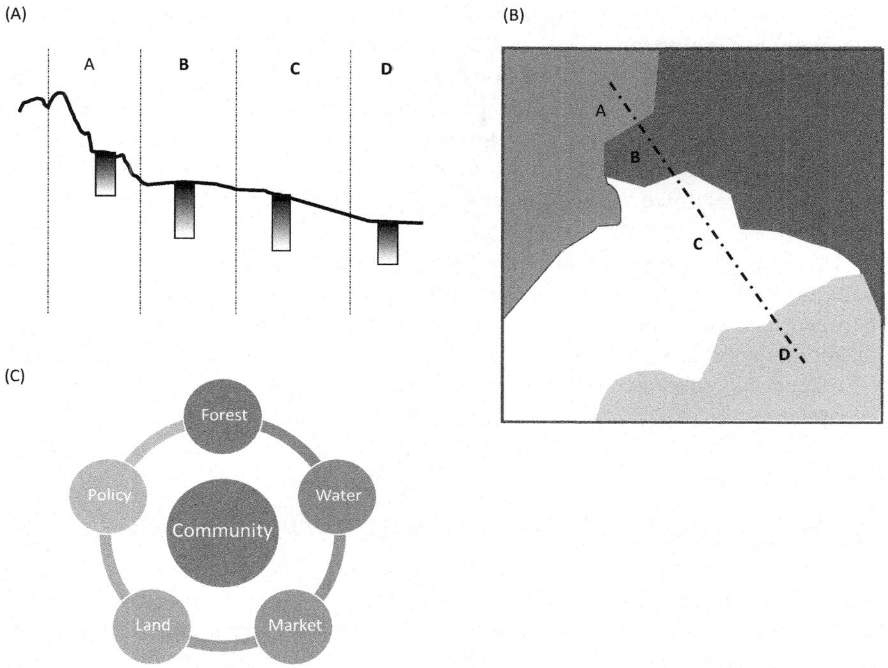

Fig. 6.1 Different disciplines use alternative ways of representing a landscape. (**a**) A 'catena' describing landscape features and the spatial distribution of soils; (**b**) A 'map' of the landscape as a mosaic, used in spatially explicit studies; (**c**) A diagram representing a 'landscape approach' to natural resource management in a community territory

development of "precision agriculture" techniques aims to address spatial heterogeneity. These techniques involve the use of sensors, indicators, and target values to adjust the application rates of fertilizers and agrochemicals based on specific needs, such as nutrient deficiencies or weed infestation levels, which are unevenly distributed in space and time. Similar principles are also applied in smallholder farming, albeit without the use of GPS-aided equipment (Tittonell et al. 2016). However, it is worth mentioning that the underlying concept of precision agriculture is essentially focused on minimizing the effect of heterogeneity. This is because, particularly in mechanized agriculture, uniformity is often perceived as a desirable attribute of agroecosystems since uniform crop stands facilitate field operations and harvesting.

There are many situations however in which spatial heterogeneity is a desirable characteristic, sometimes even *created* by farmers. For example, the roots of cassava are often 'stored' in the soil by smallholder farmers and harvested gradually for self-consumption or the market. A heterogeneous crop stand with plants at different stages of maturity is desirable in such a case. Also, in small farms with limited access to organic matter and nutrients, farmers may choose to concentrate these resources in smaller areas to ensure adequate crop productivity, thereby creating spatial heterogeneity. Plant diseases such as potato late blight caused by *Phytophthora infestans* can literally weep out a homogeneous – and genetically uniform – potato crop, especially in farming landscapes that are dominated by this crop. The spread of this disease can be limited or slowed down by introducing spatial heterogeneity (e.g., Ditzler et al. 2021). In most cases, however, farmers perceive crop variability in space as a nuisance and they try to reduce it. In agroecology, it is considered important to assess what the causes of heterogeneity are in order to either reduce, manage or reinforce spatial heterogeneity according to different objectives (cf. Chap. 7).

The following sections define terms and concepts that are often used or associated with the study of landscapes, and that are necessary to understand the relationship between landscape structural and functional characteristics, i.e., (i) synchronic and diachronic approaches; (ii) heterogeneity and variability; (iii) landscape anisotropy; (iv) landscape structure; (v) patterns and processes; (v) landscape functions and ecosystem services.

6.1.1 Synchronic and Diachronic Approaches

Spatial landscape heterogeneity and its consequences (e.g. soil and crop variability) can be assessed through synchronic and diachronic approaches. Synchronic approaches assess spatial heterogeneity and variability at one single point in time, whereas diachronic approaches aim to capture their temporal dynamics. In such cases, it is also important to define the time horizon and the time step of the analysis, depending on whether we are dealing with short versus long-term variability. Heterogeneity can be thus viewed from the perspective of an observer (a farmer, a researcher), from the perspective of its ecological functions (niches), and from static or dynamic perspectives (Kolasa et al. 1991). A recent example of a synchronic approach to landscape heterogeneity is the study of Trinco et al. (2021), who analysed the effect that the shape of grassland patches has on grassland productivity, in grassland patches that were distributed within a matrix of high forest in the Patagonian Andean region (Fig. 6.2). Although altitude had the strongest statistical effect due its inverse relationship with temperature, shape variables such as the perimeter-to-area ratio had a positive effect on grassland productivity, as strong as that of rainfall, especially during the drier late summer period.

Although the study of Trinco et al. (2021) considers measurements at three moments in time, these moments are not connected statistically, there are no carry over effects being tested in the study, and thus this is not a diachronic study, but three independent synchronic studies. An example of a diachronic approach is the study of Novotny et al. (2020, 2021), who analysed changes in landscape configuration in two municipalities of central Mexico to understand changes in food and nutritional self-sufficiency of farming families as a result of changing agrobiodiversity from 1984 to 2017. The longitudinal (diachronic) analysis of this region shows a regreening of the landscape through a reduction of bare soil (degraded land), croplands and grasslands, and an increase in forest land (Fig. 6.3a). Such a forest

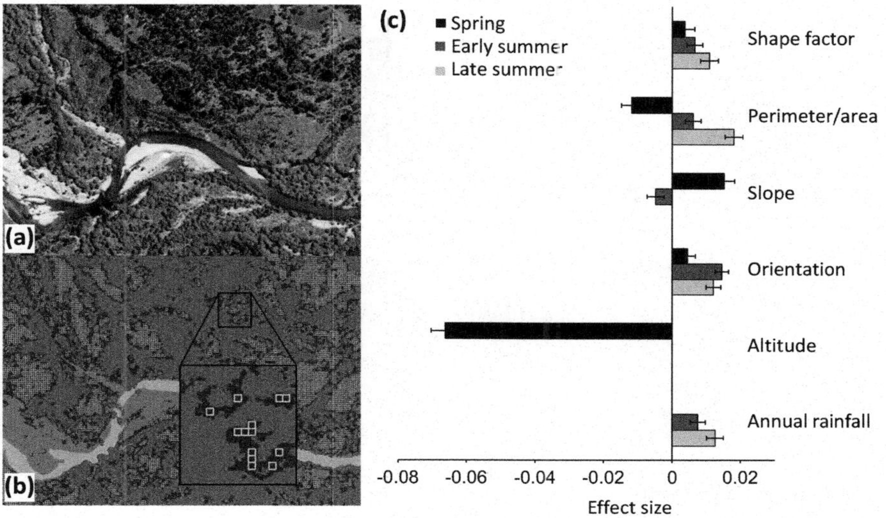

Shape of grassland patches			
Lenghtness	Low	High	Intermediate
Perimeter/area ratio	Low	Intermediate	High
Shape factor	High	Intermediate	Low
Solidity	High	Intermediate	Low
Convexity	High	Intermediate	Low
Compactness	High	Intermediate	Low

Fig. 6.2 An assessment of the effect of patch shape on native grassland productivity within patches scattered within a forest matrix in the Patagonian Andes, Argentina (Trinco, 2022, PhD Thesis Comahue University). (**a**) Detail of an image generated through pan-sharpening method from a SPOT7© image; Green areas represent dense tree cover and brown areas represent forest openings. (**b**) The same area as (**a**) after neural network classification; green = dense tree canopy, brown = grassland patch, light blue = water, grey = rock. Small yellow squares: selection of SPOT7 pixels within the grassland patches used to calculate the NDVI. CNES© 2016 and 2017, reproduced by CONAE under Spot/AIRBUS image license. (**c**) Effect size of each factor on grassland productivity (NDVI) at three moments in the year obtained through a multiple linear regression analysis (95% confidence)

encroachment is the result of several drivers, but an important one among them is the migration of young adults to Mexico City and the US that took place in the 1990s, which resulted in farmland being abandoned due to lack of labour. In southern Ethiopia, Kebede et al. (2018, 2019) studied changes in landscape configuration as a result of population growth (Fig. 6.3b). The aim of the study was to assess the resulting current potential for biodiversity-mediated biocontrol at landscape level, or

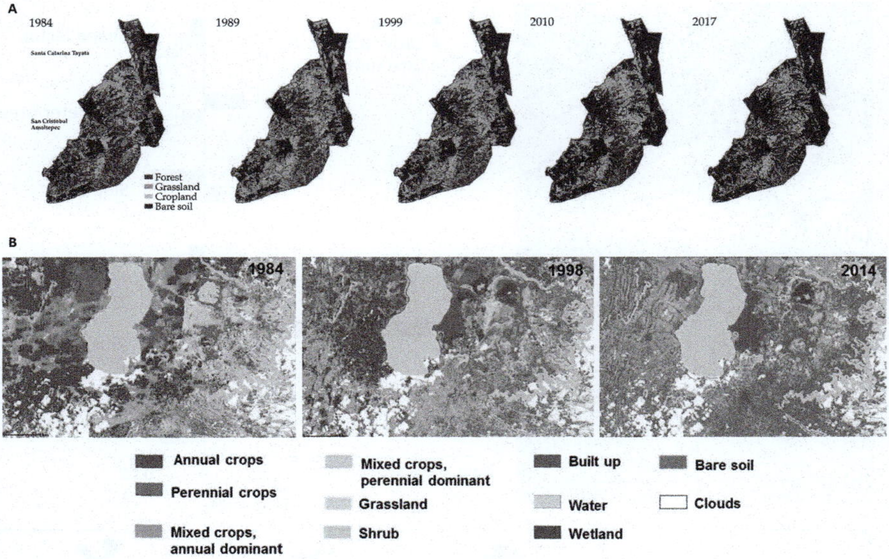

Fig. 6.3 (**a**) Changes in land use in the Mixteca Alta landscape in Oaxaca, Mexico from 1984 to 2017; two municipalities are depicted: Santa Catarina Tayata and San Cristobal Amoltepec. (Novotny, 2020, PhD Thesis Wageningen University). (**b**) Changes in land use around lake Hawassa in the southern highlands of Ethiopia from 1984 to 2014; the large grey area east of the lake corresponds to Hawassa town (Kebede, 2019, PhD Thesis Wageningen University)

for the effectiveness of agroecological measures such as push-pull cropping systems. As population increased in this region of Ethiopia, so did landscape fragmentation, with a tendency to reducing field sizes, increasing density of hedgerows, and of perennial crops around the homesteads. This resulted in more biodiverse landscapes with greater abundance of natural pest enemies.

Diachronic studies of structural or functional changes in the landscape are often done through analysing series of satellite images. The normalized difference vegetation index (NDVI), calculated from the reflective properties of the vegetation, is used as a proxy for plant biomass production. Live green vegetation gives higher NDVI values, and hence a positive slope in the NDVI trend is interpreted as plant growth or vegetation recovery. Graphical representations normally use a scale from red to green, as in Fig. 6.4c below, to represent negative (red) or positive (green) NDVI slopes over time. However, most of the studies of NDVI trends use linear regression models to assess long term NDVI trends, as illustrated in Fig. 6.4a, b. When the slope of the linear regression is negative, authors conclude that vegetation is degrading (e.g. the 'desertification' of the Sahel has been estimated like this), and when it is positive, they conclude there is a recovery (e.g., the 'regreening' of the Sahel has also been estimated like this). A limitation of linear regression models is that they do not capture cyclical, shorter term tends such as phases of recovery or relapsing. In Fig. 6.4a, it is possible to observe a recovery phase of the vegetation following a volcanic ashfall in 2011, which is not captured by the linear model. In

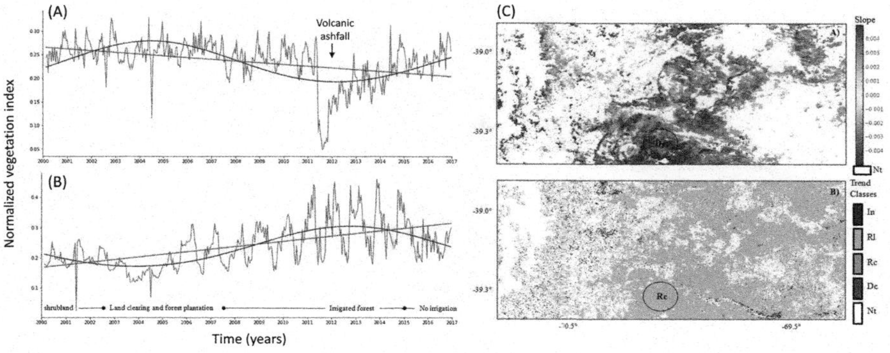

Fig. 6.4 Example of diachronic studies of landscape changes over 17 years in Patagonia, Argentina through the analysis of satellite images (Modis). The normalized vegetation index is used as a proxy for plant biomass production. (From: Easdale et al. 2018)

Fig. 6.4b, we observe a relapsing trend in vegetation as from 2014 which is also overlooked by the linear trend, which assumes that the trend is still positive. The blue, cyclical trends fited in these figures were obtained using a wavelet autoregression model (WARM), which integrates small changes in slope over shorter time paces to better reflect actual trends (Easdale et al. 2018).

Figure 6.4c compares the study of changes in vegetation using 17 years series in a portion of North Patagonia affected by the volcanic ashfall of 2011, through linear trends (above) or through wavelet autoregression (below). Most of the red areas in the image above using linear trends (degradation) appear as recovery areas in the image below using wavelets. Several green areas in the image above using linear trends (regreening) appear as relapsing areas in the image below using wavelets autoregression. In spite of these well-known shortcomings of linear models, their use is still generalised, likely due to their simplicity. The use of trend cycles is a more sensitive approach to assessing and monitoring land degradation (Easdale et al. 2019). It does require more computational capacities and mathematical knowledge, but these are resources which are not scarce nowadays. Users of linear trends are fiercely opponents of new methods that question their models. It is alarming to realise that famous global or regional assessments on land degradation or regreening, highly cited and published in top scientific journals, were done using linear trends... how valid could their conclusions be?

6.1.2 *Heterogeneity and Variability*

While in ecology heterogeneity refers to patterns that consist of dissimilar or diverse constituents, in statistics heterogeneity is the opposite of homogeneity, as in the case of homogeneity of means or variances (the latter is also known as heteroscedsaticity in statistics). The term variability, on the other hand, indicates changes in space and

Fig. 6.5 (**a**) Agricultural fields on a slope, showing a gradient of decreasing soil fertility at increasing distances from the homestead in smallholder farms of western Kenya. The yellow dot-dashed line indicates the boundary between two farms (hedgerow), one with a strong productivity gradient (left) the other one not (right). (**b**) A highly fragmented landscape in the Maragoli area, western Kenya, illustrating landscape anisotropy. Consider, for example, the study of plant species distribution or of micro-climates following the three axes depicted on the picture. It was my early work in these landscapes of the western Kenya highlands that triggered my curiosity about landscape heterogeneity and anisotropy. (Photos: P. Tittonell)

time in the value of a given property or attribute (operationalized through one or more variables in our case), and it is measured through standard parametric statistics. When such variability has a spatial structure it is termed heterogeneity, and it is quantified through the study of its patterns and scales. A distinction is necessary when dealing with spatial heterogeneity of landscape features or soil fertility, and the associated variability that may be observed in the resulting crop or grassland productivity, or forest growth, etc. In such cases, the difference between heterogeneity and variability is not merely semantic, although certainly arbitrary and decidedly linked to the agriculture jargon.

For example, crop yield variability – both in space and in time – can be the result of the interaction between soil spatial heterogeneity and other factors, such as rainfall distribution, uneven nutrient applications or the distribution of pest populations in both space and time. Soils may be relatively homogeneous and yet crop yields may vary in space (randomly or structurally). Or, especially in high input agricultural systems, yield variability in space may be small even when soils are spatially heterogeneous. The picture in Fig. 6.5a illustrates this with two contiguous farms located on a slope. While one of the farms exhibits a strong gradient of productivity down the slope, the other one, presumably with more resources to manage its crops, shows almost no productivity gradient in the same direction.

6.1.3 Landscape Anisotropy

Certain physical properties of objects or substances exhibit anisotropy, which means that the property has a different value when measured in different directions. Agricultural landscapes are typically anisotropic, as a result of horizontal and vertical gradients in combination with random or human-induced spatial heterogeneity. Spatial heterogeneity in landscapes tends to be asymmetrical, to differ when measured in different directions. Yet, an asymmetrical pattern can appear as random or *vice-versa* when heterogeneity is assessed at different spatial resolutions. Although heterogeneity is often studied on the basis of a single criterion – e.g. soil carbon, species distribution, forest cover or plant biomass – in reality it is a multi-criterial property. Heterogeneity in a landscape is the result of the uneven distribution of biological entities (flora, fauna, microbes), of geological and geomorphological features (soil texture, depth, slope), of environmental factors (temperature, wind, radiation, rainfall) and of human activities (land use, disturbance, infrastructure). Each one of these aspects will exhibit a different degree of heterogeneity, which may be characterised by gradients, uneven distributions (asymmetry) and/or randomness, and they will vary differently when measured in different directions (Fig. 6.5b).

At the same time, the various biotic and abiotic dimensions of the agroecosystem will exhibit different degrees of co-variance in space. For example, the spatial distribution of plant species and communities in natural or undisturbed environments is a good indicator of spatial heterogeneity in micro-habitats. The 'concentration' of individual species is unevenly distributed in space, often as patches, reflecting the uneven distribution of resources and environmental factors regulating plant growth. The corresponding dimension to express heterogeneity in this case would be plant species richness,[1] and this may also be a surrogate variable for e.g. soil spatial heterogeneity.

6.1.4 Landscape Structure

Most landscapes are not just randomly patchy. The various elements of the landscape do not exist in isolation from one another. Biophysical interactions in a landscape operate both *between* and *within* landscape dimensions (or 'layers', as they are termed in geographical information systems). Next to the correlation between e.g. soil spatial patterns and vegetation types, there are interactions or exchanges between the constituents of each layer, i.e. between soil units or plant communities located on different landscape positions. They interact through exchanges of energy, matter and information, and they may exhibit different degrees of spatial interdependency. Typically, above and belowground water flows connect the upper and lower parts of the landscape, carrying with them sediments, solutes (salts, nutrients), plant seeds and other organisms (Fig. 6.2). In agricultural landscapes, humans mediate many of such exchanges.

A classic example of this, from smallholder agro-pastoral systems, is the transfer of carbon and nutrients from grazing areas to croplands when farmers collect animal

[1]Animal species richness is also positively correlated with spatial heterogeneity through plant species richness, as vegetation serves as both food and habitat (cf. ecosystem productivity)

manure and apply it to their field crops. This results in human-induced soil heterogeneity, through the creation of fertile spots and large areas of degraded, overgrazed land. Human activity thus affects the functioning of the landscape not only by modifying its semi-permanent structure (e.g., deforestation, afforestation, terracing, fencing, channelling, fragmenting, levelling) but also through frequent disturbance (ploughing, harvesting, grazing). But human activities are not random. They are influenced by culture and at the same time largely determined by the pre-existing heterogeneity of the landscape. When colonising landscapes, for instance, humans allocate the best soils to annual cropping and shallow, rocky, flooded or non-workable soils to grazing or forest activities. Terracing structures are typically built in sloping areas whereas drainage ditches are common in the low laying parts of the landscape. Both terraces and drainages create spatial patterns in the landscape. In general, the presence of water and its accessibility (for irrigation, livestock or domestic use) are often structuring factors of spatial heterogeneity in human landscapes (Fig. 6.6).

6.1.5 Patterns and Processes in Landscapes

The relation between patterns and processes is central to both general ecology and agroecology, where it is also a basic requirement for the design of agroecosystems. A large diversity of the methods used in ecology are meant to help unravelling whether there is causality or interdependency in the relationship between patterns (what we perceive as diversity) and processes (which results in functions and services, but also in disservices). For example, do different levels of heterogeneity or forms of spatial arrangement in landscapes result in different types and magnitude of ecosystem services? The relationship between patterns and processes is reminiscent of the relationship between structures and functions, which is a central concept in systems analysis (cf. Chap. 2). This was illustrated with the example of a mobile phone earlier in this book: it is not the components together that make the phone work, it is the way they are organised. Could we say the same thing about landscapes? In the case of the mobile phone the relationship between structure and function, i.e., pattern and process, is a tight and unequivocal one. Remove one connection between its components and a function may be lost. Depending on which connection was removed, the phone may continue working but less effectively, or lose some functionalities. But there are some critical connections that, when removed, will make the phone stop working.

This allegory may be useful to think about and design sustainable agricultural landscapes, even if the principles might not always be demonstrable or defendable. Yet it *is* possible to imagine situations in which structures/functions may be removed from agricultural landscapes that are critical to their functioning. For example, in an irrigated fruit and vegetable production valley, stopping irreversibly the main water supply to the irrigation network will result in major changes in the agroecosystem, which may even require a complete redesign. This has been the case of dry hillsides

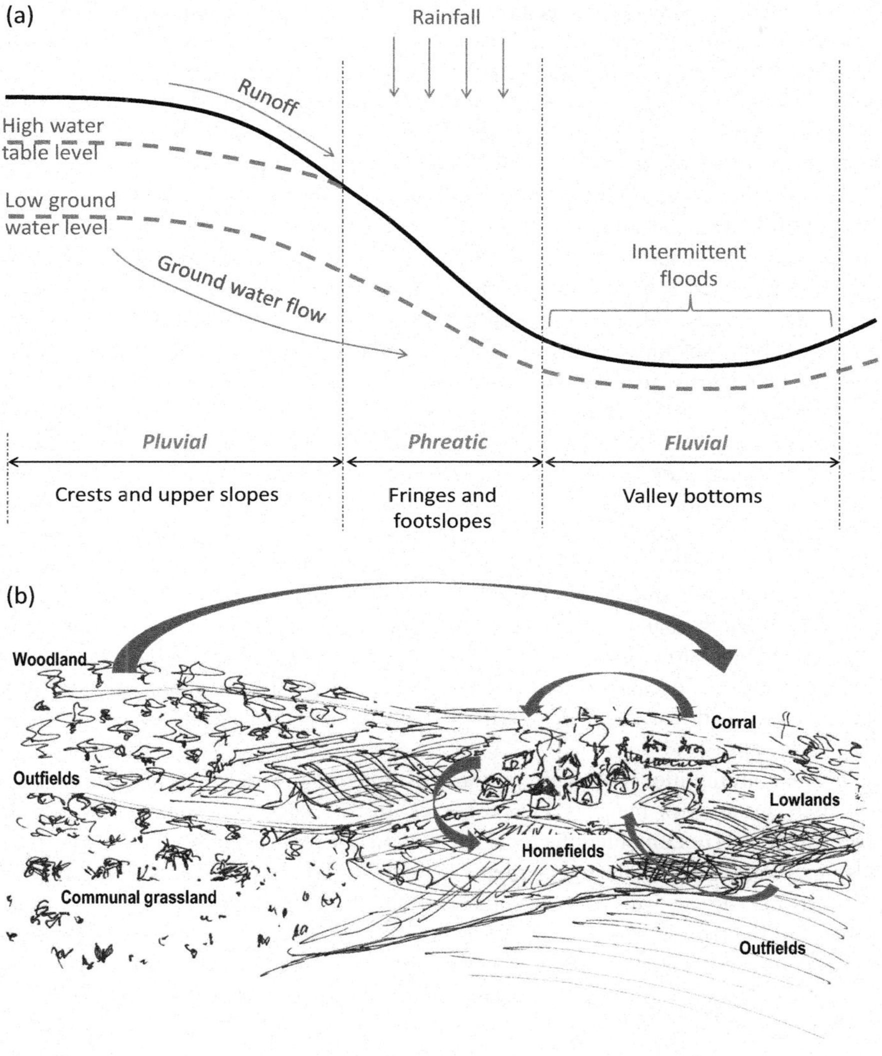

Fig. 6.6 (**a**) Scheme of a catena showing water flows above and belowground; (**b**) Scheme showing nutrient transfers from woodlands and grassland to cropland in an African village

used nowadays for sheep and goat grazing in parts of the Mediterranean region, that were formerly irrigated terraces where fruit trees, vineyards and vegetables would grow. Also, patch deforestation and the removal of trees from farmland, as in many temperate agricultural regions, has had major effects on several landscape functions and hence ecosystem services, including human habitat quality. And so on. But the allegory stops there. Because, for example, when studying or designing mechanical systems, the relationship between structure and function is formal and directional. It is first necessary to design a structure to be able to achieve the desired function(s). In biological systems, by contrast, structures and associated functions are pre-existing and we may or not know or understand them as yet. When assessing functional biodiversity and its space-time distribution on a landscape, a large part of our work consists of understanding the relationship between structures and functions, which we do through studying the relationship between patterns and processes.

In addition, landscapes are social-ecological systems that combine biophysical determinism with the complexity inherent to social processes and human agency, plus the expected causality of human physical structures such as fences, drainage, irrigation, tree patches, etc. As a result, in landscape design, structures and functions may not always be known or understood beforehand. Landscape structures are often studied as patterns, in space and time, and the identification and interpretation of such patterns depends strongly on the tools we use to assess them. Our understanding of pattern-process relationships may differ if we use satellite imagery versus on-the-ground landscape reading, of participatory mapping, or transect walks, or soil and vegetation toposequences, etc. In other words, different tools and different disciplines will result in different approximations to the relationship between patterns and processes in landscapes (or structures and functions in agroecosystems). Additionally, our study of the process-pattern relationship is generally geared towards a certain goal, and such a goal determines our methods as well. For example, pattern-process (structure-function) relationships may be studied to understand water dynamics on a landscape, or to understand biological pest control mechanisms, or to understand historical societal systems of land tenure that affect current access to resources, etc.

Rural communities have their own way of categorising and recognising landscape diversity in terms of patterns and processes. In the Barotse floodplains of Zambia, Del Río (2014) documented the local names and concepts used by farmers to refer to different landscapes units and elements (Fig. 6.7), and their variation with respect to the seasons (dry vs wet). The structuring element of this landscape is the seasonal flood. During the wet season, most of the low-lying plains known locally as *Bulozi* or *Libala* are under water. Homesteads and perennial crops are located in the uplands or in elevated patches on the landscape that appear as islands when the lowland is flooded (e.g. *lizulu, likana*). The distance between the river and the uplands vary, reaching up to 20 Km in certain parts. There are permanent ponds in the uplands. There is an intermediate zone between the uplands and lowlands known as *Saana*, where a diversity of crops are grown including rice, on the lower edges of this land unit. The different structures in the landscape deliver different functions throughout the year, during the dry and wet seasons, but also during key intermediate seasons when water tables are gradually rising or when the flood is receding. The crops grown on these structures/times of year also match the landscape patterns; a few examples are provided in Table 6.1.

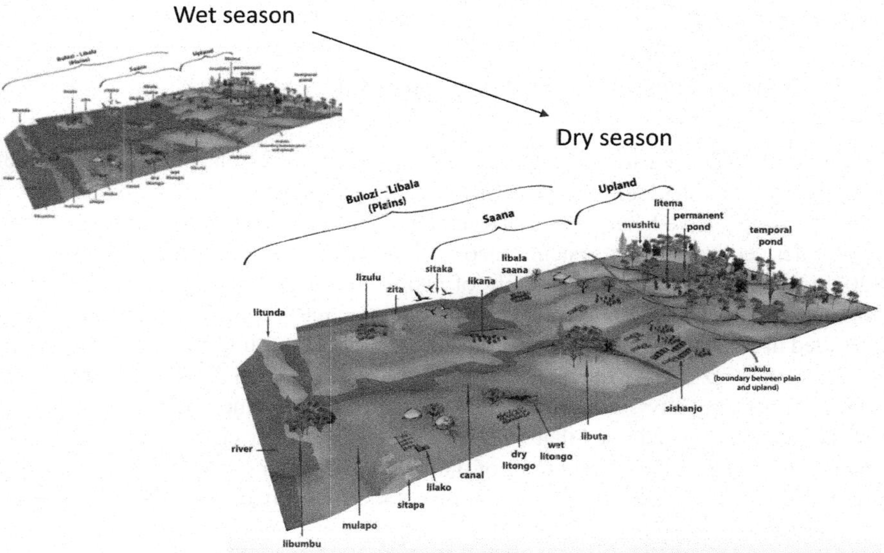

Fig. 6.7 Local names assigned to landscape units and landscape elements in the Barotse floodplain of southern Zambia, along the Zambezi river, as documented and elaborated by Trinidad del Rio and Natalia Estrada (Wageningen University, 2014)

Table 6.1 Four examples of landscape units described by Del Río (2014) in the Barotse floodplains of Zambia, indicating soil quality and main crops grown

Landscape unit	Description	Soil quality	Crops/activities
Mulapo	Low lying areas, concave shape, close to water sources in the dry season, flooded during the wet season	High soil fertility, moderate water availability	Non-cultivated
Sitapa	Similar to Mulapo but used for cultivation during short periods; albeit under risky of early floodings. Cultivation starts when the flood recedes towards the end of the dry (cold) season.	High soil fertility and water availability	Maize, Pumpkin, Yellow squash, Groundnuts, Rape, Watermelons, Potato, Okra, Rice, White squash, Local sugarcane, Orange squash, Sorghum, Carrots, Chinese cabbage, Cassava, Tomato, Hibiscus, Cowpeas
Sishanjo	It has similar characteristics as Sitapa but located in the intermediate area between the upland and floodplain (Saana), receives water from ponds (canals) or the water table, and is subject to organic matter oxidation due to drainage	High soil fertility and water availability (underground)	Rice, Vegetables, Sweet potatoes, Tomato, Maize, Rape, Cabbage, Cowpeas
Likana	Sandy areas located in the Saana area, subject to flooding; water table levels change drastically between seasons; usually borders drain water bodies.	Poor soil fertility, moderate to low water availability	Cassava, Groundnuts, Sweet potatoes, Bambara groundnuts, Squash (yellow, white), Hibiscus, Cowpeas, Rice, Watermelons

6.2 Landscapes as Complex Systems

Many ecologists refer to landscapes as complex systems, that is, systems that exhibit non-linear dynamics, self-organization, emergence, criticality, etc. Yet, while most of these concepts have been studied and modelled in disciplinary fields such as thermodynamics, economics or systems theory (Bak 1996; Auyang 1998; Thurner et al. 2018), their application to the study of ecosystems and landscapes has been more limited. Attempts to 'translate' complexity theories into ecological applications have resulted in useful but rather abstract concepts and models around the turn of the century (e.g., Ulanowicz 2004; Milne 1998; Medvinsky et al. 2001; Li 2002). At that time, the term 'bio-complexity' was coined in an attempt to examine complexity from the multiple disciplinary lenses that study landscapes and their ecology (Colwell 1998; Bruggeman et al. 2002). Definitions of bio-complexity refer to the 'complex chemical, biological, and social interactions in our planet's systems' (Colwell 1998, p. 786), or to "properties emerging from the interplay of behavioural, biological, physical, and social interactions that affect, sustain, or are modified by living organisms, including humans" (Michener et al. 2001, p. 1018), or state that "biocomplexity includes nonlinear or chaotic dynamics, unpredictable behaviour and interactions that span multiple levels of biological organisation or spatio-temporal scales" (Cottingham 2002, p. 793). The biocomplexity term and its use in the literature faded since then, but it remains a good descriptor for the properties we ascribe to diverse landscapes in agroecology.

Later on even, Cadenasso et al. (2006) proposed a framework to assess bio-complexity that represents a middle ground between complexity theory and practical ecological applications. These authors define bio-complexity as consisting of three major dimensions in landscapes: heterogeneity, connectivity and historical contingency (Fig. 6.8). These are three dimensions that (agro-)ecologists can assess in real landscapes to study their bio-complexity, as they are able to identify what patches/ entities are there, and how are they spatially arranged (heterogeneity), how are they organised and interact (connectivity), and how are they connected through time (memory or contingency). Heterogeneity has been introduced earlier (cf. Sect. 6.1.2), but the complexity in the structural heterogeneity of landscapes increases as one moves from merely describing different patches (e.g. richness, diversity) to the spatial configuration of such patches and their dynamics (shifting mosaics). Connectivity represents the organisational complexity of landscapes, in which patches may be more or less functionally connected. Organisational complexity is often proposed as a pre-requisite for resilience (e.g., Cabell and Oelofse 2012). Historical contingencies increase in complexity from contemporary direct effect through lags and legacies, to slowly emerging indirect effects. As complexity increases, the relationship between patches/ entities in a landscape shift from being direct and contemporary, to being indirect, to represent legacies, ecological memory, or lagged affects. Often such indirect and slow emerging effects can be seen in agricultural landscapes as the result of past human agency, for example through past management-induced soil heterogeneity, as will be discussed in the next chapter.

One may argue to what extent landscape bio-complexity can be assessed considering only these three dimensions. Are they enough? Do they truly reflect landscape

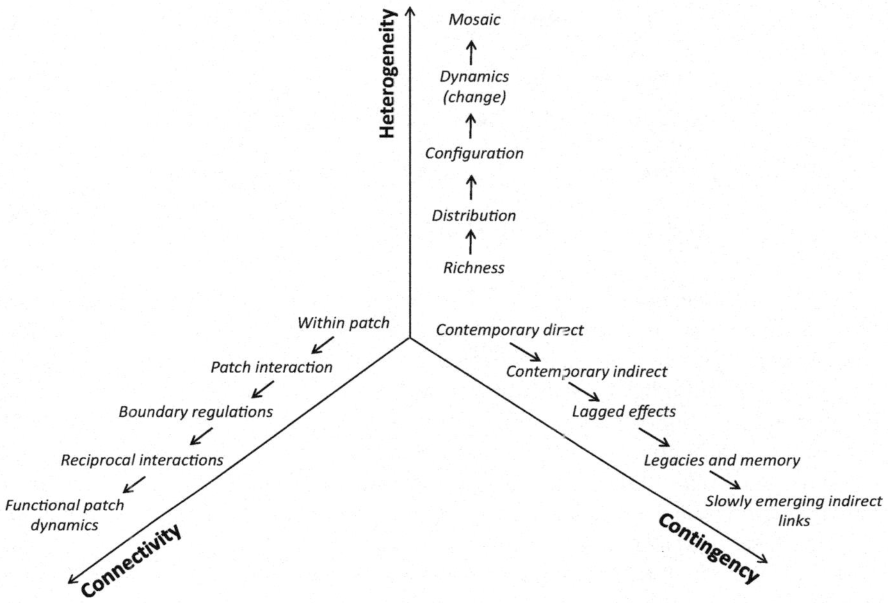

Fig. 6.8 Three dimensions of landscape bio-complexity based on the framework proposed by Cadenasso et al. (2006): spatial complexity (heterogeneity), functional complexity (connectedness) and historical complexity (contingency)

complexity? To answer these questions the reader is advised to study the original source (cf. Cadenasso et al. 2006), to examine in particular the assumptions the authors made regarding complexity theory, and the extent to which it may be approached using these three dimensions. They rely on the underlying assumptions that (i) landscape ecology is based on the interaction between structures and functions and their changes in time, and that (ii) slow and fast rates of interactions are responsible for sorting entities across scales and levels of organisation (hierarchy theory). They also illustrate the application of this framework through the analysis of a case study landscape. In agroecology, we are particularly interested in the relationship of these dimensions with human agency. In other terms, in how much are humans responsible for creating complexity, heterogeneity and diversity, and to what extent we recognise it and use it to manage agroecosystems.

6.3 Landscape Functions and Ecosystem Services

The relationship between structure and function is a core concept in systems analysis and ecology, as repeatedly stated in previous sections. Landscape structures sustain ecological processes that we humans interpret as functions. When such functions serve our purposes, we consider them as services (Chap. 3). Such is the origin of the

Table 6.2 Examples of ecosystem services classified in four main types

Ecosystem service type	Examples
Provisioning	Food, fibre and energy production, habitat provision, etc.
Support	Soil formation, nutrient availability, water storage, pollination, etc
Regulation	Pest biocontrol, carbon sequestration, watershed regulation, etc.
Cultural	Landscape recreational value, cultural heritage, food traditions, spiritual and religious values, etc.

concept of ecosystem services, to which we humans may ascribe benefits, and even a monetary value. Such a sequence of causalities is often represented as a 'cascade', from structures to functions, services and benefits (Potschin-Young and Haines-Young 2011), and it has been challenged by Diaz et al. (2018), who prefer to speak of 'nature contributions to people' (NCP) instead of ecosystem services. Such services or contributions may be defined and categorised in different ways, in the sense that they may be direct or indirect, proximal or distal, intermediate or final, or be classified by the nature of what they deliver as provisioning, support, regulation or cultural services (Table 6.2 – see also Chap. 3).

To illustrate the relationship between structures, functions, services, benefits and values, Fig. 6.9 uses the cascade model of Potschin-Young and Haines-Young (2011) exemplified with a number of services that may be provided by forest cover. A healthy forest, with its original structure, its diversity and canopy cover, a dense litter layer, and deep roots, is able to secure a number of functions. They include, among others, the reduction of water runoff through increased water infiltration and through the protection of the soil from the direct impact of the rain drops that can break the soil structure and detach sediments. The speed at which rain water falls is reduced by the canopy, and a large proportion of it penetrates the soil flowing on the surface of tree trunks and roots, reaching deep soil layers. Tree and litter cover reduce evaporation and, if sediment detachment still occurs and sediments are carried in the runoff flow, the presence of vegetation and litter slows such flow down and allows sediments to precipitate (sediment trapping). These functions together provide two major ecosystem services: watershed flow regulation and soil erosion control. These services result in a number of benefits for human population, including water capture, storage and availability, and the protection of the headwaters of streams specially when forests cover sloping land. Such water may be used to produce drinking water, to refill the aquifers and replenish the water table downstream, it may be used to produce hydroelectrical power, to be pumped for irrigation, etc. But these services also benefit humans by preventing landslides and/or floods downstream.

The economic value of such benefits may be estimated by calculating the costs of repairing the economic damage that landslides and/or floods can generate, the costs of insurances to protect infrastructures, crops, housing, etc., or the economic value of having drinking or irrigation water, or of producing hydroelectrical power. This example considers only two ecosystem services (watershed regulation and soil erosion control). But the presence and integrity of a forest may result in many

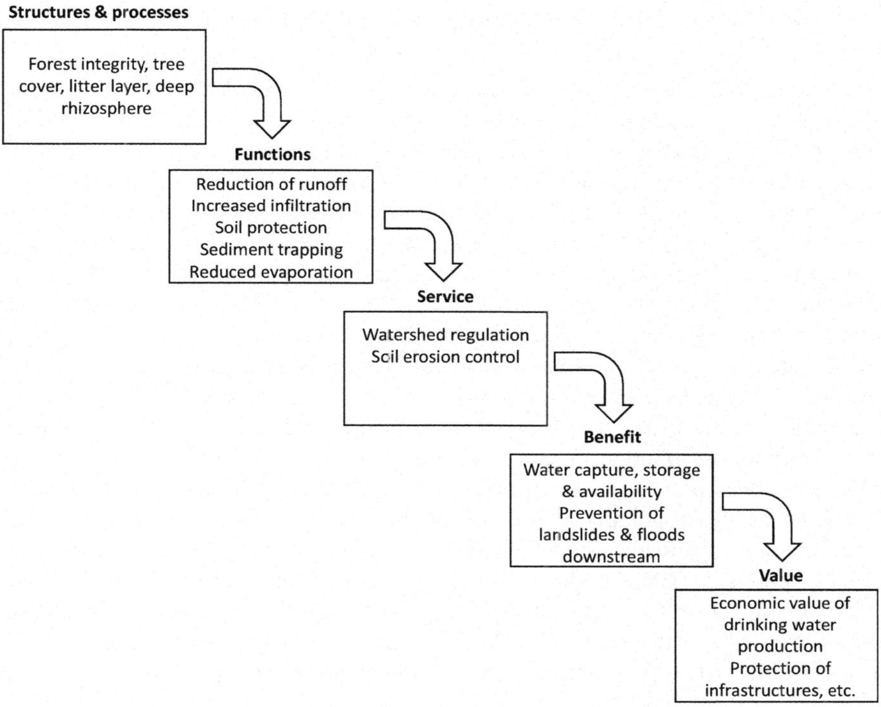

Fig. 6.9 The cascade model of Potschin-Young and Haines-Young (2011) applied to the provision of two main ecosystem services by forests: watershed flow regulation and soil erosion control. For conciseness, not all possible services, benefits and values are depicted

more ecosystem services simultaneously, such as biodiversity conservation, carbon sequestration, habitat provision, and the provision of food, fibre, fodder, timber, fuel, recreation, tourism, etc. Biodiversity conservation may be presented as an ecosystem service in itself, although it is also a 'means to an end', i.e., biodiversity is also necessary to provide other ecosystem services. This is why biodiversity conservation is often referred to as an intermediate ecosystem service.

Ecosystem services are presented here because they are nowadays broadly used as a concept, and bear a close relationship with landscape functions, mediated by biodiversity and human agency. However, in agroecology, we are interested in understanding/designing structures and functions, influencing patterns and processes, irrespective of whether they are termed ecosystem services, nature contributions to people, or whatever new terms may come in fashion in the next decades. Both 'services' and 'contributions' rely on biodiversity, and hence on the functional and structural integrity of landscapes that can be designed and managed following the principles of agroecology. Certain configurations of landscapes and their management may also lead to 'disservices', which is another way to refer to negative externalities (e.g., Tittonell et al. 2021).

6.4 Biodiversity in Agroecosystems and Landscapes

Beyond the effect of background geological and geomorphological features, spatial heterogeneity in the agricultural landscape results in and is the result of heterogeneous distribution of biodiversity in space and time, plus changes in physical attributes associated with human activities, such as infrastructure (fencing, roads, hedges, etc.), human habitat, drainage or irrigation, trees, domestic animals and their housing, crops and pastures. These modifications of the original environment create new niches for species of plants, animals and microorganisms, and purposively or not, they also facilitate the introduction of new species. Because biodiversity is essential to the provision of most ecosystem services from agricultural landscapes, understanding their nature, magnitude and space-time distribution is key for the design and management of sustainable agroecosystems. Most methods used to assess biodiversity in the landscape focus on descriptive parameters, such as richness, abundance, evenness, etc., and their distribution in space, such as patchiness, connectivity, fractality, etc. These methods provide an idea of patterns in the distribution of biodiversity in the landscape, but say nothing or little about the processes that involve the observed biodiversity. In particular, agroecologists are concerned with the functions that biodiversity provides. This is why agroecology places strong emphasis on functional biodiversity, which focusses on the processes that organisms support in communities and ecosystems, with emphasis on those that are *functional* to humans. Or, as Altieri and Nicholls (2004) proposed, on functional agrobiodiversity.

6.4.1 Functional Biodiversity

In ecology, functional biodiversity is classically defined as the number, type and distribution of functions performed by organisms in an ecosystem (e.g., Diaz and Cabido 2001). What organisms do in an ecosystem is more relevant to functional biodiversity than which species of organisms are actually there, why, or in which proportion. Species of plants, animals, microorganisms have certain traits associated with their phenotype that influence ecosystem functioning, and hence the study of functional biodiversity consists largely of studying such functional traits. Functional traits are thus components of an organism's phenotype that influence processes at ecosystem-level (Petchey and Gaston 2006). There are several indicators used in ecology to study functional biodiversity in an ecosystem. These include functional richness, evenness, or the 'distance' between phenotypes performing a similar function. But, what kind of function? As there is a large number of functional traits that could be identified in an ecosystem, a form or prioritization is needed. First of all, it is important to define the type of functions(s) for which certain phenotypic characteristics (traits) matter. Then, what is the relative contribution or relevance of a given trait to the function of interest? The treatment of such indicators and measures exceeds the scope of this book. However, some target functional traits that are of relevance to agroecology are worth mentioning.

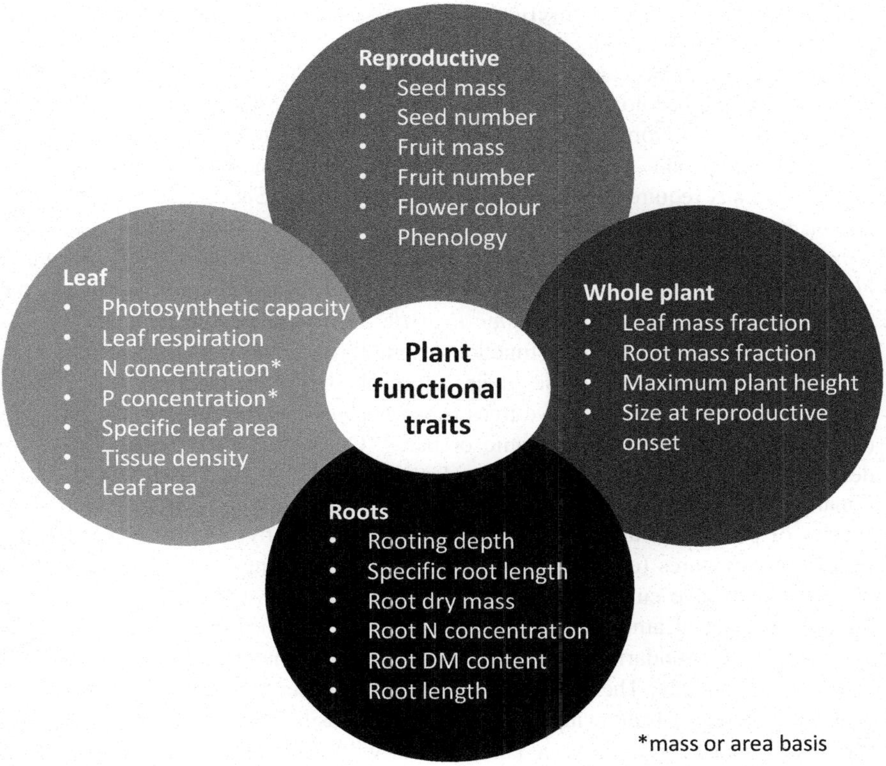

Fig. 6.10 An illustration of key plant functional traits based on Martin and Isaac (2015)

A large body of the literature on functional traits in (agro)ecosystems focuses on functional plant traits, such as leaf traits (e.g., the specific leaf area or the leaf C:N ratio) or root traits (e. g., specific root length, carboxylate and proton exudation, etc.) (Fig. 6.10). These traits have different effects on resource acquisition (radiation, water, nutrients), carbon storage, nutrient release, regulation of pathogen communities in soils, etc. The ability of plant roots to establish functional links with N_2-fixing bacteria or Mycorrhiza is variably considered by different authors as a 'plant' functional trait as well as bacterial or fungal functional traits. Some plant functional traits are relevant to more than one function simultaneously. For example, leaf-N or leaf N:lignin contents are relevant traits for soil organic matter formation or nutrient release, but also for herbivory, as leaf consumption by insects or ruminants is regulated by leaf quality. The amount and timing of flowering and the provision of nectar are key functional traits to attract pollinators and parasitoid arthropods, yet flower morphology is also an important complementary trait, as the mouth piece of different arthropod species needs to match that of the flower to access nectar. The concept of functional traits has also been used in livestock systems, particularly in the field of breeding (e.g. Egger-Danner et al. 2015). Many, in some cases most, of these functional traits or their effects are well known to farmers and constitute an intrinsic component of indigenous knowledge.

Knowledge on plant, animal or microbial functional traits may in principle inform the design of agroecological or ecologically intensive cropping systems (rotations, intercrops, cover crops, flower strips, agroforestry, etc.). However, designing agroecosystems and landscapes on the basis of knowledge on functional agrobiodiversity requires a more comprehensive approach. Such as the agroecosystem approach proposed by Moonen and Bàrberi (2008), who define functional *agro*biodiversity as "*that part of the total biodiversity composed of clusters of elements (at the gene, species, or habitat level) providing the same agroecosystem service, that is driven by within-cluster diversity*". They also postulate that the positive but also the negative and neutral functions exerted by biodiversity components should be considered. They illustrate their approach with an example of a focus agroecosystem and context: olive trees (*Olea europaea* L.) affected by the olive fly (*Bactrocera oleae* Rossi) in organic orchards. Once the target function has been defined (control of the olive fly), they identify the 'cluster' or agroecosystem functional group of interest for this target. The functional cluster in this case consists of species of parasitoids and hyper-parasitoids of the olive fly, wild plant species and vegetation structures (e.g., hedgerows, woodland) attracting natural enemies of the olive fly, and olive cultural practices known to affect olive fly (e.g., cultivar type, pruning, types and amount of natural pesticides sprayed). Finally, they define the space and time boundaries of the agroecosystem functional group and relevant indicators for their study. These include, at field and landscape scale, and over a whole year, the number of fruits with fly punctures, of parasitized fruits, and natural enemy species, presence and abundance (% cover) of wild plants and structures supporting natural enemies, etc. This idea of functional clusters of biodiversity proposed by proposed by Moonen and Bàrberi is, in my opinion, more relevant to the design of agroecosystems than the functional traits of individual components of biodiversity.

6.4.2 Agrobiodiversity and Biodiversity for Food and Agriculture

A challenge in the field of agricultural biodiversity is the multiple definitions of it that co-exist. As described in Chap. 3, and hinted in the previous paragraphs, Agrobiodiversity (ABD) entails not only the cultivated plant or domestic animal species, but also the non-cultivated ones and the diversity of habitats in the landscape. But there is also a co-exiting definition, that of Biodiversity for Food and Agriculture (BFA), which shares several components with the former but adds also wild species collected, fished or hunted, and the concept of associated biodiversity. Often ABD and BFA are simply used as synonyms – although not by their proponents. Let us refer to the definitions provided by the Food and Agriculture Organisation (FAO) of the United Nations.

The FAO (1999) defines agrobiodiversity literally as: "*The variety and variability of animals, plants and micro-organisms that are used directly or indirectly for food and agriculture, including crops, livestock, forestry and fisheries. It comprises the diversity of genetic resources (varieties, breeds) and species used for food, fodder, fibre, fuel and pharmaceuticals. It also includes the diversity of non-harvested species that support production (soil micro-organisms, predators, pollinators), and those in the wider environment that support agroecosystems (agricultural, pastoral, forest and aquatic) as well as the diversity of the agroecosystems.*" Whereas biodiversity for food and agriculture is defined as (FAO 2019): "*Biodiversity for food and agriculture is all the plants and animals – wild and domesticated – that provide food, feed, fuel and fibre. It is also the myriad of organisms that support food production through ecosystem services – called associated biodiversity. This includes all the plants, animals and micro-organisms (such as insects, bats, birds, mangroves, corals, seagrasses, earthworms, soil-dwelling fungi and bacteria) that keep soils fertile, pollinate plants, purify water and air, keep fish and trees healthy, and fight crop and livestock pests and diseases.*"

Comparing both definitions, it appears at first glance that BFA comprises ABD, yet the '*non-harvested species that support production*' may also be defined as the '*myriad of organisms that support food production through ecosystem services*'. Here, I propose to define agrobiodiversity as follows:

Agrobiodiversity is the *diversity* of cultivated and non-cultivated, harvested and non-harvested species of plants, animals and microorganisms that provide essential ecosystem services to sustain food, feed, fibre, energy and well-being (e.g. medicinal and spiritual plants), plus *the knowledge* that is necessary for their cultivation, collection, storage, processing and use by rural communities.

I believe this to be a more appropriate definition of agrobiodiversity in the realm of agroecology. A diagram corresponding to this definition is introduced in the Fig. 3.2 of Chap. 3. Agrobiodiversity is intrinsically functional. But, without the necessary knowledge for their management, all the referred species will be just genetic resources, and not agrobiodiversity.

6.4.3 Measuring Biodiversity: Alpha, Beta and Gamma Diversity

Biodiversity in landscapes is studied by considering three complementary dimensions. Alpha diversity is the diversity of species within one patch of vegetation, beta diversity is the diversity of species/communities between patches, and gamma diversity is the result of aggregating both alpha and beta diversities at landscape level. The concepts were introduced by Whittaker (1972) as an approach to

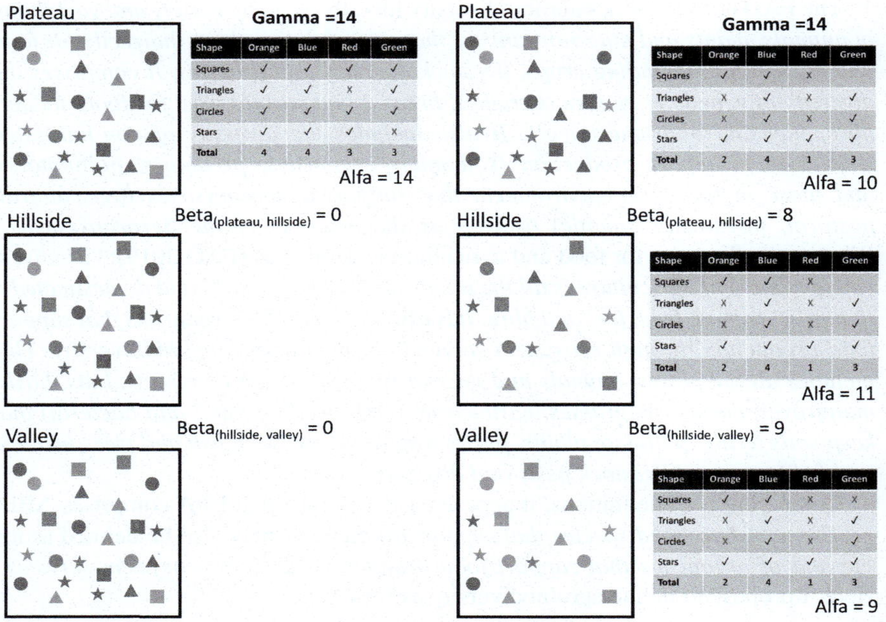

Shape	Orange	Blue	Red	Green
Squares	✓	✓	✓	✓
Triangles	✓	✓	x	✓
Circles	✓	✓	✓	✓
Stars	✓	✓	✓	x
Total	4	4	3	3

Shape	Orange	Blue	Red	Green
Squares	✓	✓	x	
Triangles	x	✓	x	✓
Circles	x	✓	x	✓
Stars	✓	✓	✓	✓
Total	2	4	1	3

Shape	Orange	Blue	Red	Green
Squares	✓	✓	x	
Triangles	x	✓	x	✓
Circles	x	✓	x	✓
Stars	✓	✓	✓	✓
Total	2	4	1	3

Shape	Orange	Blue	Red	Green
Squares	✓	✓	x	x
Triangles	x	✓	x	✓
Circles	x	✓	x	✓
Stars	✓	✓	✓	✓
Total	2	4	1	3

Fig. 6.11 An illustration of the concepts of alpha, beta and gamma diversity. The example A (left hand side patches) shows similar alpha and gamma diversity, as the three landscape units show the same levels of alpha diversity. The example B (right hand side patches) shows smaller average alpha diversity at patch level but diversity among patches, or beta diversity. The overall gamma diversity (species richness) is the same in both examples

measuring biodiversity. An example is presented in Fig. 6.11. Alpha diversity is basically the species richness and it may be similar to gamma diversity in spatially homogeneous landscapes (Fig. 6.11a). In some cases, landscapes may exhibit patch heterogeneity, or beta diversity, but little alpha diversity within patches. In agroecology, we use these concepts to characterise biodiversity across scales, in agricultural landscapes, land use systems, agroecosystems, down to the levels of individual farms and fields. Farms with large numbers of different crop species, grasslands and/or land uses may exhibit high beta diversity, where the different fields and paddocks are considered as patches. But also, and specially in smallholder agriculture, one may encounter high species diversity within fields, or alpha diversity, in the form of intercrops, agroforestry or polycultures (cf. Chap. 3).

Plate 6.1 illustrates how alpha and beta diversity may look like in smallholder agriculture, as determined by farmers' management decisions. In the Pictures A and B we see cassava intercropped with groundnuts and with maize. In both cases, if we do not consider weed species, the *planned* alpha diversity is 2. The farmer planned to grow two plant species together, simultaneously. In Pictures C and D we can see small patches (beta diversity) of different single-species crops, such as Napier grass,

Plate 6.1 Examples of alpha (above) and beta (below) diversity in smallholder agroecosystems of East Africa. Cassava plants intermingled with plants of groundnuts (**a**) or maize (**b**) in Eastern Uganda. Fields close to the homesteads, showing a plot of maize in the front and banana plants and trees in the background (**c**), and fields of food (maize) and fodder (Napier grass) crops divided by living fences, with trees growing near the homesteads in Western Kenya (**d**). (Photos: P. Tittonell)

maize, bananas, etc., plus living fences (*Euphorbia* hedges in both cases), and sparse fruit and timber trees. Each plot of a different crop species exhibits alpha diversity = 1, but there is beta diversity (between plots) and gamma diversity (adding also the trees and hedges). At higher levels of analysis, or low levels of resolution, as when analysing land uses or large areas of the landscape, the patch diversity within a small farm of 1 or 2 ha may end up being considered as alpha diversity, especially when the 'patches' are sufficiently large to comprise several farms with their area.

Thus, at higher levels of analysis (or lower resolution), it is not possible to detect differences in the way biodiversity is organised in space and time (structure), and hence it becomes more difficult to infer their functions. For example, consider the Pictures A and B in Plate 6.2. The diversity in these pictures may be the same when working at landscape or regional scales, in both cases it would be considered alpha diversity and have perhaps a similar value. Picture A shows a polyculture on an agroecological farm in Cuba, whereas Picture B shows an organic horticultural farm in temperate Spain. When analysing the agroecosystem at farm scale, as seen in the pictures, it is clear that in Picture A we are seeing alpha diversity and in Picture B beta diversity. Interactions between different species (e.g. facilitation, mutualism, competition, etc.) are more likely to happen in A than in B. Over time, however, we may see a similar structure in the polyculture of Picture A year after year, or perhaps

Plate 6.2 Examples of alpha, beta and gamma diversity. (**a**) An example of a polyculture on an agroecological farm in Cuba, exhibiting alpha diversity within a single field plot. (**b**) Organic horticulture in the Basque Country, Spain, showing fine-grain beta diversity. (**c**) A silvo-pastoral system in Colombia in which high species diversity is found along the fences but not within or between the paddocks. (**d**) Native grasslands in Uruguay (*Campos* region) which exhibit high levels of alpha diversity but relatively low beta diversity, due to homogeneous grazing management over large areas and sparse subdivision of paddocks (Photos: P. Tittonell)

with small changes in the herbaceous component of the polyculture. In the cropping pattern of Picture B we expect to see frequent crop rotation, up to three subsequent crops grown per year in a given area. Moreover, as crop rotations in horticulture are largely dictated by the market, the size of the plot of land grown to a certain crop may vary from year to year, the plots are not fenced and hence the 'identity' of a field plot is lost over time. Additionally, both types of space-time arrangements in these two examples have huge consequences for the biodiversity one-level below: soil biodiversity.

Diversity in livestock systems can also de described in terms of alpha, beta and gamma diversity. This applies specially, but not exclusively, to graze-based livestock systems. In the examples of Plate 6.2, Pictures C and D we see two contrasting situations. Picture C shows a relatively small paddock of Napier grass surrounded by living fences of native trees and leguminous fodder trees in a silvo-pastoral system in the Cauca valley, Colombia. Many spontaneous species colonise the areas under the tree canopy, contributing diversity. Both alpha and beta diversity are low, because only Napier grass is grown in every small paddock. Yet, gamma diversity at landscape level may be high, due to the diversity of species to be found in the tree hedges. These spontaneous species contribute not only to livestock diets but also to ecosystem regulation. In Picture D, we see native grasslands in the plains known as *Campos* in Uruguay, in which alpha diversity is very high. Yet the landscape is quite homogeneous and the farms and paddocks are large, and sparsely fragmented. The result is high alpha and gamma diversity but relatively low beta diversity. Beta diversity emerges as the result of natural heterogeneity in the landscape, such as by the presence of marshes and wetlands, rock outcrops or slopes.

Trees on farm represent an important element of biodiversity in the landscape generally appearing as alpha rather than beta diversity. Trees can be organised in the landscape following patterns such as rows, patches or isolated. Tree windrows are common in many farming systems, in which they also often demarcate fields or farms, and may serve as shade for livestock. In smallholder farms, trees are most common in home gardens, where they provide shade and shelter, firewood, timber, fruits, nuts. They are normally arranged in such a way that different vertical strata are used by different tree species. Trees may also be scattered in the landscape, growing isolated within cropping fields, or even as parklands or savannahs that comprise both trees and crops. Trees on farm may also be planted as monospecific woodlots, such as *Eucalyptus* or *Tephrosia* grooves in the tropics, that provide important amounts of timber and fuelwood to smallholder farmers when they are harvested.

To assess the diversity and carbon storage potential of trees and other perennial vegetation on smallholder farms, Henry et al. (2009) conducted an exhaustive research in western Kenya where they established allometric models to estimate biomass of most perennial species grown by farmers, either as woodlots, hedge rows, windrows, home gardens or individual trees (Fig. 6.12a). They observed that, beyond the cases of monospecific woodlots or windrows in which diversity was practically nil, in all other variegation units more biodiversity was associated with greater carbon stocks aboveground. Unfortunately, carbon credits pay for C stored in vegetation irrespective of whether it is stored in highly diverse stands or in a monospecific tree formations. For these smallholder farmers to be able to access the carbon market, conclude Henry et al. (2009), they will need to apply together, joining forces over hundreds of households (each owning about 1 ha of land), and over 20 years, generating large transaction costs. Under current conditions, the carbon market is not accessible to smallholder farmers, in spite of their large contribution to carbon storage in their diverse tree stands.

6.4.4 Landscape Biodiversity and Human Nutrition

In the context of smallholder, family agriculture, landscape biodiversity has a major impact on nutrient self-sufficiency for animal and human populations (Chap. 1). This entails not only the diversity of crops grown or animals raised on every single farm, but also the diversity of associated and wild species (see above: Biodiversity for

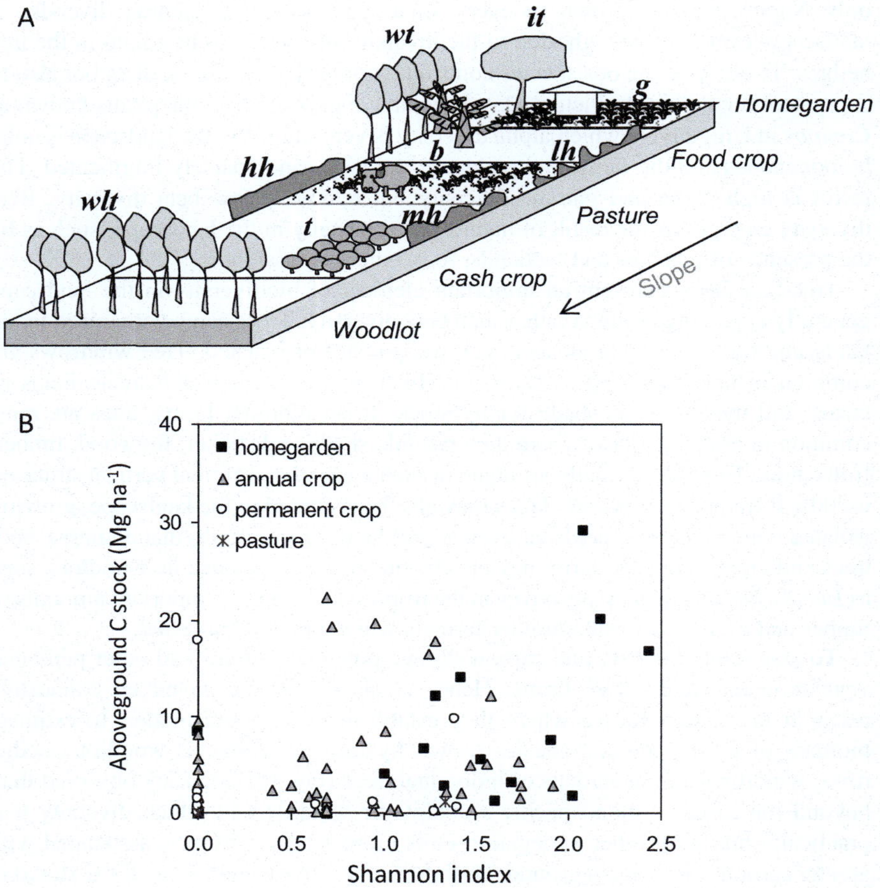

Fig. 6.12 Biodiversity of perennial species and their carbon storage in their aboveground biomass. (**a**) A categorisation of perennial vegetation types to assess carbon storage though allometric models; (**b**) Higher plant diversity (Shannon index) was often associated with high carbon storage. (From: Henry (2009) MSc Thesis AgroPariTech)

Food and Agriculture). At local level, such diversity of food products can be found not necessarily on each farm but on the local market, which sells a combination of food produced locally and imported from other regions. A survey done by Félix et al. (2018) in central Burkina Faso (Sahelian region) illustrates this by showing the proportion of different nutrients in the diet that local people derive from their own farm, from the local market, or from the surrounding landscape (associated and wild biodiversity) (Fig. 6.13). During the wet season, when food is available, the farm is the main source of energy, iron and zinc, but not of vitamin A, which is mostly derived from food bought on the market. During the dry season, when families suffer food shortages, they derive important amounts of iron (25%) or vitamin A (37%)

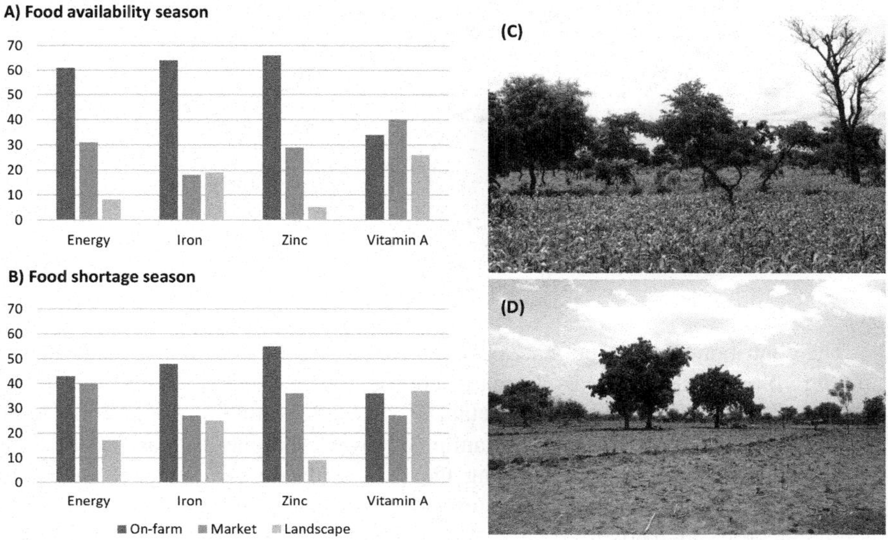

Fig. 6.13 Relative sourcing of food energy (calories), iron, zinc and Vitamin A from on-farm production, the local market or the landscape (wild biodiversity) by rural households in Sahelian Burkina Faso, during times of food availability (**a**) and shortage (**b**), modified from Félix et al. (2018). Pictures of the area during the wet (**c**) and dry (**d**) season. (Photo credit: G. Felix)

from the 'landscape'. In this particular case, the landscape exhibits scattered patches of forest, shrubland and grazing land, all of them managed communally, and hosting a diversity of edible wild plants (leaves, roots, fruits) and game.

On-farm biodiversity contributes to diverse diets not only through the different species of crops grown or animals raised on the farm, but also through the presence and management of spontaneous plants, often indigenous, that people use as food. These include a diversity of plants that an outsider would consider to be 'weeds' (indeed, some of these species are listed as weeds in agricultural extension material), plus other traditionally cultivated plants known sometimes as minor or 'underutilized' crop species in the literature (Li et al. 2020). Others prefer to refer to them as traditional or indigenous vegetables. Some species of traditional vegetables may be cultivated by farmers (or kept, when emerging spontaneously) but also act as weeds when in large numbers. Sometimes crop species grown for their grain or fruit present other edible organs that rural people use in their diet, such as pumpkin leaves, seeds and flowers, or cowpea leaves. In the home gardens of smallholder farms, such species may be often seen grown for their leaves, while they are grown for their main economic plant organ in the open fields of the farm. Figueroa et al. (2008) made an inventory of traditional vegetables identified by and with farmers in two villages of western Kenya, in Vihiga and Migori districts, and calculated their contribution to household nutrition as compared to maize, the main staple crop in the region (Table 6.3).

Table 6.3 Percent contribution of maize and traditional vegetables to household daily nutritional requirements in two rural areas of Kenya (Figueroa et al. 2008)

	Vihiga		Migori	
	Maize	Traditional vegetables	Maize	Traditional vegetables
Energy	59.0	4.0	71.0	3.0
Protein	10.9	44.8	12.5	32.6
Calcium	7.5	48.3	7.4	27.5
Iron	9.2	29.7	11.2	22.8
Vitamin A	Traces	16.5	Traces	11.9

The contribution of traditional vegetables to household nutrition was substantial in both villages, but greater in Vihiga where population density is higher and farms smaller (Table 6.3). Traditional vegetables contributed protein, calcium and iron, and between 10% and 20% of the Vitamin A. However, Figueroa et al. (2008) show that these figures varied widely along the year, with the greatest contribution of traditional vegetables taking place during times of food scarcity (e.g. the months before the crops are ready to be harvested and when the household grain reserves are already exhausted). Thus, traditional vegetables should not only be valued for their role on the overall diet composition of the family over the year, but specially for the strategic role they play during times of food scarcity.

6.5 Summary and Concluding Remarks

The agroecosystem is the core object of agroecology. It is a special type of system that combines multiple dimensions, it is governed by natural laws and steered by humans through design and management. Agroecosystems are dynamic and spatially heterogeneous, and their properties can be used to assess their sustainability through evaluation criteria and indicators. One way of studying the agroecosystem is from a landscape perspective, considering its horizontal structure and mosaic patterns in space and time. Landscapes present spatial heterogeneity, which may be described based on patterns, that may reveal also different processes taking place on the ground. Landscapes are studied considering their anisotropy and the spatial autocorrelation of their constituent elements, as will be further explained in the following chapter. A landscape, be it a portion of a territory delineated within our visual range, or following physical or administrative boundaries, may be considered an agroecosystem. But a landscape may also be larger than an agroecosystem, and include it or not, or be a ‘natural’ landscape, or an urban landscape, etc. An agroecosystem, on the other hand, may be defined at different scales, from a single farm to a region, and hence be smaller or larger than a landscape. In agroecology, both concepts tend to converge at a scale that considers communities and their territories, communal decision-making and ecological interactions. While the agroecosystem concept brings implicitly in the notion of systems, the landscape brings in the spatial dimension and the tools to study spatial diversity.

Landscapes host biodiversity, which is the basis for the provision of ecosystem services or nature contributions to society. Biodiversity is often measured by classifying it into alpha, beta and gamma diversity, respectively within patches, between patches and total landscape biodiversity. But most important in agroecology is the notion of functional diversity, that is, the functions associated with clusters of biodiversity within the landscape. Identifying such clusters requires first to define the objective for which functions are studied. Yet there are also generic characteristics known as functional traits (of plants, animals and organisms) that are generally known to participate in several ecosystem functions. Some authors propose to see landscapes as complex systems, and define several dimensions to describe such complexity. These include heterogeneity, connectivity, and contingency. In agroecology, these dimensions are useful to understand landscape complexity so long as they can be linked to human agency, as farmers are the key source of knowledge and the architects of landscape complexity. They not only understand such complexity, they manage it to achieve their production and reproduction goals, amongst them the production of sufficient and nutritionally diverse diets year round.

References

Altieri MA, Nicholls CI (2004) Biodiversity and pest management in agroecosystems, 2nd edn. The Haworth Press, New York

Auyang SY (1998) Foundations of complex-systems theories in economics, evolutionary biology, and statistical physics. Cambridge University Press, Cambridge, p 404

Bak P (1996) How nature works: the science of self-organized criticality. Springer, New York, p 212

Bruggeman FJ, Westerhoff HV, Boogerd FC (2002) BioComplexity: a pluralist research strategy is necessary for a mechanistic explanation of the "live" state. Philos Psychol 15:411–440

Cabell J, Oelcfse M (2012) An indicator framework for assessing agroecosystem resilience. Ecol Soc 17. https://doi.org/10.5751/ES-04666-170118

Cadenasso M, Pickett STA, Grove M (2006) Dimensions of ecosystem complexity: heterogeneity, connectivity, and history. Ecol Complex 3:1–12. https://doi.org/10.1016/j.ecocom.2005.07.002

Colwell R (1998) Balancing the biocomplexity of the planet's living systems: a 21st century task for science. Bioscience 48:786–787

Cottingham K (2002) Tackling biocomplexity: the role of people, tools, and scale. Bioscience 52: 793–799

Del Río T (2014) Farming systems characterization in three communities from the Barotse floodplains, Zambia: Relating landscape with production and diversity. MSc Thesis Wageningen University

Diaz S, Cabido M (2001) Vive la difference: plant functional diversity matters to ecosystem processes. Trends Ecol Evol 16:646–655. https://doi.org/10.1016/S0169-5347(01)02283-2

Diaz S, Pascual U, Stenseke M, Martín-López B, Watson R, Molnár Z, Hill R, Chan K, Baste I, Brauman K, Polasky S, Church A, Lonsdale M, Larigauderie A, Leadley P, van Oudenhoven A, Plaat F, Schröter M, Lavorel S, Shirayama Y (2018) Assessing nature's contributions to people. Science 359:270–272. https://doi.org/10.1126/science.aap8826

Ditzler L, van Apeldoorn DF, Schulte RPO, Tittonell P, Rossing WAH (2021) Redefining the field to mobilize three-dimensional diversity and ecosystem services on the arable farm. Eur J Agron 122:126197. https://doi.org/10.1016/j.eja.2020.126197

Easdale MH, Bruzzone O, Mapfumo P, Tittonell P (2018) Phases or regimes? Revisiting NDVI trends as proxies for land degradation. Land Degrad Dev 29:433–445. https://doi.org/10.1002/ldr.2871

Easdale MH, Fariña C, Hara S, Pérez León N, Umaña F, Tittonell P, Bruzzone O (2019) Trend-cycles of vegetation dynamics as a tool for land degradation assessment and monitoring. Ecol Indic 107. https://doi.org/10.1016/j.ecolind.2019.105545

Egger-Danner C, Cole JB, Pryce JE, Gengler N, Heringstad B, Bradley A, Stock KF (2015) Invited review: overview of new traits and phenotyping strategies in dairy cattle with a focus on functional traits. Animal 9(2):191–207

FAO (2019) The state of the World's biodiversity for food and agriculture. In: Bélanger J, Pilling D (eds) FAO Commission on genetic resources for food and agriculture assessments. Rome. 572 pp. http://www.fao.org/3/CA3129EN/CA3129EN.pdf

FAO (Food and Agriculture Organization of the United Nations) (1999) Agricultural biodiversity, multifunctional character of agriculture and land conference. Background Paper 1. Maastricht, Netherlands. September

Félix GF, Diedhiou I, Le Garff M, Timmermann C, Clermont-Dauphin C, Cournac L, Groot JCJ, Tittonell P (2018) Use and management of biodiversity by smallholder farmers in semi-arid West Africa. Global Food Secur 18:76–85. https://doi.org/10.1016/j.gfs.2018.08.005

Figueroa BM, Tittonell P, Ohiokpehai O, Giller KE (2008) The contribution of traditional vegetables to household food security in two communities of Vihiga and Migori Districts, Kenya. In: International symposium: underutilized plants for food, nutrition, income and sustainable development, 3–7 March, Arusha, Tanzania

Henry M, Tittonell P, Manlay R, Bernoux M, Albrecht A, Vanlauwe B (2009) Biodiversity, C stocks and sequestration potential in aboveground biomass in smallholder farming systems of western Kenya. Agric Ecosyst Environ 129:238–252

Kebede Y, Bianchi F, Baudron F, Abraham K, de Valença A, Tittonell P (2018) Implications of changes in land cover and landscape structure for the biocontrol potential of stemborers in Ethiopia. Biol Control 122:1–10

Kebede Y, Baudron F, Bianchi FJJA, Tittonell P (2019) Drivers, farmers' responses and landscape consequences of smallholder farming systems changes in Southern Ethiopia. Int J Agric Sustain. https://doi.org/10.1080/14735903.2019.1679000

Kolasa J, Pickett S, Rollo DC (1991) The hertegogeneity of heterogeneity – a glossary. In: Kolasa, Pickett (eds) Ecological heterogeneity. Springer, New York, pp 1–23

Li B-L (2002) A theoretical framework of ecological phase transitions for characterizing tree-grass dynamics. Acta Biotheor 50:141–154

Li X, Yadav R, Siddique KHM (2020) Neglected and underutilized crop species: the key to improving dietary diversity and fighting hunger and malnutrition in Asia and the Pacific. Front Nutr 7. https://www.frontiersin.org/articles/10.3389/fnut.2020.593711

Martin AR, Isaac ME (2015) Plant functional traits in agroecosystems: a blueprint for research. J Appl Ecol 2015(52):1425–1435. https://doi.org/10.1111/1365-2664.12526

Medvinsky AB, Tikhonova IA, Aliev RR, Li B-L, Lin Z-S, Malchow H (2001) Patchy environment as a factor of complex plankton dynamics. Phys Rev E 64:021915-1–021915-7

Michener WK, Baerwald TJ, Firth P, Palmer MA, Rosenberger JL, Sandlin EA, Zimmerman H (2001) Defining and unraveling biocomplexity. Bioscience 51:1018–1023

Milne BT (1998) Motivation and benefits of complex system approaches in ecology. Ecosystems 1: 449–456

Moonen AC, Bàrberi P (2008) Functional biodiversity: an agroecosystem approach. Agric Ecosyst Environ 127:7–21

Novotny IP, Fuentes-Ponce MH, Tittonell P, Lopez-Ridaura S, Rossing WAH (2020) Back to the people: the role of community-based responses in shaping landscape trajectories in Oaxaca, Mexico. Land Use Policy 100:104912. https://doi.org/10.1016/j.landusepol.2020.104912

Novotny IP, Fuentes-Ponce MH, Lopez-Ridaura S, Tittonell P, Rossing WAH (2021) Longitudinal analysis of household types and livelihood trajectories in Oaxaca, Mexico. J Rural Stud 81:170–181. https://doi.org/10.1016/j.jrurstud.2020.10.022

Petchey OL, Gaston KJ (2006) Functional diversity: back to basics and looking forward. Ecol Lett 9:741–758. https://doi.org/10.1111/j.1461-0248.2006.00924.x

Potschin-Young M, Haines-Young R (2011) Ecosystem services. Prog Phys Geogr 35:575–594. https://doi.org/10.1177/0309133311423172

Thurner S, Klimek P, Hanel R (2018) Introduction to the theory of complex systems. Print ISBN-13: 9780198821939

Tittonell P, van Dis R, Vanlauwe B, Shepherd K (2016) Managing soil heterogeneity in smallholder African landscapes requires a new form of precision agriculture. In: Lal R, Stewart BA (eds) Soil specific farming, advances in soil science. CRC Press – Taylor & Francis Group NW, Boca Raton, pp 199–224

Tittonell P, Hara SM, Álvarez VE, Aramayo VM, Bruzzone OA, Easdale MH, Enriquez AS, Laborda L, Trinco FD, Villagra SE, El Mujtar V (2021) Ecosystem services and disservices associated with pastoral systems from Patagonia, Argentina – a review. Cah Agric 30:43. https://doi.org/10.1051/cagri/2021029

Trinco FD, Rusch VE, Howison RA, Garibaldi LA, Tittonell PA (2021) Implications of landscape configuration on understory forage productivity: a remote sensing assessment of native forests openings. Agrofor Syst 95:1675–1688. https://doi.org/10.1007/s10457-021-00676-w

Ulanowicz RE (2004) On the nature of ecodynamics. Ecol Complex 1:341–354

Whittaker RH (1972) Evolution and measurement of species diversity. Taxon 21:213–251. https://doi.org/10.2307/1218190

Chapter 7
Spatial Heterogeneity in Agroecosystems

Abstract As explained in the previous chapter when describing landscapes, spatial heterogeneity refers to patterns that consist of dissimilar or diverse constituents, it is a form of variability has a spatial structure, and it is quantified through the study of patterns and scales. This chapter builds on the previous one to address the spatial properties of landscapes in an agroecological sense, that is, as influenced or shaped through human agency. It proposes a number of approaches to study, describe and categorise spatial heterogeneity in landscapes and soils. Spatial heterogeneity and variability are studied mostly from the perspective of the land manager, at the scales at which farmers normally perceive them and consider them in their management decisions. There is a large palette of methods to assess spatial heterogeneity from an ecological perspective in the specialised literature, focusing on natural ecosystems. This Chapter focuses on the characteristics and sources of spatial heterogeneity in the *agro*ecosystem, with a strong emphasis on soil properties and crop variability, two aspects of agroecosystems with which farmers and practicing agroecologists are usually concerned. Methods are proposed to categorise heterogeneity by means of observation, simple statistics, local soil knowledge and indicators.

7.1 Sources of Variability in Agroecosystems

Before examining patterns of spatial heterogeneity at different scales and delving specifically into soil spatial variability or heterogeneity, let us classify the various types and sources of variability in both space and time that can be found in agroecosystems (Fig. 7.1). Climate, geology, geomorphology, and biome, along with their interactions, are the major sources of macro-variability. These factors also translate into large spatial patterns of heterogeneity, such as variations in soil types, vegetation, and land uses. Although these factors also operate at lower hierarchical levels, influencing meso- and micro-variability, their effects are less obvious at these levels compared to those of human agency, such as land use and management decisions. Land use shapes landscapes and farms, generating meso-variability within agroecosystems. Additionally, differences in land use across watersheds or regions, which may arise from disparities in cultures, policies,

P. Tittonell, *A Systems Approach to Agroecology*,
https://doi.org/10.1007/978-3-031-42939-2_7

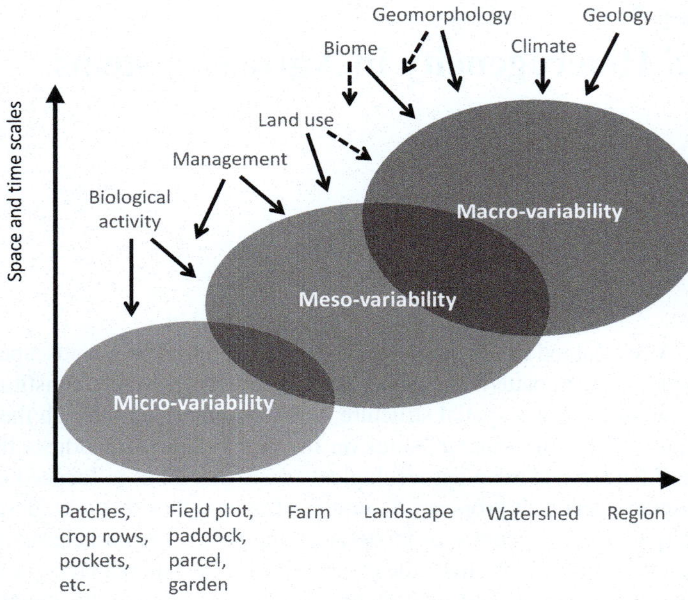

Fig. 7.1 Sources and levels of variability in agroecosystems. Dotted arrows indicate secondary influences. The intersection between the sets indicates interactions between levels of variability

regulations, markets, or socio-economic contexts, contribute to reinforcing macro-variability as well. Management decisions at the farm scale wield significant influence over meso-variability, while management at the field scale also impacts micro-variability. Biological activity, including that of domestic animals such as grazing, browsing, or trampling, affects both micro- and meso-variability. The activities of soil biota (microorganisms, arthropods, earthworms, rodents, plant roots, etc.) have a profound effect on spatial and temporal micro-variability.

In Fig. 7.1, the intersections between the sets that depict each level of variability represent their interactions. Alternatively, the diagram could have been drawn as nested sets of variability, with the macro level containing the other two. However, this would have made the graph less intelligible. These levels and sources of variability will shape the agroecosystem differently, influencing human agency and biodiversity. In some cases, they will result in spatial patterns that can be described as heterogeneity. The extent to which these different sources of variability shape spatial heterogeneity patterns also depends on the type of agroecosystem under consideration, especially when comparing smallholder farming to large-scale farming (Plate 7.1). Apart from the differences in topography between the images in Plate 7.1, they illustrate that while large-scale farming strives to minimize heterogeneity in order to implement uniform management practices across extensive areas, smallholder farming deliberately utilizes heterogeneity to allocate different activities, including human habitat, in space and time. These various sources and

A

B

Plate 7.1 (**A**) A smallholder farm in Bella Vista, Valle del Cauca, Colombia; (**B**) A recently sown cereal field in Tres Arroyos, Buenos Aires, Argentina. (Photos: P. Tittonell). Both models of agriculture co-exist in Latin America, sometimes even within the same landscape

levels of variability can also give rise to different temporal patterns in agroecosystems. For example, soil cover in the agroecosystem of Picture A will be more or less constant throughout the year, while in that of Picture B will vary abruptly. The same could be said about the carbon exchange balance, or the habitat quality for soil biodiversity, which will exhibit more variable temporal patterns in B than in A.

7.2 Spatial Interdependence

The degree of spatial dependence between landscape constituents is quantified through their spatial autocorrelation, which measures the resemblance between 'neighbours' as a function of the distance that separates them. Autocorrelation is the jargon term for heterogeneity when dealing with spatial data. When data are autocorrelated they violate the assumption of randomness necessary for standard parametric statistical analysis. That is, a high degree of spatial autocorrelation means that near neighbours are more similar than far away neighbours, which points to the existence of patchy patterns. Autocorrelation is computed by means of semi-variances (γ):

$$\gamma(h) = \frac{1}{2N(h)} \sum_{i=1}^{n} (Z(xi) - Z(xi + h)) \tag{7.2}$$

Where, N is the number of observation pairs separated by distance h, $Z_{(x_i)}$ is the value of the variable of interest at location x_i and $Z_{(x_i + h)}$ is its value at a distance h from location x_i. When there is spatial autocorrelation the value of γ will increase with distance, because near neighbouring observations are more similar than far away ones (again, this is highly sensitive to the scale at which spatial correlation is analysed). Graphically, this is represented by semi-variograms. In the example of Fig. 7.2 the semi-variance is low between proximate neighbours and increases with distance, typically reaching an asymptotic value as distant neighbours become spatially independent. The parameters of the model fitted through the semi-variance estimates are used to quantify the spatial dependency of the studied property as well

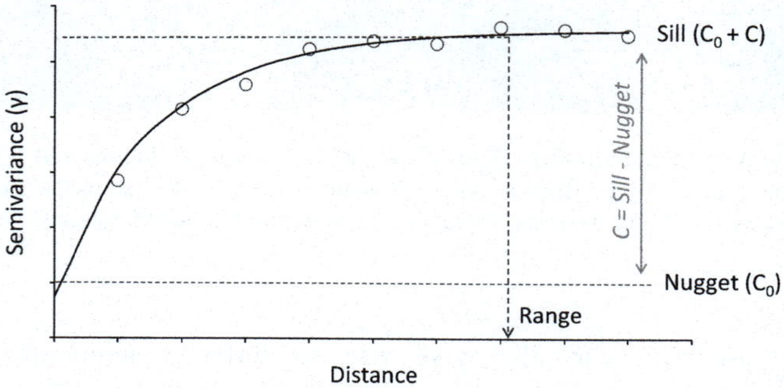

Fig. 7.2 A hypothetical semi-variogram. Hollow circles indicate semi-variance estimates (at constant distances between them) and the solid line is a model fitted through these points

as the degree of heterogeneity or patchiness. In the spherical model[1] fitted in Fig. 7.2 the 'sill' (asymptote) provides a measure of total population variance, whereas the 'nugget' (intercept) represents the level of 'unexplained' variance due to either sampling errors or to spatial dependency at scales not explicitly measured. The range indicates the distance at which the asymptote is reached. Beyond this range, data are considered to be spatially independent. The difference between sill and nugget indicates the degree of spatial structure of the variance, or the degree of spatially dependent predictability of the property. The smaller the ratio (sill - nugget)/nugget (C/C_0) the greater is the degree of noise or unpredictable error.

The ability to detect spatial patterns by means of semi-variance calculations depends first of all on the nature of the variable under study, whether biotic or abiotic for example, but also on the characteristics of the area and, most importantly, on sampling design, specifically on the minimum and maximum distances between samples. Nested sampling schemes may be very useful to reveal spatial variability at multiple scales. Fig. 7.3 shows hypothetical semi-variograms and surface maps. In Fig. 7.3a spatial variability is characterised by gradients across large areas, with smoothly continuous patches (the range is reached at a long distance). In Fig. 7.3b patches are small and more sharply discontinuous, characteristic of small-scale heterogeneity (the range is reached at a shorter distance). Fig. 7.3c illustrates a case of nested heterogeneity, when factors shaping heterogeneity operate at more than one single scale. It is hard to design sampling schemes when these patterns are not known a priori and, especially in the case of nested heterogeneity, the results are often hard to interpret. Sampling designs differ in their ability to deal with landscape anisotropy (cf. Chap. 6), when measurements in different directions may yield different semi-variogram models; but they will also vary in terms of their costs, time requirements and logistics.

The diagrams presented in Fig. 7.3 are, in principle, scale-agnostic or independent of scale. However, in agricultural landscapes, each of these patterns is most commonly observed at specific spatial scales. Gradients, such as the one depicted in Fig. 7.3a, are typically associated with the inherent heterogeneity of soils, vegetation, and often land use in a landscape. Variations in soil depth or texture, related to different landscape positions (e.g., soils formed on alluvial sediments in valley bottoms, shallow soils on ridge crests, etc.), or changes in water regimes (water table depth, water infiltration rates, slope, exposure, etc.) significantly impact the distribution of vegetation and land use types in agricultural landscapes. In such cases, soil maps, if available, provide a useful initial approximation of the spatial heterogeneity extent in the landscape. At lower hierarchical levels, such as the farm or field level, heterogeneity is greatly influenced by community and population processes (e.g., tree distribution, faunal activity, etc.) as well as human activity (e.g., differential allocation of nutrient resources in space, cultivation intensity, etc.). This type of heterogeneity often exhibits a patchy nature, with patches varying

[1] For more details on different types of models used in spatial correlation analysis see e.g. Fortin and Dale (2006)

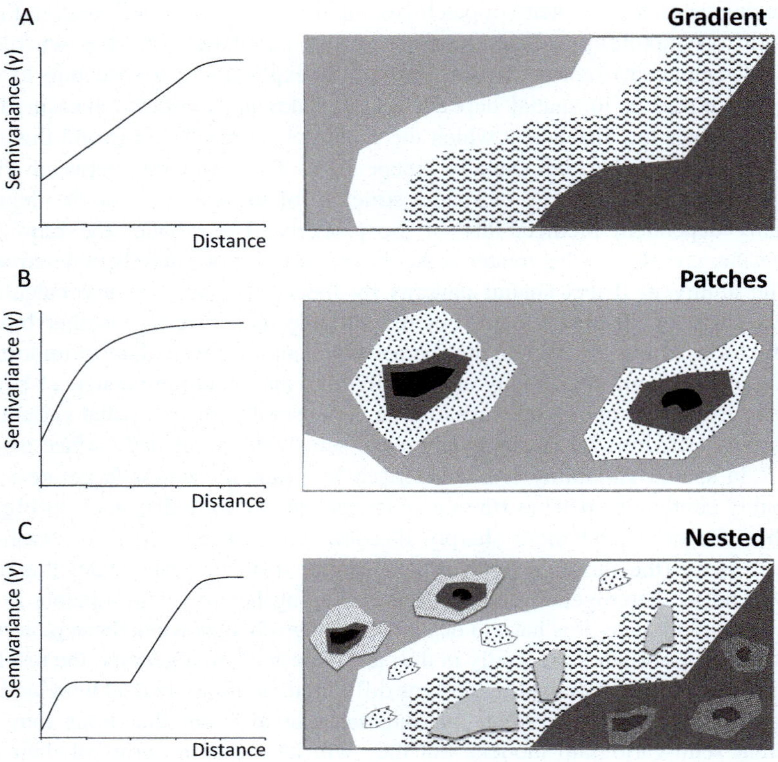

Fig. 7.3 examples of semi-variograms corresponding to different spatial patterns of heterogeneity. (Inspired by the schemes of Ettema and Wardle 2002)

in size, shape, and distribution depending on the underlying processes or the spatial resolution of the analysis (refer to Fig. 7.1).

Patchiness can originate when different fields of a farm exhibit different land uses or current soil fertility levels, or within such fields, often at levels that we define as micro-variability. A semi-variogram model with a good level of spatial predictability can be used for interpolation ('*kriging*') to estimate non-sampled points and their estimation error, and to produce an accurate spatial representation of heterogeneity. An example of micro-variability is shown in the soil carbon interpolation map in Fig. 7.4, which was done after intensive sampling of one hectare of land prior to the establishment of a field experiment (Kintché et al. 2010). The concentration of carbon in certain spots corresponds to the previous presence of trees in this former savannah soil.

Since soil heterogeneity is an essential component of overall heterogeneity in agricultural landscapes, the following sections will deal with two major sources of variability with which agroecologists need to deal in their practice: inherent (natural) and human-induced soil heterogeneity.

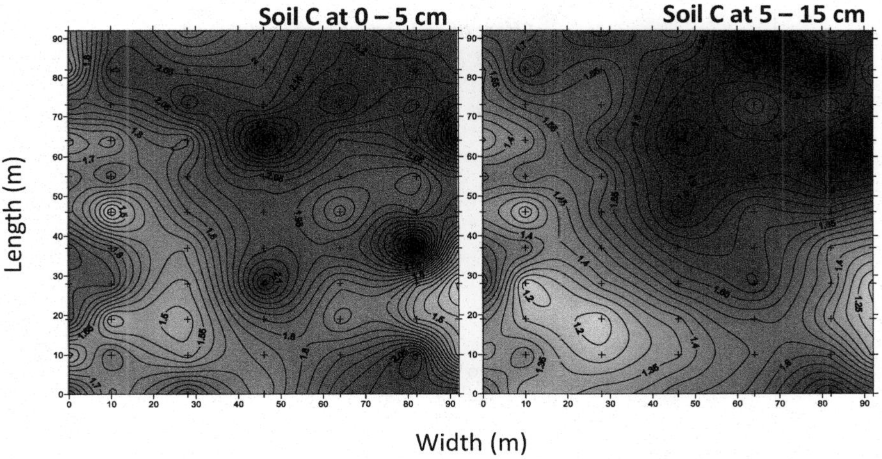

Fig. 7.4 Example of graphical representations of soil organic carbon content obtained through interpolation (kriging) after high density sampling (77 samples per hectare at each depth) at Kolokopé experimental station, Togo. The pattern shows higher concentration of soil C at both depths in places where trees had grown before the land was cleared for cultivation. The objective of this study was to obtain a detailed characterisation of spatial soil C heterogeneity before establishing a long-term field trial in which changes in soil C were to be assessed every year. (Source: Kintché et al. 2010, PhD Thesis Université de Dijon)

7.3 Inherent Soil Heterogeneity

As described above, soils are inherently heterogeneous in space as a result of interacting geological, geomorphological, climatic and biotic processes. Depending on the soil indicator being considered as well as on the spatial resolution adopted, such heterogeneity may present itself in the form of gradients, patchiness or in nested patterns, or just randomly (no spatial effect). For example, soil texture typically varies at greater spatial intervals than soil microbial communities. Soil organic carbon is a commonly used indicator of soil heterogeneity, as it is often positively associated with desirable soil physical, chemical and biological properties. At the same time, soil carbon contents reflect the effects of soil parent material, topography, landscape position, climate, soil hydrological properties, biome, ecosystem primary productivity and agricultural management. The interacting effects of these factors can be illustrated by the data presented in Fig. 7.5, showing total organic carbon measured in soils that differ in texture class (expressed here as the sum of clay plus silt fractions in the soil), from climatically different environments, and from sites under different vegetation and land use types in western Kenya.

The effect of soil texture, both directly through greater protection of soil organic matter and indirectly through increasing water holding capacity and hence ecosystem primary productivity as soils get finer, corresponds to the horizontal variation in

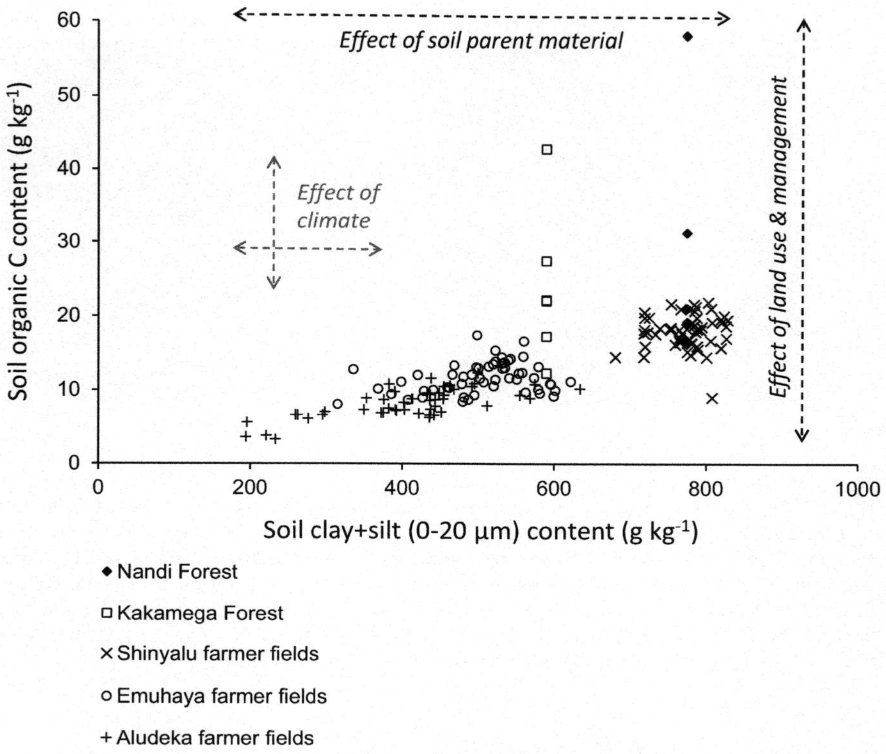

Fig. 7.5 Soil carbon content (0–20 cm, mass-corrected) measured in farmer fields and in the forests reserves of Kakamega and Nandi in western Kenya

Fig. 7.5. The effects of vegetation type, land use and/or agricultural management correspond to the vertical variation; i.e., for a given clay plus silt content, soil carbon can vary enormously depending on whether the original vegetation is still present (e.g. Nandi forest in Fig. 7.5) or not (Shinyalu farmer fields). The differences among farmer fields for a given clay plus silt content correspond to land use (e.g. perennial versus annual cropping) and management effects (e.g. application of animal manure), but also to climatic conditions (cf. farmer fields at Aludeka vs. Emuhaia at clay plus silt contents from 400 to 600 g kg^{-1}). As these data come from a range of agroecological conditions, the effect of climate is confounded with management and soil parent material. Climatic conditions impact on both the horizontal and vertical variation in Fig. 7.5 through their influence on ecosystem primary productivity and soil organic matter decomposition rates. In addition, in this particular example, there is co-variation between soil texture and climate; i.e., soils are finer, temperatures lower and rainfall more abundant in the highlands around Nandi forest.

7.3.1 Macro-Variation: Climate, Biome and Geology

The analysis of the example above points to the fact that soil indicators always need to be referenced with respect to climatic conditions, type of biome and type of soil. In Fig. 7.5, for example, a soil carbon content of 1% (or 10 g kg^{-1}) will correspond to the average soil C level observed in fields with clay plus silt contents around 400 g kg^{-1}, or to the lowest level that could be measured in fields of 800 g kg^{-1} clay plus silt. A soil with 1% organic carbon will be within range in the first case and very poor in the second (or relatively fertile when clay plus silt contents are about 200 g kg^{-1}). To capture the actual impact of soil texture on soil organic carbon, the data in Fig. 7.5 should be expressed as organic carbon associated with the finer soil physical fraction, with average diameters smaller than 20 μm, instead of total soil carbon. This is because the finer fraction participates in the formation of complex organo-mineral structures that bind organic matter in stable forms in the soil. Other forms of 'protection' of organic matter in the soil are micro aggregates, which constitutes a physical rather than physicochemical form of protection. Figure 7.6 illustrates another case of large-scale heterogeneity with topsoil carbon data from central Argentina, where different land uses and vegetation types co-existed (Fig. 7.6a) before the rapid expansion of agriculture that took place during the last two decades. In this region, altitudes change quite abruptly within a distance of one to two hundred kilometres, creating strong ecological gradients that follow changes in temperature (mean averages ranging from 12 to 20 °C) and rainfall (from 500 to 1200 mm $year^{-1}$).

7.3.2 Meso-Variation: Landscape

The organic carbon data potted in Figs. 7.5 and 7.6 correspond to soils located in well-drained positions within the landscape. Most of the variation observed is the result of large gradients of soil texture, climate and vegetation/land use type across locations. But within a single landscape, these drivers will also vary across different topographic positions (landscape units), which may also be dominated by different soil types (cf. Chap. 6). An example of such meso-variability within landscapes is presented in Table 7.1 using data from the highlands of Madagascar (Alvarez 2012), illustrated with pictures in Plate 7.2. In these heavily dissected landscapes, in which rainfed crop and livestock production in the hills and high plains coexists with paddy rice cultivation in the lowlands, soil types and their properties may vary quite substantially within short distances (within hundreds of meters or less). In this case, most soils are found within the same texture class, except for those in drained lowlands (formed on a mix of alluvial and colluvial sediments), but they differ in the type of parent material and soil formation process, largely influenced by their topographic position. The brown soils formed on volcanic ashes are typically located on hills and high plateaus; they present good physicochemical properties (except for

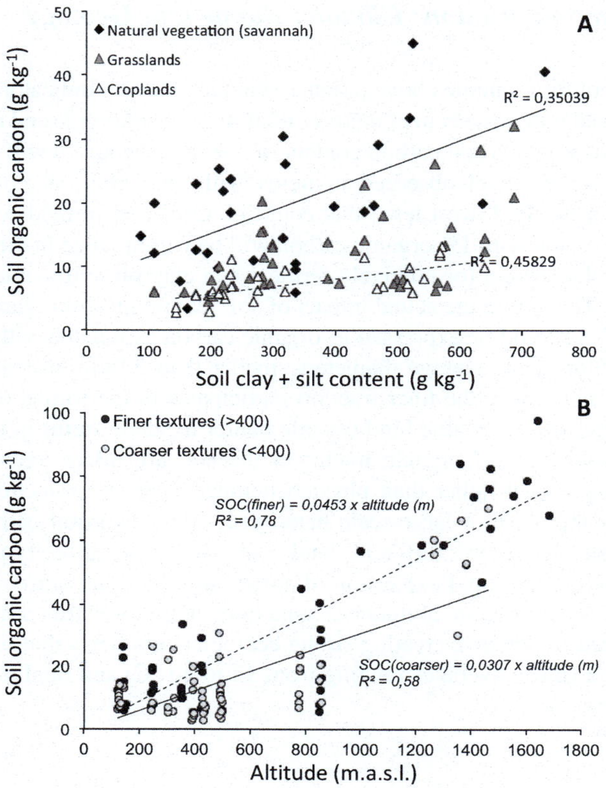

Fig. 7.6 Soil carbon content (0–20 cm) in central Argentina. (**a**) Soil C as a function of clay plus silt content, per type of land use; regression lines fitted through the natural vegetation (solid) and cropland (dotted) points. (**b**) Soil C as a function of altitude (meters above sea level), for soils with clay plus silt contents smaller (finer) or greater (coarser) than 400 g kg^{-1}. Both regression lines were forced through zero. Note that the observations below 900 m correspond to a subset of the ones presented in panel A

their acidity) and are used for rainfed cultivation of annual crops and pastures. Flooded lowland soils are in most cases yellow ferrallitic soils used for rice cultivation and receiving every season more inputs of carbon and nutrients than the other two types. Again, the confounded effects of inherent soil fertility and soil management are difficult to unravel.

Soil spatial heterogeneity within landscapes can be characterised through the analysis of the spatial autocorrelation of soil properties, using the semi-variogram models described earlier in this chapter (cf. Figs. 7.2 and 7.3). An example from two landscapes in Eastern Uganda is presented in Fig. 7.7, in which semi-variogram models were fitted to the variation in soil organic carbon and available (Olsen-extractable) phosphorus (see Tittonell et al. 2013 for further details). The case study landscape in Tororo district is characterised by a fairly flat topography dominated by

Table 7.1 Variation in soil properties (0–20 cm) in 487 fields sampled across three soil types and four major landscape units in the highlands of Antsirabe, Madagascar (from Alvarez 2012, PhD Thesis Montpellier SupAgro)

	n	Bulk density	Clay + silt	SOC	TSN	P_{Extr}	pH_{water}	CEC
		(mg m^{-3})	(g kg^{-1})	(g kg^{-1})	(g kg^{-1})	(mg kg^{-1})		(cmol kg^{-1})
Soil type								
Brown volcanic	*106*	0.83	602	32.8	2.28	84.1	5.22	–
Red ferrallitic	*130*	0.93	569	17.1	1.24	62.7	5.80	5.70
Yellow ferrallitic	*251*	0.98	568	28.7	2.01	68.8	5.30	5.63
Landscape unit								
Hills and high plains	*293*	0.85	585	32.7	2.28	72.5	5.15	4.58
Terraced foothills	*76*	0.93	561	18.0	1.24	74.3	5.91	5.84
Flooded lowlands	*101*	0.94	577	18.0	1.34	64.0	5.70	6.32
Drained lowlands	*17*	0.93	474	8.0	0.52	58.7	6.00	6.02

SOC Soil Organic Carbon, *TSN* Total Soil Nitrogen, P_{Extr} Extractable Phosphorus with the Olsen-Dabin method, *CEC* Cation Exchange Capacity (The method used to determine the CEC does not perform well for soils formed on volcanic ash)

Plate 7.2 Images from rural areas of Antsirabe, in the highlands of Madagascar, illustrating the meso-variability in the landscape, showing in the different pictures the units characterised in Table 7.1: Hills and high plains; Terraced foothills; Flooded lowlands; and Drained lowlands. (Photos: P. Tittonell)

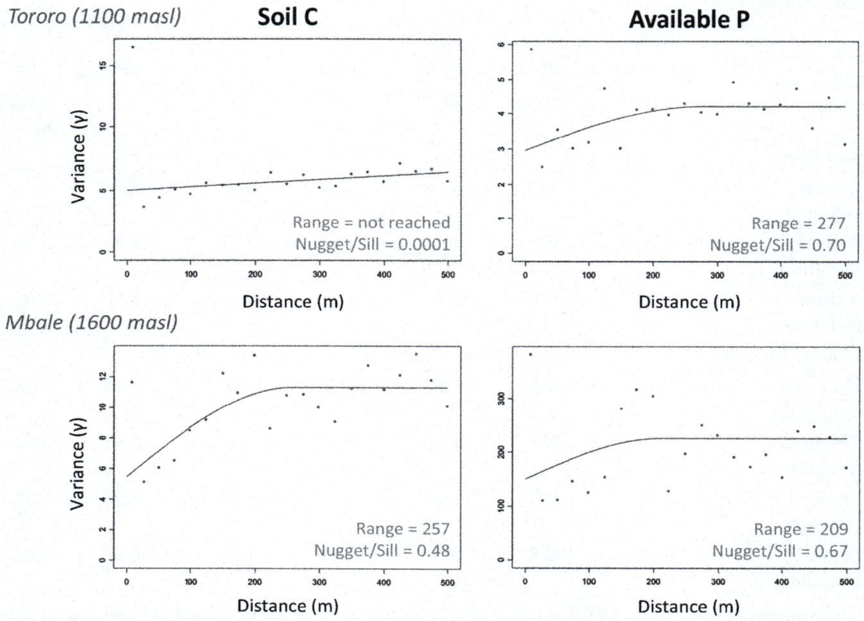

Fig. 7.7 Semivariograms fitted to the spatial variation in soil carbon and extractable phosphorus in two landscapes from Eastern Uganda (Tororo and Mbale districts), using data from Tittonell et al. (2013)

coarse textured soils, and annual crops (cassava, millet) alternating with fallow areas used for extensive grazing. The landscape in Mbale district presents an undulating topography, with steep slopes and soils formed on volcanic ashes, which are more intensively cultivated with annual and perennial crops (banana, coffee) and intensive livestock production through cut and carry feeding systems. Soil carbon did not exhibit spatial autocorrelation in Tororo, where a pure nugget effect can be observed in the semi-variogram. In Mbale, soil carbon exhibited less variance between contiguous fields with respect to fields located far away from each other. The maximum variance (range) was reached at 257 m, and 48% of this variance was associated with differences between adjacent soils (nugget/sill). While average soil carbon levels were comparable between both sites, although generally higher in Mbale (Table 7.2), whereas the average levels of available phosphorus were one order of magnitude higher in Mbale. Yet the spatial autocorrelation models for available P were quite similar at both sites (Fig. 7.6). About 70% of the variation was not explained by distance in both cases, and the sill was reached at a somewhat farther distance in Tororo.

A closer look at the average soil properties presented for these two landscapes in Table 7.2 indicates that soils located in different landscape units did not vary much in their physicochemical properties in Tororo, while they varied in terms of slope, soil carbon and available phosphorus in Mbale. The spatial variation in soil carbon, as discussed earlier in this chapter, is largely associated with variations in topographic

Table 7.2 Average slope, distance to the homesteads and soil properties of fields located in different positions of the landscape in two villages located in Tororo and Mbale districts of Eastern Uganda (from Tittonell et al. 2013)

Site	Slope	Distance[a]	Particle size (%)			SOC	Av. P	pH water
Landscape position	(%)	(m)	Clay	Silt	Sand	(g kg^{-1})	(mg kg^{-1})	(1:2.5)
Tororo								
Upslope	1.2	76	18	17	65	10.2	2.9	6.3
Midslope	3.0	62	19	16	65	10.1	3.5	6.5
Marshland border	2.1	155	23	16	61	10.3	3.5	6.7
Mbale								
Upslope	9.9	16	18	20	62	16.9	55.2	6.4
Midslope	20.8	48	23	22	55	13.8	25.9	6.3
Footslope	2.6	78	24	22	54	10.9	12.3	6.5
Valley bottom	3.4	82	21	22	57	14.5	25.6	5.8

[a] Average distance from the centre of the field to the homestead within each farm

positions and soil types occurring within the landscape. The variation in phosphorus is also determined by parent material, topography (erosion) and vegetation/land use type, but in the particular case of smallholder farmers it is largely influenced by agricultural management practices. Localised application of ashes, mineral fertilisers or animal manure can lead to strong variability in available P at short distances, which explains why this indicator is more susceptible to management in soils in which it is inherently present at low levels. In the example of Table 7.2, fields located in the higher topographic positions of the landscape in Mbale are those in which perennial crops such as coffee and banana are located, crops that receive frequent nutrient inputs. Here again, the example shows that it is often difficult to unravel the effect of inherent and management-induced soil heterogeneity.

7.4 Farmer-Induced Soil Heterogeneity

Farmers can induce soil heterogeneity through land use choices as well as through management practices. In the first case, soil heterogeneity can be the result of co-existing annual or perennial crops, pastures, woodlands, grazing areas, housing areas, etc. Each of these land uses will affect soil properties in a different way. Under perennial crops, soils are often less physically disturbed by ploughing, or at least less frequently than under annual cropping. Pastures and grazing areas include the effect of grazing animals and of other elements of the fauna that are frequent in grasslands (e.g., birds, molls, etc.), both of which can affect soil heterogeneity through preferential grazing, dejections, trampling or digging. Soils under woody vegetation, such as fruit orchards, tree plantations or natural woodlands receive annual inputs of organic matter and nutrients in the form of leaf litter, or in the form of fertilisers, which tend to favour soil carbon and nutrient concentration around the tree trunks. Heavy trafficking in perennial crops and plantations can also be a cause of soil heterogeneity. Soil levelling, ditching, terracing or contouring can all affect soil heterogeneity, although these operations often propend to soil homogeneity. Poor drainage or irrigation control systems in irrigated agriculture can also lead to soil heterogeneity.

Soil heterogeneity caused by farmer management decisions is most conspicuous in smallholder agriculture, where farming practices are not mechanised and where organic matter and nutrient resources are often scarce and thus preferentially allocated to certain fields within the farm, often the ones that are closest to the homesteads as will be explained later. However, even in large scale mixed-farming, it is also common to see that fields located in the proximity of livestock facilities tend to receive more inputs of animal manure than fields located far away, creating gradients of decreasing soil fertility (e.g., available P) at increasing distances from the stalls. Soil heterogeneity in large-scale farming is however often masked by the high rates of fertiliser application employed. In smallholder agricultural landscapes, farmer-induced soil heterogeneity is often nested within the other patterns of heterogeneity described before, such as soil type or vegetation gradients, as illustrated in Fig. 7.8. The greatest influence of human agency on soil heterogeneity in smallholder agriculture is often observed at the scale of individual farms, between fields located on different topographic positions, or at increasing distances from the homesteads. In many smallholder systems, the fields around the homestead that may be termed generically home gardens (e.g., *transpatios* in Central America, *champs de case* in west Africa, *kibanja* or *shamba* in different parts of East Africa, or *pekarangan* in Indonesia), are forms of agroforestry that often receive most of the organic matter and nutrient inputs, in some cases also most of the water inputs, and host a diversity of annual and perennial crops grown in mixtures. Such fields tend to exhibit more favourable soil quality attributes than the rest of the fields of the farm (cf. Chap. 6).

When farmers use animal manure or compost to manage their soils, and when such materials need to be transported by hand, the labour required for such operations may result in a burden that not all farmers are able to afford. In such cases, and certainly when farmers are unable to hire in additional labour force, manure or compost will tend to be more frequently applied in nearby fields, in detriment of the soil fertility of fields located farther away. This may result in positive nutrient balances (nutrient inputs minus nutrient outputs) in the home gardens, at the expense of negative nutrient balances in the outfields, as has been broadly documented (e.g., Tittonell et al. 2005; Ramisch 2005; Samaké et al. 2005; Zingore et al. 2006; Ebanyat 2009). In some cases, the concentration of nutrient and labour resources around the homesteads may respond to merely socio-cultural factors, such as aesthetic considerations regarding the looks of the fields that surround the homestead, or security, since crops located far away from the homestead risk to be damaged by marauding livestock, by wildlife, or by thieves. It is evident that the stylised patterns of soil heterogeneity at different scales illustrated in Fig. 7.8 correspond well with the patterns illustrated earlier in Fig. 7.3. This is not a general

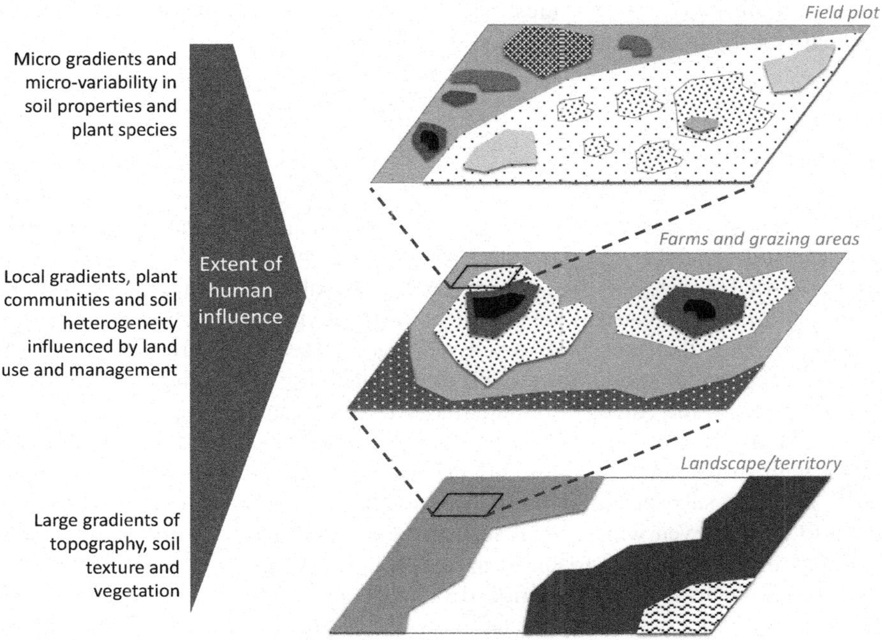

Fig. 7.8 Schematic representation of possible patterns of soil variability at different scales, with reference to smallholder agroecosystems

rule, but rather the result of frequent observation of these patterns in practice. Heterogeneity tends to be patchy at the lowest scales, and to exhibit gradients at higher scales.

At lower scales, notably within fields, heterogeneity may be caused by different management-related factors, but also largely affected by biological ones. Heterogeneity may be reinforced by the distribution and activity of termites, ants, earthworms, etc. Typically, the presence of old termite mounds in tropical soils creates 'islands' of fertility that farmers use for their benefit by growing nutrient-demanding crops on them (e.g., Félix et al. 2018). In the Sahel, shrubs of the natural flora often act as traps for sediments and organic matter that are transported by wind or water, and because they stay green during the dry season, they favour greater biological activity in their rhizosphere as compared with the surrounding soil. Farmers make use of such heterogeneity to grow crops in combination with these shrubs (Lahmar et al. 2012). Still in the Sahel, farmers management of micro-variability in soil topography to reduce risks of crop failure has also been well documented in the past (e.g., Brouwer et al. 1993). Recent evidence is pointing to the fact that soil fertility management can induce heterogeneity in soil microbiological communities, in their diversity and richness, as well as in the integrity and functioning of their trophic networks (El Mujtar et al. 2019; Lupatini et al. 2019). This is an incipient field in agroecology that requires much research as yet, but that offers promising insights for the future management of soil fertility.

7.5 Spatial Patterns

Due to the mechanisms described above, the combination of inherent and human-induced soil variability results in spatial patterns, and this is why we can generally speak of soil heterogeneity (i.e., heterogeneity is variability with a spatial structure – cf. Chap. 6). Spatial patterns are conspicuous in all types of agricultural landscapes worldwide, but they are accentuated where resources are scarce, or where the inherent geomorphology influences differential land use and management practices across space. In the Andes of South America, for example, agroecosystems developed that make use of the pedoclimatic diversity created by altitude to grow different crops with varying intensity, resulting in strong zoning of soil fertility (cf. Caulfield et al. 2020). In sub-Saharan Africa, where resources are scarce, such spatial patterns tend to emerge even within small surface areas, often less than one hectare. Such patterns of within-farm variability in sub-Saharan Africa have been termed soil fertility gradients, as they describe declining soil fertility levels as one moves away from the homesteads (Tittonell et al. 2005). Strictly speaking, these are not always gradients but variations of soil fertility attributes in space that may sometimes appear as gradual or continuous gradients, and other times as sharp differences between contiguous or distant fields, or as discrete gradients (Fig. 7.9). One of the first publications describing such patters in Africa is the one of Prudencio (1993), documenting ring management patterns as the result of resource allocation practices by smallholder farmers in Burkina Faso. As a consequence of more intensive management of the fields near the village (*champs de case*), with more frequent inputs of animal manure, and less intensive management of the bush fields (*champs de brousse*) from where the crop residues are harvested and/or fed to livestock, soils patterns describe concentric 'rings' of declining fertility at increasing distances from the homesteads (Fig. 7.9a). In such type of agroecosystem, communal herds of livestock are managed by herders who 'park' their animals in certain fields overnight to enrich the soil with their dejections. These are also generally the fields near the homesteads. There may be economic transactions associated with this practice, especially when transhumant livestock herds pass through a village along their herding route (Diarisso et al. 2015). Farmers may pay or offer something in exchange to nomadic herders to park the animals in their fields at night. In these systems, that still subsist specially in West Africa, households are located in a village and families own or usufruct fields scattered in both the village and bush zones. A single family may own or manage two or more fields but these are not necessarily next to each other.

In highly populated regions (e.g., 500–1000 inhabitants per Km^2), and especially where households own or usufruct a well-delimited farm consisting of a number of fields or subdivisions, fencing is common and soil fertility variations tend to describe

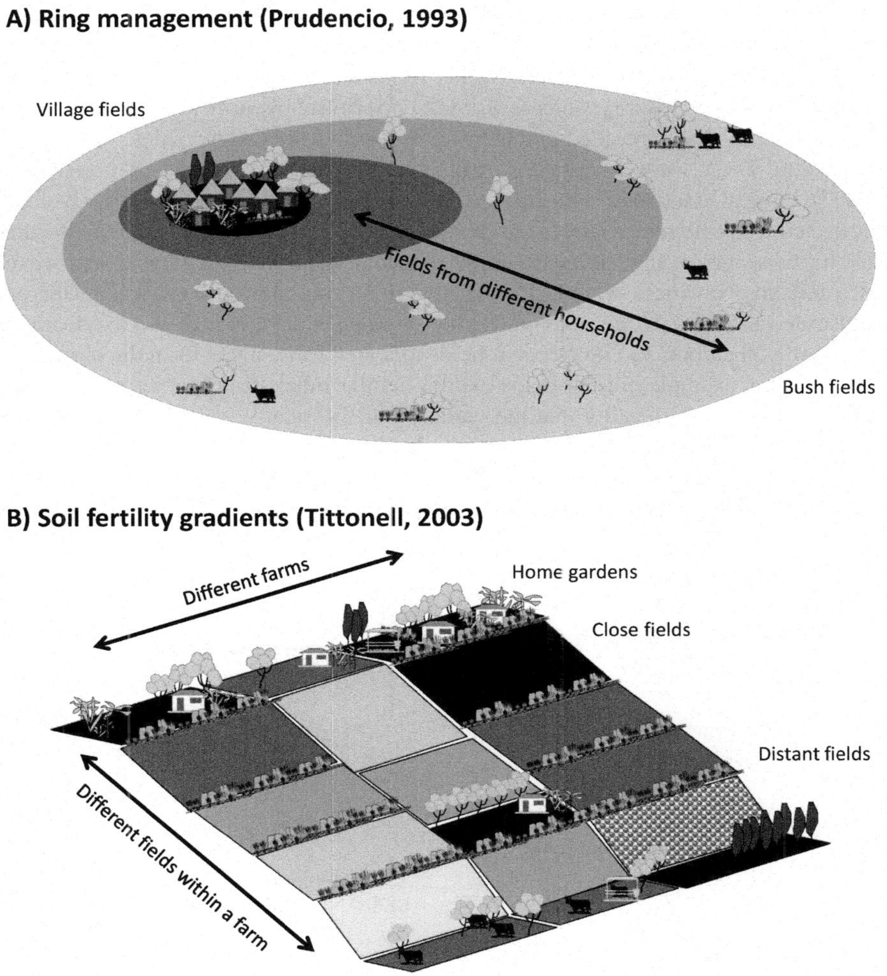

Fig. 7.9 Representation of two types of soil fertility gradients in smallholder agriculture. These examples were documented in sub-Saharan Africa by Prudencio (1993) in Burkina Faso and by Tittonell (2003) in Kenya. The intensity of colour brown indicates the level of soil fertility, generally higher at short distanced from the homesteads or from the village

discrete spatial patterns (Fig. 7.9b). This pattern was first described as a soil fertility gradient by Tittonell (2003) in Kenya, but it is not restricted to sub-Saharan Africa. In fact, it is one of the most common spatial patterns observed in smallholder agriculture worldwide, provided that individual fields are delimited as different management units. Every field receives a certain amount of nutrient and carbon inputs, resulting in different nutrient and carbon balances. When balances are positive there is nutrient concentration; when they are negative there is nutrient depletion. Fields that are close to the homestead receive most of the nutrient inputs, especially the home gardens, where crops are grown for family self-consumption. Fields that are far from the homesteads receive less inputs, specially of manure which is laborious to carry, and are also least prioritised by farmers in terms of labour allocation; i.e., they are tilled and planted late, with less seed densities, weeded less often, etc. These were described as management intensity gradients (Tittonell et al. 2007), closely associated with soil fertility gradients. In addition, the crop residues collected from the remote fields are fed to animals, and the manure is then allocated to the close fields at planting, reinforcing the soil fertility gradient every season (Castellanos-Navarrete et al. 2015). The resulting spatial pattern may be as the one illustrated in Fig. 7.10 for a single farm in western Kenya, where we see declining levels of soil carbon and nutrients in fields that are farther away from the homestead (F1). In this example, all the fields exhibit similar inherent properties such as clay plus silt content, indicating that the variation in soil quality is due to land use and management decisions, and only partly to landscape variability (e.g. slope).

In less populated regions, agricultural activities tend to be organised in the landscape following topographic or hydrogeological features such as soil textural

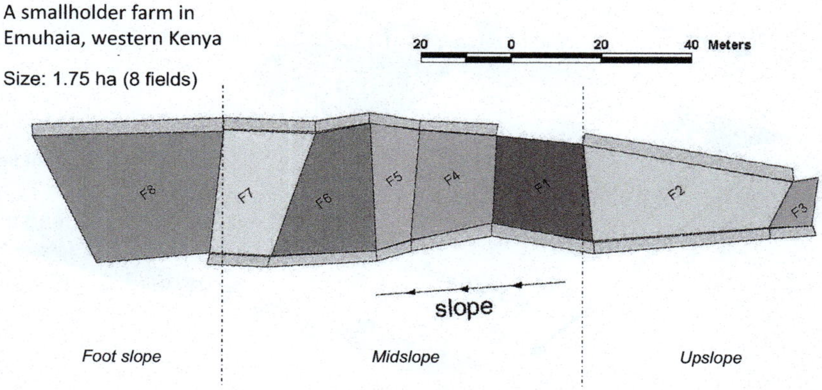

Field code	Land use	Area (ha)	Slope (%)	Clay + Silt (%)	SOC (g kg^{-1})	Total N (g kg^{-1})	Av. P (mg kg^{-1})	Ex. K (cm$_+$ kg^{-1})	Ex. Ca (cm$_+$ kg^{-1})	Ex. Mg (cm$_+$ kg^{-1})	pH water (1:2.5)
F1	MZ/VG/BN/FT	0.056	11	50	14.3	1.1	1.9	0.27	3.6	1.3	5.8
F2	MZ/BS/CP	0.075	4	48	16.0	1.3	3.4	0.34	5.1	1.1	5.8
F3	WDT	0.008	1	49	15.2	1.3	2.6	0.29	4.3	1.6	5.6
F4	MZ/BN/VG	0.036	10	55	11.6	0.9	1.1	0.25	2.7	1.2	5.5
F5	MZ/CV/SP/NG	0.026	12	58	12.6	1.0	0.9	0.27	3.0	1.5	5.4
F6	NG	0.028	22	56	6.9	0.4	0.5	0.22	1.8	1.0	5.2
F7	FLW	0.043	7	54	8.8	0.7	2.9	0.14	0.2	0.2	4.9
F8	WDT	0.112	10	50	8.6	0.7	0.9	0.13	1.9	0.5	4.9

Fig. 7.10 Characterization of soil spatial variability in a smallholder farm in Emuhaia, Kenya. F1 to F8 indicate the eight fields of the farm that totals 1,75 ha. F1 indicates the field where the homestead is located, surrounded by a home garden. Land use legend: MZ is maize, VG vegetables, BN bananas, FT fruit trees, BS beans, NG Napier grass, WDT woodlot, FLW fallow land, SP sweet potato, CV cassava, CP cowpea

classes, soil depth, slopes, depth of the water table, rockiness, proximity of surface water, etc. For example, deeper and better drained soils are used for cropping while low lying, water logged areas, sloping surfaces or rocky soils are used for livestock or forest. As landscapes become more densely populated, land subdivision disfigures such original patterns and activities are present often in less suitable land units (e.g. cropping fields in rocky or shallow soils). Sometimes the spatial organisation of a landscape responds to strict governmental rules allocating land to different uses, including human habitation or industries. The designation of each portion of land is not easily modifiable unless through changes in the official land use plan. This is the most common system in Europe, for example, where every squared meter of land has a certain designation. But several countries in sub-Saharan Africa have undergone processes of similar land use planning under colonial rule. In Zimbabwe, for example, the more fertile deep clayey soils were allocated to colonising white Rhodesian farmers, while the local people – also often the land labourers of those large farms – were gathered in villages or 'native reserves' where each family had access to cultivable land on coarse sandy soils of poor inherent fertility (Bessant 1992). This organisation, and limited access to resources, led to the spatial patterns of soil fertility represented in Fig. 7.11a, with nutrient concentration in homefields at the expense of the outfields (Carter and Murwira 1995). Differences in productivity between homefields and outfields (or 'contours') may be as wide as ten-fold (cf. Zingore et al. 2006).

In even less populated regions, such as in pioneering agricultural frontiers that can still be observed in the Amazon, Congo or Borneo forest, but also in many other forest regions of the world, shifting cultivation is gradually evolving towards sedentary and spatially stable forms of farming. Rural families often own or usufruct a plot of land around their homesteads and one or more plots of land in newly opened forest or savannah soils which are still fertile, resulting in 'inverse' gradients of soil fertility as shown in Fig. 7.11b (i.e., the outfields or bush fields are more fertile than the homefields – Adjei-Nsiah et al. 2004). These outfields may or not be subject to shifting cultivation, depending on population density, forest protection laws or labour availability, often resulting in hybrid agroecosystems that combine sedentary (homefields) and shifting cultivation (outfields) systems. Although regulations in several countries prevent shifting cultivation, there are still a large number of communities that live in forest fringes and rely strongly on forests as part of their agroecosystem. In the mid-hills of Nepal, for example, Alomia-Hinojosa et al. (2020) described how livestock marauding in the forest during the day and being stalled at night represent the main input of carbon and nutrient to smallholder farms. Livestock generate a net flow of nutrients – a nutrient subsidy – from forest to agricultural land which is vital for the functioning of these agroecosystems. A similar pattern has been described by Duriaux Chavarría et al. (2020) in Ethiopia, where farms closer to the forest exhibited higher levels of soil fertility indicators than those located farther away, thanks to the nutrient transfers generated through livestock grazing in the forest.

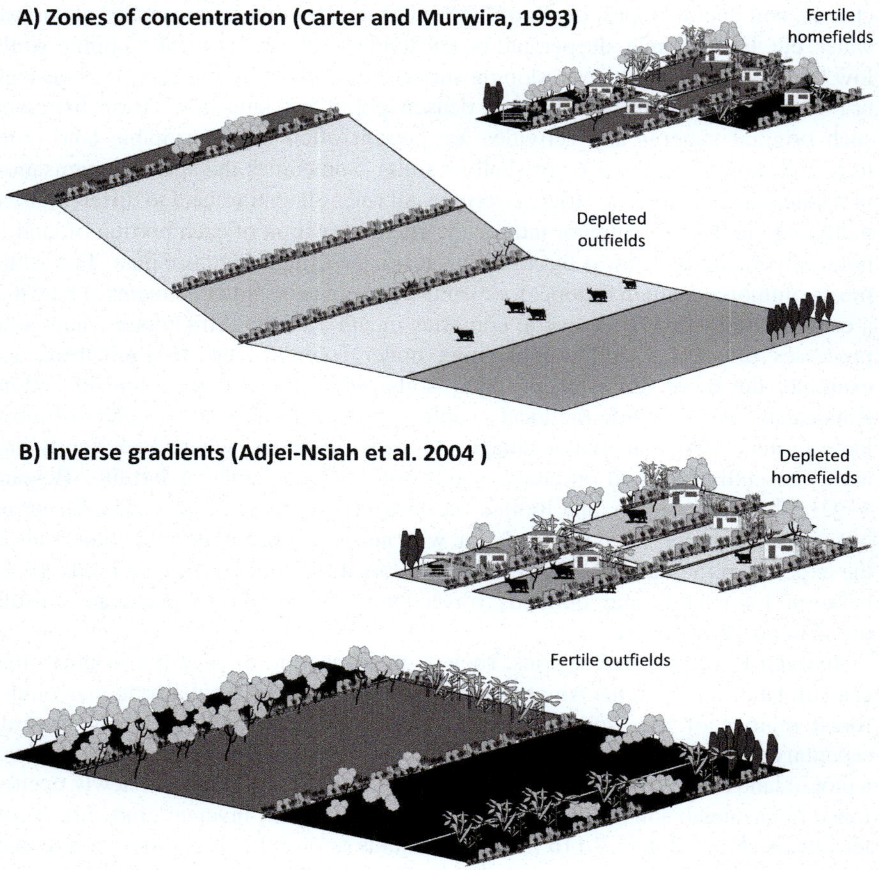

Fig. 7.11 Representation of two types of soil fertility zoning when homefields and outfields are not contiguous. These patters were described in sub-Saharan Africa by Carter and Murwira (1995) in Zimbabwe and by Adjei-Nsiah et al. (2004) in Ghana. The intensity of colour brown indicates the level of soil fertility, higher at short distanced from the homesteads when farmers concentrate nutrients in the homefields (**a**) or higher in bush fields located away from the villages and where forms of shifting cultivation are still being practiced (**b**)

7.6 Assessing Heterogeneity

Assessing soil heterogeneity poses a serious challenge to soil scientists, as much as dealing with it poses challenges to agroecologists. The in depth treatment of statistical theory to deal with spatial heterogeneity exceeds the scope of this book. The aim of this section is just to introduce a couple of practical examples to be considered when designing methodological approaches to study soil heterogeneity in agroecosystems. I will focus on sampling designs, statistical models (parametric and non-parametric), methods for soil analysis and the use of farmer's knowledge to categorise spatial variability.

7.6.1 Sampling Design

Assessing heterogeneity requires careful sampling designs, and clear objectives regarding the type of heterogeneity that is being targeted, the purpose that leads to studying it, the indicators to be assessed, and the techniques and logistics available. One way of assessing soil heterogeneity is to quantify the proportion of the variance that is associated with each level/pattern of heterogeneity, at the scales of landscapes, farms or individual fields (cf. Fig. 7.7). If heterogeneity at higher scales is well-known and documented, for instance by means of soil or vegetation maps, then sampling can be best stratified within each major unit. This, however, is not always the case. When the exact sources and patterns of soil heterogeneity are not well-known a priori, it may be at least possible to put forward some hypothesis to explain what could be the major drivers of heterogeneity. For example, in a study of soil heterogeneity at different scales in the East African highlands, Tittonell et al. (2013) hypothesised that soil heterogeneity was affected by human population density, distance to markets and agricultural potential (including the inherent fertility of dominant local soils). Population density affects farm sizes, markets affect access to inputs, livestock activities and land use patterns, the agricultural potential of each site may affect the cropping frequency (e.g., bi-modal rainfall allows harvesting two crops per year), while the inherent fertility of soils affects land use, agricultural productivity and soil resilience. Then, six sites were selected based on these factors, with respectively dense and sparse population densities, far from or close to markets, and with high or low agricultural potential (Table 7.3). Needless to say, often these factors were associated among each other across the region (e.g., high potential areas were often highly populated and close to markets).

Spatially randomised Y-sampling frames were used in this case to select 250 case study farms to study farm- and field-level soil heterogeneity, and to relate these with household resource endowment and agricultural management decisions (Tittonell et al. 2010, 2013). The Y-frames consisted of three transects of 1.3 Km each departing from one point in three directions, at 120° apart from one another, as indicated in Fig. 7.12. Farms were selected along these transects at 100, 400 and 1300 m from the centre, resulting in 10 farms sampled per Y-frame (the 25 Y-frames were located in different districts of Kenya and Uganda, as explained in the source publications – Table 7.3). Such spatial distribution allowed a maximum of possible distances between sampling points, which is highly convenient when fitting semi-variograms (cf. Fig. 7.7), with the minimum number of samples. That is, if we consider all the possible distances between the 10 sampling points in the Y-frame we can measure up to 12 different distances (Fig. 7.12a). Using a grid, for example, we would need to sample 16 farms to be able to obtain 9 different possible distances (Fig. 7.12b).

Table 7.3 Criteria used to select six sites to study *soil heterogeneity*, *considering demography*, *access to market and production potential* in the highlands of East Africa (from: Tittonell et al. 2013)

Region	Site	Altitude (masl)	Annual rainfall (mm)	Dominant soil types (FAO)	Topography	Population density (km^{-2})	Farm sizes (ha)	Major food and cash crops	Livestock
Central Kenya	Meru south	1500	1600	Nitosols, Ferralsols	Strongly undulating, slopes up to 45%	800	0.5–3	Maize, beans, coffee, tea	Intensive dairy systems; cultivation of fodder crops; improved cattle
	Mbeere	1100	700	Lixisols, Arenosols	Fairly flat to gently undulating, slopes <5%	400	1–10	Sorghum, cowpea, khat, groundnuts	Free ranging local zebu (traction) and goats; night corralling
Western Kenya	Vihiga	1600	1800	Nitosols, Ferralsols	Gently undulating, slopes 5 to 20%	1000	0.3–2	Maize, beans, tea coffee	Tethered cattle grazing in compound fields, zero grazing
	Siaya	1200	1400	Ferralsols, Acrisols	Fairly flat, slopes <3%	350	0.5–5	Maize, cassava, sugarcane, cotton	Free grazing and tethered local cattle, used for traction
Eastern Uganda	Tororo	1100	1100	Acrisols, Vertisols	Fairly flat, slopes <3%	250	1–8	Cassava, sorghum, cotton, tobacco	Free grazing in communal grasslands; local zebu used for traction
	Mbale	1600	1200	Ferralsols, Acrisols	Gently undulating, slopes up to 30%	350	0.5–5	Bananas, beans, coffee	Zero grazing of cattle (traction) and goats; donkey used for transport

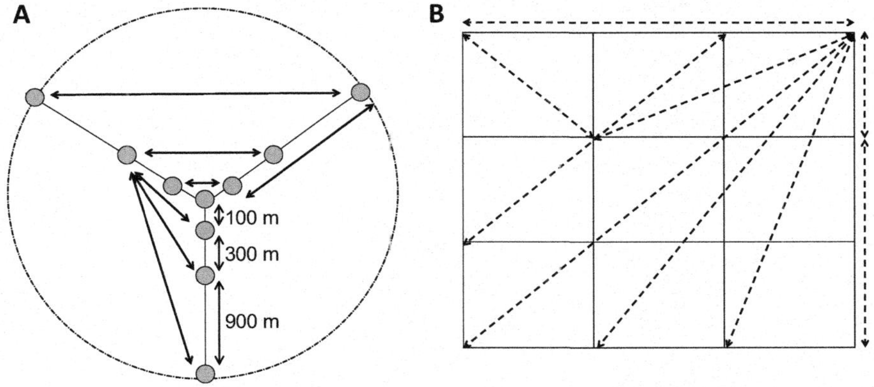

Fig. 7.12 Y-sampling frames used to characterise soil variability in East African smallholder farms (**a**), as compared with a regular grid sampling (**b**)

7.6.2 Mixed-Effect Models

When soil diversity is nested within so many levels of variability, it may be of intertest to assess what is the set of factors that has the strongest influence on the various indicators of soil fertility. In the case described in the previous paragraph, soil variability may be ascribed to differences across regions, sites, sampling frames, farms and fields (and also to micro-variability within fields) (Table 7.3). One approach to study the structure of the variance across these levels is to use mixed-effects models, a statistical method to analyse data that are not independent, multilevel or hierarchical, longitudinal (time series), or correlated. In other words, when it is expected that the variance is partitioned within and between groups of observations, as in hierarchical sampling designs (within and between sites or farms, in our case). Soil observations at field level are not independent, since they tend to be more similar within a farm or a sampling frame, whereas sample units at higher level, such as region or locality are independent. There are two possible ways to deal with such hierarchy: (i) to average the data per farm or sampling frame and then run the analysis using averages, which will be independent, or (ii) to run one analysis per farm or sampling frame, since data within them are independent. There are several disadvantages of these two approaches, one of them being that models will not be making use of the information of the entire dataset. Mixed-effect linear models are a sort of combination of these two approaches. Mixed models are a special type of linear models, which allow for fix and random effects. In a linear regression, the data are random values, but the parameters are fixed effects. In a mixed model, the parameters are fixed at a higher level, but random at lower levels.

Playing with different mixed effect models, that is, setting higher level factors as fixed or random allows redistributing variances to study effects at lower levels. For example, the relationship between field slope and soil carbon may be negative at

Table 7.4 Relative proportion (%) of the total variance in different soil properties that is explained by the random terms of a mixed model, and the same when the factor Site is set as fixed in the model (from Tittonell et al. 2013)

Variance term	Soil organic C	Total N	Available P	Exchangeable K+	Exchangeable Ca++	Exchangeable Mg++
All random terms						
Site	50.7	48.9	57.6	26.4	12.0	25.5
Locality	16.8	19.2	11.5	10.8	19.7	18.1
Farm type	8.7	11.2	7.2	14.0	27.0	23.7
Field	23.8	20.8	23.6	48.9	41.3	32.6
Total variance	0.268	0.342	1.724	0.418	0.361	0.307
Site set as fixed						
Locality	34.1	37.5	27.2	14.6	22.4	24.3
Farm type	17.7	21.9	17.1	19.0	30.7	31.9
Field	48.2	40.7	55.8	66.4	46.9	43.8
Total variance	0.132	0.174	0.731	0.308	0.318	0.229

local level (i.e., the higher the slope the lower the C content) but positive when comparing across sites (i.e. higher average soil C in high altitude sites with higher average slopes). Mixed-effect models may help disentangling such nested patterns of variability. Table 7.4 presents the distribution of the variance (%) between factors for six different soil fertility indicators using two different models, one that assumes Site to be a random effect and one that assumes it to be a fixed effect. In the first case, about half of the variability in soil C, total N and extractable P is associated with differences across sites (50.7, 48.9 and 57.6%, respectively), whereas about 20% is associated with variability between fields within farms (23.8, 20.8, 23.6%). In the second case, when Site is set as a fixed effect, then most of the variance is associated with differences between fields, which is the variability of interest at farm and landscape level. In this study, Tittonell et al. (2013) went on adding fix terms to the model as a way to characterise and understand spatial heterogeneity at farm level, linking this up with farmer management decisions and their level of resource endowment (see source article for further details). Mixed effect models are not the only method for statistical assessment of soil variability, but they are recommended when the data allows their use. When data are more complex and include different type of variables then other methods are needed, especially non-parametric ones, as in the example that follows.

7.6.3 Classification and Regression Trees

Beyond soil quality indicators, agroecologists are often concerned with variability in crop growth and productivity, which is strongly linked with soil spatial heterogeneity but also with many other factors. Management factors such as choice of cultivars, sowing dates and densities, weed infestation level and frequency of weeding (and method), amounts and timing of organic or mineral fertiliser applications, soil tillage management, amount and type of irrigation if present, preceding and accompanying crops, presence and incidence of pests and diseases, etc., are some key management factors affecting crop variability next to (and in interaction with) soil heterogeneity. Unravelling such complexity requires collecting information about these different variables plus climatic and location specific data and information. One challenge that arises immediately is that not all data are normally distributed or quantitative or continuous. For example, a dataset to characterise maize yield variability in smallholder farms in Kenya (n = 151 fields) included soil indicators (e.g. carbon, nitrogen, pH, depth, etc.), level of weed infestation (visual scoring: 1 = low; 5 = high), planting date (Julian day), cultivar (local names and hybrid names), presence or absence of the parasitic weed *Striga* sp. (yes/no), soil fertility class used by local farmers (local names), plant density (# ha^{-1}) and so on (cf. Tittonell et al. 2008). In other words, a complex dataset that mixes quantitative continuous, discrete, binary and nominal data, and exhibits non-linearity, is highly skewed or multimodal, and has large numbers of missing observations.

One method to deal with such complex data is the use of classification and regression tree (CART) analysis. Inherently non-parametric, no assumptions are made in CART analysis regarding the underlying distribution of values of the predictor variable (Tittonell et al. 2008). The method consists of recursive binary partitioning, and can be used to identify clusters of observations – crop yields in this example – that respond to similar conditions in terms of the candidate independent variables – i.e. the above-mentioned explanatory variables. In simple terms, at each stage the method identifies the explanatory variable that best partitions the data in two groups, such that the variance between groups is maximised and that within groups is minimised. In a subsequent step, each subgroup is partitioned in two by a new variable. The result is a ramifying binary tree that can be allowed to 'grow' its branches as much as the data allows it, or that can be 'pruned' down to a satisfactory clustering of the data. Figure 7.13 illustrates the use of CART to the analysis of crop yield variability in two different socio-ecological contexts.

The variability in maize grain yields measured in 151 smallholder farmers' fields in western Kenya was first partitioned in two groups according to the level of resource use intensity, no use or low use (left) and medium to high (right) (Fig. 7.13a). Since smallholders use a diversity of resources to 'feed' their crops, from animal manure to mineral fertilisers, but also kitchen waste, compost, chicken manure, dead leaves and crop residues, etc., and since they are in most cases unable to quantify the exact amounts used of different resources, a score was proposed to categorise fields according to the intensity of resource use: no use (0), low (1),

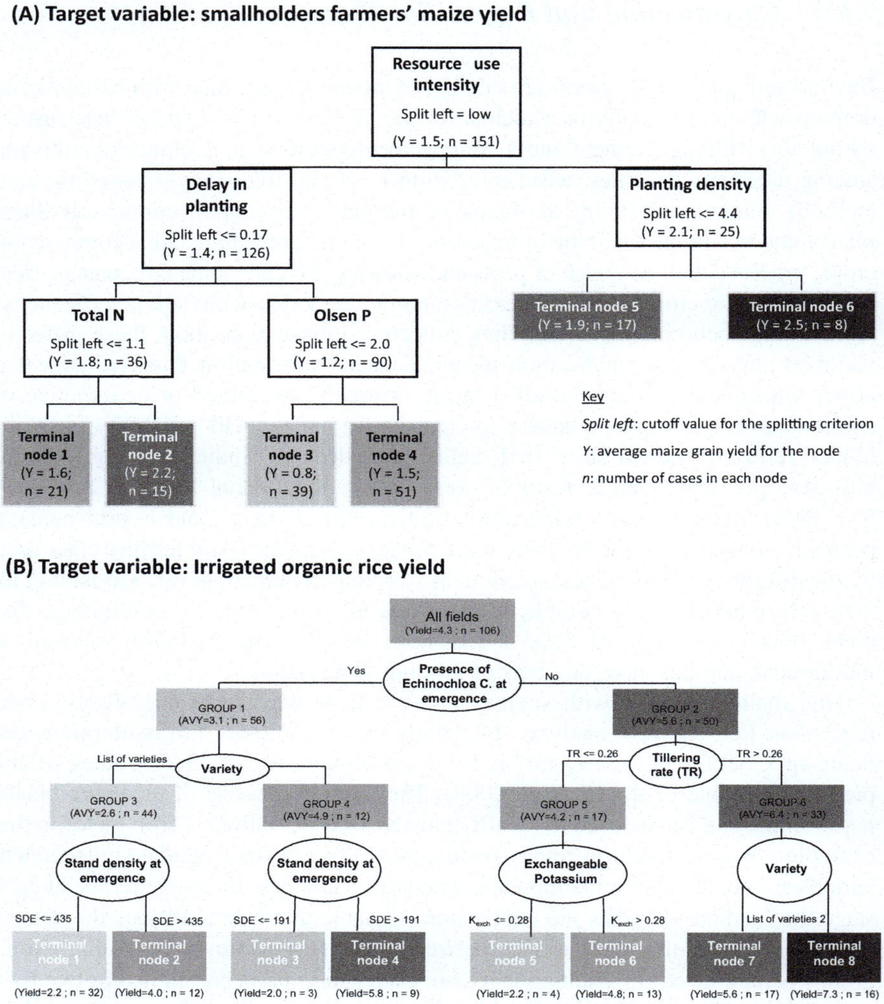

Fig. 7.13 Applications of CART to the analysis of (**a**) maize yield variability in farmers' field in western Kenya (adapted from Tittonell et al. 2008) and (**b**) organic rice yields in farms of the Camargue region of France. (Adapted from Delmotte et al. 2011)

medium (2) and high (3) – the level 'high' is only relative to this context, as the amount of nutrients applied in this category of fields may still be far from sufficient for proper crop yield. The subgroup of fields on the right-hand side of the tree, receiving medium to high level of resource, comprises 25 fields that exhibited an average maize yield of 2.2 t ha^{-1}. This group is further split in two subgroups by plant density, with a threshold value of 4.4 plants m^{-2}; 17 fields with lower plant density and average yields of 1.9 t ha^{-1} and 8 fields with higher plant density and average yield of 2.5 t ha^{-1}. These subgroups are considered terminal nodes, which means that there is not gain in model fitness by further splitting them. The terminal node 6 in Fig. 7.13a represents the best yielding farmer fields in the region of study and an important benchmark against which other fields can be compared, for example, by calculating yield gaps. Indeed, CART is an interesting methodology to complement yield gap analysis, as done by Delmotte et al. (2011) – see later.

Still in Fig. 7.13a, the left-hand side splitting groups all the fields (n = 126) that received no or low level of resources, with an average yield of 1.4 t ha^{-1}. These fields are further split by the delay in planting, with early planted fields splitting left (n = 36, average yield 1.8 t ha^{-1}) and late planted fields to the right (n = 90, average yield 1.2 t ha^{-1}). Even before any soil variable emerged as explaining yield variability, this analysis highlighted that there were 90 fields (out of a total of 151) that were planted late and receiving no or low resources. Thus, management-related factors were strongly determining yield variability in this context. Of course, there is often a certain degree of correlation between management and soil fertility factors, since farmers tend to prioritise those fields they perceive to be more fertile, by planting them earlier, with more seeds per unit area and using more resources (cf. Tittonell et al. 2005). The subgroup of early planted fields is further split by the level of total nitrogen in soil, with a cut-off value of 1.1 g N kg $soil^{-1}$, resulting in two terminal nodes. Terminal node 1 groups 21 fields that receive no to low resources, are planted early but have low levels of soil N, yielding 1.6 t ha^{-1} on average, while terminal node 2 includes 15 early planted fields that yield on average 2.2 t ha^{-1}. Note that the threshold of 1.1 g N kg $soil^{-1}$ is just a value found through statistics but not a physiologically meaningful threshold for maize productivity, and indeed quite a low value. Finally, the group of late planted fields (n = 90) is split according to the level of extractable phosphorus in the soil (using the Olsen method), with a cut-off value of 2 mg P kg $soil^{-1}$. Here again, this threshold value distinguishes between soils with poor and very poor P levels, as the threshold used to decide on fertiliser applications for maize is often in the order of 10 to 15 mg kg^{-1} in different regions. Terminal node 3 groups the lowest yielding fields (n = 39, average yield 0.8 t ha^{-1}), which are planted late and with no or low resource use, and exhibit very low levels of soil P.

Let us move to another region of the world, the Camargue region in southern France, where Delmotte et al. (2011) applied the method to explain irrigated rice yield variability under organic and conventional crop management (Fig. 7.13b). There is no need to repeat the detailed analysis done for the case of maize in Fig. 7.13a, since the reader may have already understood how to 'read' a CART. What is important to highlight in this new example, is that weed management emerges as a main factor determining rice yields, especially in organic systems where chemical herbicides are not used. In fact, one of the reasons to produce rice in flooded fields is to control weeds. But *Echinocloa* sp. is a plant that shares the same ecological niche as rice, striving under the same conditions in which rice does, and hence it is a major competitor for resources in rice fields. In fields that exhibit high levels of infestation with *Echinocloa* at the stage of rice emergence (n = 56) yields are on average lower and then the right choice of the rice variety becomes the next most important factor determining yield variability. The highest yields (7.3 t ha^{-1} on average) were attained in 16 fields with low weed infestation levels and high tillering rates (which is also indicative of high N availability in soil), as multiple tillers per plant allow increasing rice density and outcompete weeds (Terminal node 8). As can be seen in these two examples, this way of analysing yield variability provides deeper insights into the causes of variability than using linear models as classically done in such studies. It provides a better diagnosis of the processes underlying spatial heterogeneity.

7.6.4 *Infrared Spectroscopy*

Soil sampling and analysis to capture heterogeneity within farms may be quite costly and time-consuming, as many samples need to be collected per unit area to properly capture variability. One way of reducing costs and save time considerably, is the use of soil spectral analysis or spectroscopy. Soil samples are air-dried and crushed as usually done for wet chemistry analysis, and then placed in a petri dish under a spectrometer that projects a beam of light through them, which may be near- or mid-infrared (NIRS or MIRS, respectively). What is measured is the degree at which the soil sample reflects the light, and the reflectance has a different pattern for each soil sample over the entire spectrum considered. Figure 7.14 illustrates this with the relative reflectance curves of soil samples from six farms in western Kenya (three localities: Aludeka, Emuhaia and Shinyalu, two types of farm per locality: low and high resource endowment) using near infrared light (wavelengths between 350 and 2500 nm). The shape of each curve – each sample – can be analysed to characterise the shape of the peaks and valleys (in mathematical terms, this means taking the derivatives of the curve at each point) or spectral 'signature' of the soil sample. In Fig. 7.14 we can see how the differences in shape of the spectra for each field of the farm are wider in certain contexts than in others; they are widest in the high resource endowment farm at *Shinyalu*, a relatively large farm located on a heavily dissected slope. In Aludeka, where soils are coarse and sandy, the spectra produced by the soils of the fields of the two farms differ less in their shape. The shapes of infrared spectra depend on the molecular structure of mineral and organic constituents of soils. Since the organo-mineral composition determines soil functions, such as the ability to retain and supply nutrients and water, the distribution and protection of soil organic carbon in different pools, nitrogen mineralization, cation exchange, structural stability, etc., the infrared spectra can be used to predict different soil properties across a wide range of soil types (e.g., Tittonell et al. 2008; Janik et al. 2007; Terhoeven-Urselmans et al. 2010; Bellon-Maurel and McBratney 2011; Soriano-Disla et al. 2014).

Figure 7.14 also illustrates the relationship between measured (i.e. wet-chemistry analysis) versus NIRS-predicted values for carbon, nitrogen and phosphorus on 159 fields of smallholder farms in the same region, showing a relatively good fit between both methods (with levels of unexplained variance ranging between 17 and

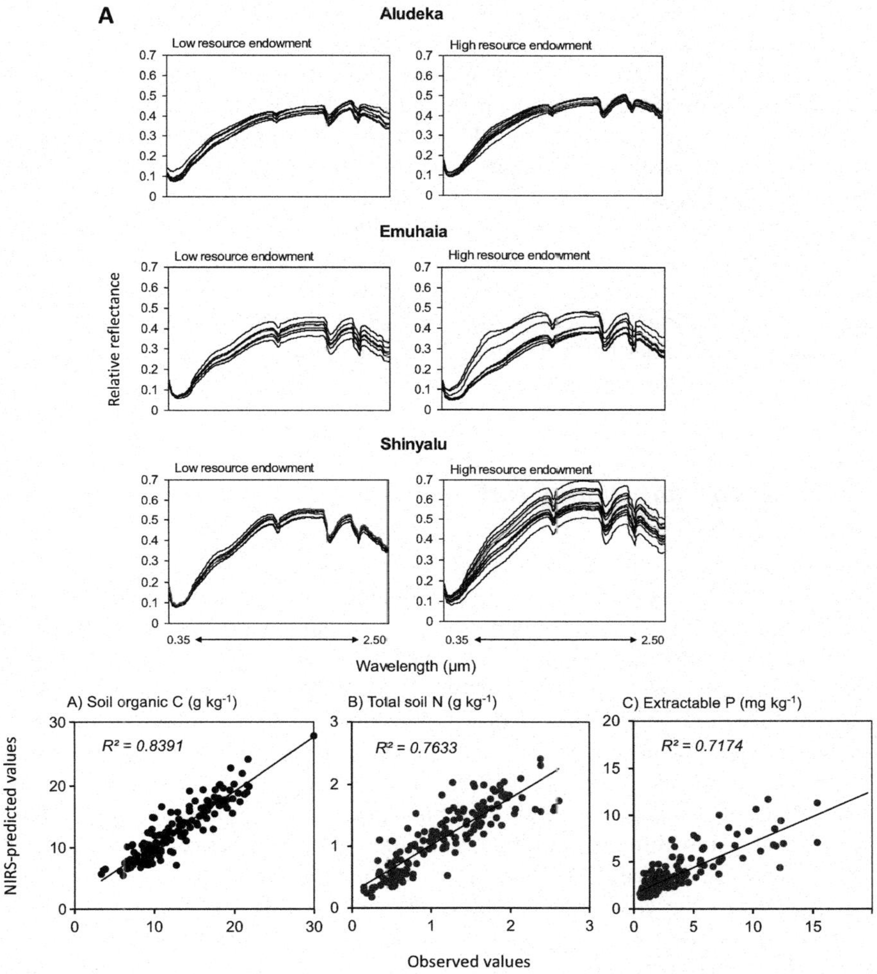

Fig. 7.14 *Above*: Examples of soil near infrared spectra taken on each field of high and low resource endowment farms, at three localities in western Kenya (n = 6 farms). *Below*: The relationship between observed (measured) values of soil organic carbon, total nitrogen and extractable P and those predicted trough near-infrared spectroscopy (NIRS) from 159 smallholder farmer fields in western Kenya. (Modified from Tittonell et al. 2008)

29%). It is important to stress that, to be able to obtain good predictions of soil properties, the spectral method needs to be calibrated against conventionally measured data. Calibrations require that at least 20 to 30% of the soil samples be analysed through both methods, spectral and chemical. In a lab that uses spectroscopy, calibrations are done at the beginning but also monitored and improved regularly, e.g. once a year, and often also regularly contrasted against the results of other labs through crosschecks. NIR spectroscopy is also used to analyse vegetation samples, and it has been used to assess forage quality for more than 40 years. More recently, the use of medium infrared spectroscopy (MIRS) showed even better performance at predicting soil quality (Tittonell et al. 2016). The cost of spectral infrared soil analysis is much lower than conventional wet chemistry, but they should not be underestimated: samples still need to be taken, processed, dried, crushed, and a fifth to a third of them analysed for calibration purposes. Attempts have also be made to use the spectra as predictors of soil fertility straight away, without calibrating against classical soil properties. One way to do this is by analysing the relationship between soil spectra and crop yields, and then computing a spectral 'index' of soil fertility based on crop productivity (Tittonell 2003). But since crop yields depend on so many other variables beyond soils (e.g. sowing density, weed infestation, etc.), the predictive capacity of such indexes tends to be low.

7.6.5 Local Soil Quality Indicators

Last but not least, soil quality and soil heterogeneity can be studied through the knowledge of local people about the extent and sources of soil variability. Before embarking in describing some methods available to do this, let us examine the theory behind the use of local knowledge. The study of the knowledge of local people about their environment is known as ethnoecology. A good definition of ethnoecology was proposed by Prof. Victor Toledo, one of the 'fathers' of agroecological thinking in Latin America: "Ethnoecology is an interdisciplinary approach exploring how nature is viewed by human groups through a screen of beliefs and knowledge, and how humans use their images to acquire and manage natural resources" (Toledo 2002: 514). What distils from this definition is that there is a nexus between knowledge and beliefs systems. Indeed, Toledo refers to this as the *Kosmos-Corpus-Praxis* triad (Toledo 1992), the *kosmos* representing the cosmovision and beliefs system, the *corpus* the knowledge system, and the *praxis* the set of practical operations that results from that knowledge system (Fig. 7.15). *Kosmos*, *corpus* and *praxis* delimit an 'ethnoscape' as proposed by Pauli et al. (2016) in their assessment of local farmers' knowledge on soil fauna across a wide range of agroecosystems.

Ethnopedology is a sub-discipline of ethnoecology that focuses on local soil knowledge, local perceptions and management practices (Barrera-Bassols and Zinck 2003). The term 'pedology' has been in use for a long time in soil science to refer to the study of soil genesis, morphology, survey and interpretation (Buol et al. 1997), with a focus on its dynamics and history (Duchaufour 1998). Edaphology, on the other hand, studies the properties of soils and their relationship with plant growth (Ulery 2005). Pedology studies the soil as a natural body in the landscape, while edaphology studies soil from an utilitarian perspective, as a medium for plant growth and other life forms, but also for uses such as buildings or waste disposal. Its relationship with soil formation dynamics and history through visual indicators and landscape positions places pedology closer to local soil

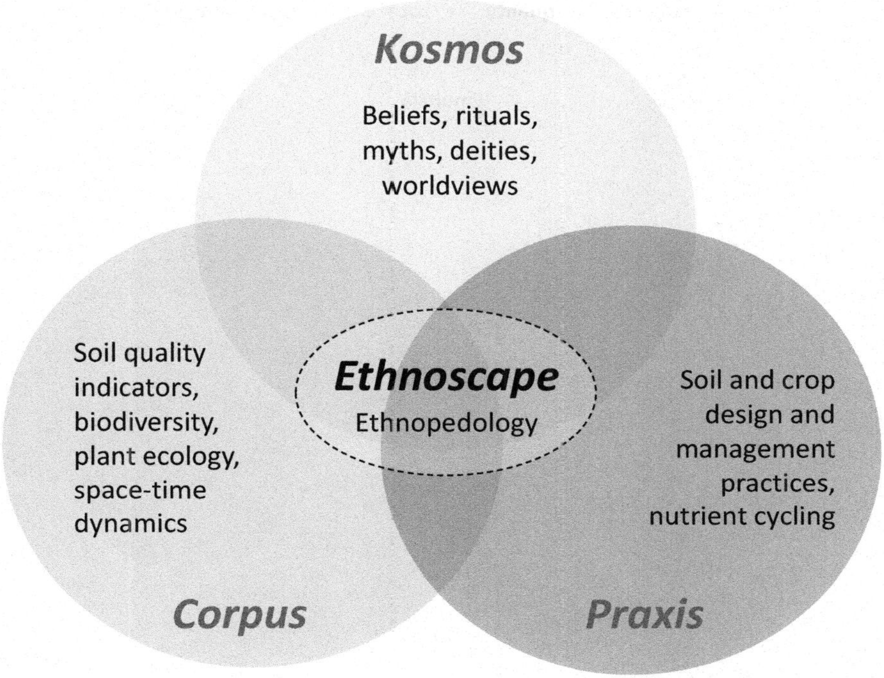

Fig. 7.15 A representation of the triad *kosmos-corpus-praxis* proposed by Victor Toledo (2002) delimiting an ethnoscape, following Pauli et al. (2016), applied in this case to soil knowledge (ethnopedology)

knowledge than edaphology, hence the use of the term ethnopedology (WinklerPrins and Barrera-Bassols 2004).

There have been various attempts to link local soil knowledge to indicators of soil quality through universal guidelines (e.g. Barrios et al. 2006). One limitation of this approach is that, as represented in Fig. 7.15, local soil indicators are strongly associated with local beliefs and knowledge systems, and these are not always extrapolatable across regions. This led others to propose building on local soil knowledge to develop indicators that are only locally relevant, such as the occurrence of spontaneous plants indicating soil degradation processes (Nezomba et al. 2017). Local knowledge has also been the basis of decision guides aimed at selecting organic materials as soil amendments (e.g. Mapfumo et al. 2005), soil restorative cultivation systems (e.g. Lahmar et al. 2012) or for the design of on-farm soil fertility 'try-outs' together with farmers (Misiko and Tittonell 2011).

Local indicators of soil quality may vary widely even within the same region, as illustrated for three locations in western Kenya located within a circle of 120 km in diameter (Fig. 7.16). The characteristics of the local landscape influence greatly the type of soil quality indicators used by farmers. For example, the slope of the field is an important indicator of soil quality for farmers in Shinyalu but not in Aludeka; the

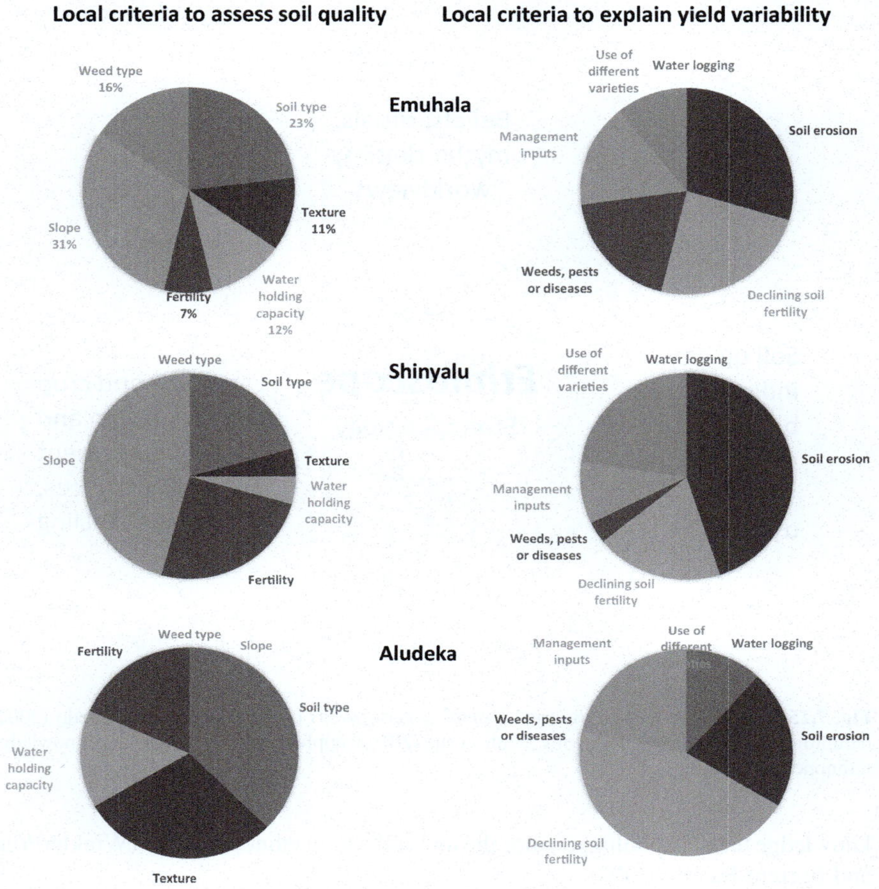

Fig. 7.16 Criteria used by farmers to classify soils according to their perceived quality (left) and to explain differences in crop (maize) yield variability (right) in three localities of western Kenya (n = 45 participating farmers). Emuhaia is located at 41 km from Shinyalu, and 123 km from Alludeka. Aludeka lies at 100 km from Shinyalu. Farmers indicated more than one possible criteria. (Adapted from Tittonell et al. 2005)

opposite can be said about soil texture. Weeds are used as indicators of soil quality in Emuhaia, but not in the other two locations, even though they are not so far away and the same types of weeds tend to occur in the three sites. When asked about the reasons behind crop yield differences within their farms and landscapes, the answers tended to be more homogeneous across the three localities in the example of Fig. 7.16 (right hand side). Soil erosion and declining soil fertility, two indicators of soil degradation, were the main reasons put forward by farmers in all three sites, followed by input use.

Table 7.5 Soil properties for the two main soil types identified by farmers in Aludeka, western Kenya (from Tittonell et al. 2007)

	Clay	SOC	N_{total}	$P_{extr.}$	Exchangeable cations ($cmol_{(+)}$ kg^{-1})				pH
Local soil name	(%)	$g\ kg^{-1}$	$g\ kg^{-1}$	$mg\ kg^{-1}$	Ca^{++}	K^{+}	Mg^{++}	H^{+}	(1:2.5)
Apokkor	25.9	10.3	0.71	6.5	3.9	0.51	1.1	0.16	5.7
Assing'e	16.5	7.6	0.45	4.9	1.9	0.18	0.6	0.28	5.2
SED	*2.3*	*0.9*	*0.04*	*2.3*	*0.6*	*0.18*	*0.2*	*0.17*	*0.2*

SOC soil organic C; Standard errors of the differences (SED) are indicated in *Italics*

Local soil quality indicators are not generalisable across localities. Yet within localities, local soil classifications may be strongly associated with formal physico-chemical indicators measured in the laboratory, as illustrated in Table 7.5 with the rsults of the laboratory analysis of the main two soil types identified by farmers in Aludeka.[2] But of course the indicators presented in Table 7.5 are generic features of two distinct soil types, located in different position in the landscape, that exhibit inherently different quality attributes. Both soil types could also be found in farmers' fields in situations of either good management and fertility, or poor management and degradation. Thus local classifications of soil quality, health or fertility status should not be confused with local identification of soil types, as is the case in the example of Table 7.5.

Another, complementary approach to ethnopedology consists of farmers classifying fields or areas of the farm according to their perceived 'fertility', based on the productivity of the crops. Strictly speaking, this should be termed ethno-edaphology (Aguilar Orea et al. 2019; Morales-Espinoza et al. 2021; Delgado Vargas et al. 2022) because of the more utilitarian view on soil. The classification of their fields into fertile, average or intermediate, and poor by 159 farmers from the same thee localities in western Kenya resulted in the average maize yield, management and soil fertility indicators presented in Table 7.6. Obviously, famers did not use the same indicators, and the classification of their own fields was based on their history of use as well. An interesting observation from this study was how farmers adapted their management to the perceived fertility of the fields, prioritising the soils that were perceived as more fertile, which were planted earlier, with denser crop stands, and weeded more often, resulting in significantly higher maize yields. This is why I have sustained for some time that what smallholder farmers practice is a form of precision agriculture, albeit without GPS-aided equipment (Tittonell et al. 2016). Among the soil chemical indicators, available phosphorus was more sensitive and reflected much better farmers' soil quality categorisations across localities than nitrogen or potassium. This match between variability in soil fertility indicators and management intensity, prompted me to speak of management intensity

[2]The location of these soils in the landscape, locally known as Apokor and Assing'e, is represented in the diagram of the Fig. 7.4a of Chap. 4

Table 7.6 Local classification of fields based on their perceived fertility level in three localities of western Kenya, and the corresponding maize yields, planting dates, plant densities, weed infestation level, and average levels of N, P and K measured in soils (n = 159 fields), from Tittonell et al. 2005

Locality	Land quality class	Maize yield (t ha^{-1})	Delay in planting (days)	Plant density (pl m^{-2})	Weed infestation score	Total soil nitrogen (g kg^{-1})	Extract. phosphorus (ppm)	Exch. Potassium ($cmol_+$ kg^{-1})
Emuhaia	Fertile	2.4	17.9	2.43	0.70	1.25	14.0	0.58
	Average	1.8	18.6	2.40	1.19	1.05	2.2	0.28
	Poor	0.9	27.3	2.16	1.65	1.10	1.8	0.25
Shinyalu	Fertile	2.5	15.6	2.62	0.85	1.52	9.8	0.40
	Average	1.5	16.9	2.09	1.45	1.65	3.9	0.33
	Poor	1.0	22.1	1.93	1.69	1.47	2.4	0.29
Aludeka	Fertile	1.5	18.0	1.73	0.91	0.73	8.4	0.66
	Average	1.1	21.6	1.76	1.38	0.50	3.7	0.34
	Poor	0.5	33.6	1.57	1.72	0.41	1.7	0.13

gradients, that emerge alongside the soil fertility gradients shown in Fig. 7.9 and reinforce them (Tittonell et al. 2007).

Together with my former colleague, anthropologist Michel Misiko, we worked extensively in western Kenya some 20 years ago to untangle the relationship between local soil quality indicators, soil types, physicochemical soil analysis, local soil management and crop productivity (Plate 7.3, pictures A–F). Farmers developed their own try-outs, which were participatory on-farm experiments they co-designed and implemented to investigate questions they prioritised collectively. We supported those farmers' experiments through a.o. the monitoring of changes in soil indicators as measured in the soil laboratory. We developed ways to 'translate' the results of the analysis of soil samples from farmer's fields into familiar images for them to be able to target management interventions. When explaining soil indicators such as carbon, nitrogen, phosphorus and potassium, we used the analogy of a typical local meal: a plate of ugali, beef and kales (Picture D). Ugali (corn meal) represented N, beef represented P, kales represented K and the size of the plate represented C. When a soil sample presented high C, low N, low P and high K, for example, we represented that soil graphically as a big plate containing small amounts

Plate 7.3 Integrating local and scientific soil knowledge and participatory experimentation (try-outs) in farmers' fields in Emuhaia, western Kenya. Pictures A and B, Michael Misiko and Pablo Tittonell in the local Farmer Field School in 2002. Picture C, a field of maize classified as fertile (*rotuba sana* in Swahili) and a proud farmer showing her maize to my former colleague Isaack Ekise in western Kenya. Pictures D and E, using the analogy of a typical local meal to explain the result of soil analysis (Nitrogen = maize ugali, Phosphorus = beef, Potassium = kales, Carbon = size of the plate). Picture F, farmers installing a try-out (participatory on-farm experiment) co-designed among them

of ugali and beef, and a large portion of kales (Picture E). This worked out very well even with participants who were not able to read or write, but recognised the analogy between soil nutrients and meal ingredients.

Although local soil quality classifications and indicators are not generalisable, there are however certain visual or tactile indicators that could be generalised to assess soil fertility in the field, such as colour, shape of aggregate, friability, depth, etc. Table 7.7 illustrates the use of some visual indicators of soil quality and degradation in 1988 fields that were previously classified by farmers as poor, average or fertile in eastern Uganda, western and central Kenya. Soil erosion signs, hard settings and stoniness had almost no relationship with farmer's soil quality classification. Slope and position in the landscape were more closely associated, with the majority of the 'fertile' fields according to famers located in the upper positions of the landscape with slopes of less than 5%. Visual soil quality indicators are universal, not local ones, but they may be useful when 'calibrated' locally with the community.

Table 7.7 Visual indicators of soil quality and degradation and their frequency of occurrence in fields classified by farmers according to their perceived fertility as Poor, Average and Fertile in 1988 farmer's fields in East Africa (eastern Uganda, western and central Kenya)

Indicator	Category/	Fields per category		Occurrence within soil qualiy classes (%)		
	type	*n*	(%)	Poor	Average	Fertile
Soil erosion	Sheet	340	17	19	18	13
	Rill	431	22	29	20	13
	Mass	16	1	1	1	1
Hard settings	Temporary	227	92	13	11	9
	Permanent	19	8	1	1	2
Stoniness	0–5%	1855	93	93	94	94
	5–25%	72	4	3	4	4
	25–50%	39	2	2	2	2
	50–75%	12	1	1	1	0
	> 75%	10	1	1	0	0
Slope class	0–5%	919	46	37	49	55
	5–10%	442	22	22	22	22
	10–20%	317	16	16	16	15
	20–40%	247	12	18	11	7
	> 40%	63	3	6	2	1
Landscape	Upslope	371	19	12	17	33
	Midslope	1423	72	77	74	58
	Footslope	158	8	10	8	5
	Bottomland	36	2	1	1	3
Flooding (occasional/regular)		60	3	3	2	5
Total number of fields per farmers' soil fertility class:				646	934	408

7.7 Summary and Concluding Remarks

Heterogeneity is a form of variability that has a spatial pattern. It is an important property of landscapes, as it reflects the interacting effects of natural and human processes, and underlies the functions that result in key ecosystem services. While management practices in classical agronomy tend to push agroecosystems towards uniformity, agroecology sees opportunity in spatial heterogeneity and hence methods are needed to understand it and categorise it. In this chapter we described methods and approaches to deal with landscape heterogeneity, in particular those that consider spatial autocorrelation, and focused on soil heterogeneity as an important driver of agroecosystem functioning. Soil heterogeneity can be inherent, as determined by geology, geomorphology, climate and biome, resulting from variations in soil depth, texture, soil water regimes, biological activity, etc. Soil heterogeneity can also be induced by human agency, through land use and management decisions. Both types of variability coexist in landscapes, they can be nested within each other, and they often interact, since farmers make management choices based on their perception of soil quality or 'niches' for their production activities. The resulting spatial patterns have been generically termed soil fertility gradients, as changes in soil quality tend to follow a certain directionality (e.g., decreasing at increasing distances from the homesteads in smallholder agriculture). There is a diversity of methods to study heterogeneity, especially in the fields of geographic information systems or geostatistics. This chapter presented a few methods that are useful in agroecology, and not often seen in the specialised literature. These methods, which were illustrated with examples from smallholder agriculture, described sampling design, statistical modelling (parametric and non-parametric) and rapid soil characterisation through spectroscopy or local soil quality indicators. Agroecological design and management practices need to embrace landscape spatial heterogeneity as a key attribute of the agroecosystem that forms the basis for positive biological interactions and (social-ecological) resilience mechanisms.

References

Adjei-Nsiah S, Leeuwis C, Giller KE, Sakyi-Dawson O, Cobbina J, Kuyper TW, Abekoe M, Van der Werf W (2004) Land tenure and differential soil fertility manage- ment practices among native and migrant farmers in Wenchi, Ghana: implications for interdisciplinary action research. NJAS Wageningen J Life Sci 52:331–348

Aguilar Orea GF, Ruiz Rosado O, Ortiz Solorio CA, Alcudia LA (2019) La etnoedafología como instrumento para la caracterización de agroecosistemas a nivel local, el caso de un ejido cafetalero del centro de Veracruz. Investigaciones geográficas, ISSN 0188-4611, ISSN-e 2448-7279, N°. 99, 2019

Alomia-Hinojosa V, Groot JCJ, Speelman EN, Bettinelli C, McDonald AJ, Alvarez S, Tittonell P (2020) Operationalizing the concept of robustness of nitrogen networks in mixed smallholder systems: a pilot study in the mid-hills and lowlands of Nepal. Ecol Indic 110:105883

Alvarez S (2012) Pratiques de gestion de la biomasse au sein des exploitations familiales d'agriculture-élevage des hauts plateaux de Madagascar: conséquences sur la durabilité des systems, PhD Thesis Montpellier SupAgro, France

Barrera-Bassols N, Zinck JA (2003) Ethnopedology: a worldwide view on the soil knowledge of local people. Geoderma 111:171–195. https://doi.org/10.1016/S0016-7061(02)00263-X

Barrios E, Delve RJ, Bekunda M, Mowo J, Agunda J, Ramisch J, Trejo MT, Thomas RJ (2006) Indicators of soil quality: a South–South development of a methodological guide for linking local and technical knowledge. Geoderma 135:248–259

Bellon-Maurel V, McBratney A (2011) Near-infrared (NIR) and mid-infrared (MIR) spectroscopic techniques for assessing the amount of carbon stock in soils—critical review and research perspectives. Soil Biol Biochem 43:1398–1410

Bessant LL (1992) Coercive development: land shortage, forced labor, and colonial development in the Chiweshe Reserve, colonial Zimbabwe, 1938–1946. Int J Afr Hist Stud 25(1):39–65

Brouwer J, Fusell LK, Herrmann L (1993) Soil and crop growth micro-variability in the West African semi-arid tropics: a possible risk-reducing factor for subsistence farmers. Agric Ecosyst Environ 45:229–238

Buol SW, Hole FD, McCracken RJ, Southard RJ (1997) Soil genesis and classification. Iowa State University Press, Ames

Carter S, Murwira H (1995) Spatial variability in soil fertility management and crop response in Mutoko communal area, Zimbabwe. Ambio 24:77–84

Castellanos-Navarrete A, Tittonell PA, Rufino MC, Giller KE (2015) Feeding, crop residue and manure management for integrated soil fertility management – a case study from Kenya. Agric Syst 134:24–35

Caulfield ME, Fonte SJ, Groot JCJ, Vanek SJ, Sherwood S, Oyarzun P, Borja RM, Dumble S, Tittonell P (2020) Agroecosystem patterns and land management co-develop through environment, management, and land-use interactions. Ecosphere 11(4):e03113. https://doi.org/10.1002/ecs2.3113

Delgado Vargas I, Leonel H, Moreno-Molina A, Chávez N, Paz P (2022) Ethnoedaphology as an integrating process between academy and peasant knowledge in the productive system of coffee (Coffea arabica L.) of southwestern Colombia. Agric Sci 13:1013–1030. https://doi.org/10.4236/as.2022.1310062

Delmotte S, Tittonell P, Mouret J-C, Hammond R, Lopez-Ridaura S (2011) On farm assessment of rice yield variability and productivity gaps between organic and conventional cropping systems under Mediterranean climate. Eur J Agron 35:2232–2236

Diarisso T, Corbeels M, Andrieu N, Djamen P, Tittonell P (2015) Biomass transfers and nutrient budgets of the agro-pastoral systems in a village territory in South-Western Burkina Faso. Nutr Cycl Agroecosyst 101:295–315

Duchaufour P (1998) Handbook of pedology: soils, vegetation and environment. A. A. Balkema, Rotterdam

Duriaux Chavarría J-Y, Baudron F, Gergel S, Yang K, Eddy I, Sunderland T (2020) More people, more trees: a reversal of deforestation trends in Southern Ethiopia. Land Degrad Dev 1. https://doi.org/10.1002/ldr.3806

Ebanyat P (2009) A road to food? : Efficacy of nutrient management options targeted to heterogeneous soilscapes in the Teso farming system. Uganda Proefschrift Wageningen

El Mujtar N, Muñoz BP, Mc Cormick M, Pulleman PT (2019) Role and management of soil biodiversity for food security and nutrition; where do we stand? Glob Food Sec 20:132–144

Ettema CH, Wardle DA (2002) Spatial soil ecology. Trends Ecol Evol 17(4):177–183. https://doi.org/10.1016/S0169-5347(02)02496-5

Félix GF, Scholberg JMS, Clermont-Dauphin C, Cournac L, Tittonell P (2018) Enhancing agroecosystem productivity with woody perennials in semi-arid West Africa: a meta-analysis. Agron Sustain Dev. https://doi.org/10.1007/s13593-018-0533-3

Fortin M-J, Dale MRT (2006) Spatial analysis. A guide for ecologists. Cambridge University Press, Cambridge, UK

Janik LJ, Skjemstad JO, Shepherd KD, Spouncer LR (2007) The prediction of soil carbon fractions using mid-infrared-partial least square analysis. J Aust Soil Res 45(2):73–81

Kintché K, Guibert H, Sogbedji JM, Levêque J, Tittonell P (2010) Carbon losses from savannah soils under cotton-cereal rotations in semiarid Togo. Plant Soil 336:469–484

Lahmar R, Bationo BA, Lamso ND, Guéro Y, Tittonell P (2012) Tailoring conservation agriculture technologies to West Africa semi-arid zones: building on traditional local practices for soil restoration. Field Crop Res 132:158–167

Lupatini M, Korthals GW, Roesch LFW, Kuramae EE (2019) Long-term farming systems modulate multi-trophic responses. Sci Total Environ 646:480–490. https://doi.org/10.1016/j.scitotenv.2018.07.323

Mapfumo P, Mtambanengwe F, Giller KE, Mpepereki S (2005) Tapping indigenous herbaceous legumes forsoil fertility management by resource-poor farmer in Zimbabwe. Agric Ecosyst Environ 109:221–233

Misiko M, Tittonell P (2011) Counting eggs? Smallholder experiments and try-outs as success indicators of adoption of soil fertility technologies. In: Bationo A et al (eds) Innovations as key to the green revolution in Africa. Springer Science + Business Media B.V, pp 1137–1144. https://doi.org/10.1007/978-90-481-2543-2_116

Morales-Espinoza I d C, Ortiz-Solorio CA, Gutiérrez-Castorena MDC, Gutiérrez-Castorena EV (2021) Estudio Etnoedafológico para el reconocimiento de tipos de usos asociados con cadenas productivas en el ejido de Santa Cruz, Durango. Terra Latinoamericana 39:e853. Epub 17 de junio de 2021. https://doi.org/10.28940/terra.v39i0.853

Nezomba H, Mtambanengwe F, Tittonell P, Mapfumo P (2017) Practical assessment of soil degradation on smallholder farmers' fields in Zimbabwe: integrating local knowledge and scientific diagnostic indicators. Catena 156:216–227

Pauli N, Abbott LK, Negrete-Yankelevich S, Andrés P (2016) Farmers' knowledge and use of soil fauna in agriculture: a worldwide review. Ecol Soc 21(3):19. https://doi.org/10.5751/ES-08597-210319

Prudencio CY (1993) Ring management of soils and crops in the west African semi-arid tropics: the case of the mossi farming system in Burkina Faso. Agric Ecosyst Environ 47(3):237–264

Ramisch JJ (2005) Inequality, agro-pastoral exchanges, and soil fertility gradients in southern Mali. Agric Ecosyst Environ 105(1–2):353–372

Samaké O, Smaling EMA, Kropff MJ, Stomph TJ, Kodio A (2005) Effects of cultivation practices on spatial variation of soil fertility and millet yields in the Sahel of Mali. Agric Ecosyst Environ 109(3–4):335–345

Soriano-Disla JM, Janik LJ, Viscarra Rossel RA, Macdonald LM, McLaughlin MJ (2014) The performance of visible, near-, and mid-infrared reflectance spectroscopy for prediction of soil physical, chemical, and biological properties. Appl Spectrosc Rev 49:139–186

Terhoeven-Urselmans T, Vagen T-G, Spaargaren O, Shepherd KD (2010) Prediction of soil fertility properties from a globally distributed soil mid-infrared spectral library. Soil Sci Soc Am J 74: 1792–1799

Tittonell P (2003) Soil fertility gradients in smallholder farms of western Kenya. Their origin, magnitude and importance. Quantitative approaches in systems analysis no. 25, ISBN 90–6754–713-1, Wageningen, The Netherlands, pp. 233

Tittonell P, Vanlauwe B, Leffelaar PA, Shepherd KD, Giller KE (2005) Exploring diversity in soil fertility management of smallholder farms in western Kenya. II. Within-farm variability in resource allocation, nutrient flows and soil fertility status. Agric Ecosyst Environ 110:166–184

Tittonell P, Vanlauwe B, de Ridder N, Giller KE (2007) Heterogeneity of crop productivity and resource use efficiency within smallholder Kenyan farms: soil fertility gradients or management intensity gradients? Agric Syst 94:376–390

Tittonell P, Shepherd KD, Vanlauwe B, Giller KE (2008) Unravelling the effects of soil and crop management on maize productivity in smallholder agricultural systems of western Kenya – an application of classification and regression tree analysis. Agric Ecosyst Environ 123:137–150

Tittonell P, Muriuki AW, Shepherd KD, Mugendi D, Kaizzi KC, Okeyo J, Verchot L, Coe R, Vanlauwe B (2010) The diversity of rural livelihoods and their influence on soil fertility in agricultural systems of East Africa – a typology of smallholder farms. Agric Syst 103:83–97

Tittonell P, Muriuki A, Klapwijk CJ, Shepherd KD, Coe R, Vanlauwe B (2013) Soil heterogeneity and soil fertility gradients in smallholder agricultural systems of the East African highlands. Soil Sci Soc Am J 77:525–538

Tittonell P, van Dis R, Vanlauwe B, Shepherd K (2016) Managing soil heterogeneity in smallholder African landscapes requires a new form of precision agriculture. In: Lal R, Stewart BA (eds) Soil specific farming, advances in soil science. CRC Press/Taylor & Francis Group NW, Boca Raton, pp 199–224

Toledo VM (1992) What is ethnoecology? Origins, scope and implications of a rising discipline. Etnoecológica 1(1):5–21

Toledo VM (2002) Ethnoecology: a conceptual framework for the study of indigenous knowledge on nature. In: Stepp JR, Wyndham FS, Zarger R (eds) Ethnobiology and biocultural diversity: proceedings of the seventh international congress of ethnobiology. University of Georgia Press, Athens, pp 511–522

Ulery AL (2005) Encyclopedia of soils in the environment. https://www.sciencedirect.com/referencework/9780123485304/encyclopedia-of-soils-in-the-environment

WinklerPrins A, Barrera-Bassols N (2004) Latin American ethnopedology: a vision of its past, present, and future. Agric Hum Values 21:139–156. https://doi.org/10.1023/B:AHUM.0000029405.37237.c8

Zingore S, Murwira HK, Delve RJ, Giller KE (2006) Soil type, management history and current resource allocation: three dimensions regulating variability in crop productivity on African smallholder farms. Field Crop Res 101(3):296–305

Part III
Design-Oriented Approaches

Chapter 8
Evaluation and Indicators in the Design of Agroecosystems

Abstract Agroecosystem design is not a one-time exercise; it requires constant evaluation and readjustment. This chapter provides a non-exhaustive overview of evaluation methods and indicators used in the (re)design of agroecosystems. Indicators are the central tool of any evaluation process. Choosing indicators based on objectives and system properties is a first step in agroecosystems evaluation, and several features of indicators must be considered. Since agroecology is a "dialogue of wisdoms" (cf. Chap. 1), wherein knowledge from multiple stakeholders merges in the redesign of agroecosystems, participatory methods for co-creating knowledge, prototyping, and evaluation are interwoven throughout the sections of this chapter and the following two chapters. A special case of comprehensive and participatory systems design, called co-innovation, is introduced here and illustrated with examples towards the end of this chapter.

8.1 Design and Evaluation

"Design is the first signal of human intention." This sentence was coined by William Macdonough, one of the proponents of the 'Cradle to Cradle' approach to industrial design and architecture. This approach relies on three major principles that are also highly applicable in agroecology: (i) waste is food, (ii) use current solar income, and (iii) celebrate diversity. The first principle refers to recycling and reusing materials (nutrients, carbon, water) in various production processes. The second principle emphasizes maximizing the capture and utilization efficiencies of solar radiation. The third principle highlights the importance of diversity in multiple aspects. In the case of agroecology, diversity can be directly associated with biodiversity in space and time, the integration of diverse knowledge systems (scientific, practical, and traditional knowledge), and the inclusion of different perspectives and stakeholders in the design process. Agroecology, and to some extent conservation or regenerative agriculture, were originally built on these principles, namely on recycling, efficiency and diversity (Tittonell 2014). In other words, these principles can be utilized to guide the design of sustainable agroecosystems.

P. Tittonell, *A Systems Approach to Agroecology*,
https://doi.org/10.1007/978-3-031-42939-2_8

Research for the design of agroecosystems differs from agroecosystem analysis, although some of the tools and methods used in the process may be the same. To recap some of the concepts discussed in Chap. 2 (refer to Fig. 2.3 and related text), the design process begins with the identification of a "purpose" and then proceeds by determining the necessary functions and structures to achieve that purpose. In the agricultural design literature, often referred to as prototyping in English or as *conception* in French, these elements are given slightly different names. For instance, the well-established framework by Vereijken (1987) defines objectives or criteria (purpose), theoretical prototypes (functions), and pilot prototypes (structures). However, to determine whether a new design is suitable or preferable compared to the current situation, an evaluation is also essential. Thus, design and evaluation are inherently interconnected throughout any iterative and gradual process of agroecosystem (re)design. Without an evaluation framework, it is not possible to design effectively.

In most cases, a method is devised to 'conceive' the design, and one or more methods are used to evaluate it. Evaluation can take place either *a priori* or *a posteriori*, using indicators to simulate the behaviour, test or monitor the new system. Such indicators should reflect the degree of fulfilment of the objectives or purpose. In systems analysis terms, recapping once again the concepts introduced in Chap. 2, such indicators should also be able to reflect system properties. For example, Hashimoto et al. (1982) proposed three criteria to evaluate the performance of water resource systems: how likely a system is to fail (reliability), how quickly it recovers from failure (resilience), and how severe the consequences of failure are (vulnerability). Recently, Laborda et al. (2023) studied vulnerability patterns of rural households in the face of environmental hazards by examining their capacity to withstand a shock (resistance) and their ability to recompose their livelihood after it (recovery), and defined resilience as the integration of resistance and recovery. Using systems properties as criteria has the advantage of making the assessment generic, irrespective of the actual indicator to be measured, therefore allowing for comparative analysis across systems of different nature.

In practical situations, the design or, more commonly, the redesign of agroecosystems involves a diverse range of stakeholders,[1] including farmers, policy makers, and sometimes agricultural researchers, extension agents, or private sector actors who may or may not provide technical advice. Each stakeholder brings their own perspectives and priorities to the table. They consider different types of indicators or similar indicators with varying thresholds when assessing a new system design. Often, stakeholders prioritize indicators that reflect short to mid-term system performance, with a strong focus on economic factors such as income, costs, employment generation, or return on investments. Typically, stakeholders may not

[1] Stakeholder is a term first coined by R.E. Freeman in his book: 'Strategic Management: A Stakeholder Approach", to refer to those who can affect or be affected by the activities of a company. Other terms used in the literature to refer to stakeholders are actors, social actors or social subjects.

explicitly consider system properties but rather focus on objectives and expected outcomes and impacts. As a result, they may overlook or neglect emerging and unexpected but plausible consequences of a certain design. This highlights the importance of identifying objectives and indicators based on system properties, as well as conducting ex-ante assessments and field testing of new designs in the agroecosystem design process. These steps help ensure a comprehensive evaluation of potential outcomes and minimize the risk of unintended consequences.

8.2 Methods for Agroecosystem Design and Evaluation

Much of what has been written around design in agriculture concerns the cropping system, chiefly the allocation of crops in space and time through species sequencing and rotation (e.g., Meynard et al. 2006), and through association of annual and perennial crops in agroforestry (e.g. Van Noordwijk et al. 2022). In animal production systems, examples of agroecosystem design range from rotational grazing schemes (Vosin 1993) to the design of animal production facilities (e.g. Bos and Groot Koerkamp 2008). Concepts of design are also used in both plant and animal breeding (i.e., ideotypes). Whole-farm designs including the livestock and crop sub-systems have received less attention, although some remarkable work has been done in some tropical agroecosystems (e.g. Funes-Monzote 2008).

Design takes very different meanings in other fields of knowledge and practice, such as architecture, information technology or management science. In such cases, design is defined as an active process that generates not only new plans, products or processes, but also concepts and knowledge. In the design of agricultural systems, Meynard et al. (2012) make also a distinction with regards to the actual design process; they differentiate between:

Pre-set or rule-based design: In which objectives and methods are decided and established from the start of a standardised process, in which there is a division of tasks between researchers, developers and farmers in a continuum that is known as technology transfer; this process has been inherent to much of the development of modern agriculture over the second part of the twentieth Century.

Innovative design: In which some of the objectives or criteria may be known from the start while others are added throughout the process, and the necessary methods and knowledge to arrive at the new design are not known a priori; this non-standardised, explorative pathway towards agroecosystem design challenges the classical division of tasks between research, extension and farming practice, and necessitates participatory methodologies to support the process of co-creation and knowledge exchange. The authors further differentiate between:

- Gradual design, step-by-step or ‘progress loops’, in which changes to the agroecosystem are included gradually, accompanying a process of incremental learning by different stakeholders (particularly farmers), minimising risks, and responding to a limited set of objectives; this is typically the pathway followed in co-innovation approaches (see later in this chapter);

- *De Novo* design, or design by rupture, in which the process of design should be almost independent from the actual constraints (resources, policies, institutions) that condition the structure and functioning of the current agroecosystem, in order to generate a broader set of alternative designs not restricted by the current reality; this approach is less adapted to conciliate the expectation of farmers and other stakeholders with the exploration of alternatives, and is the approach classically used in model-based design (cf. Chap. 10);

Considering the purpose, the actual methods employed and the nature of the research outcomes Le Gal et al. (2010) reviewed a large volume of scientific literature and proposed a decision tree to classify approaches used in research around agroecosystem design (Fig. 8.1). The first criterion is whether the researchers aim (i) to create and test a new design *per se*, e.g., an innovative cropping system to be tested by researchers and/or farmers in pilot, experimental or demonstration fields, or (ii) to support the design done by farmers exploring new sets of technologies, management practices or contextual scenarios. In the first case, it is possible to differentiate methods that employ simulation modelling in the process of design, or

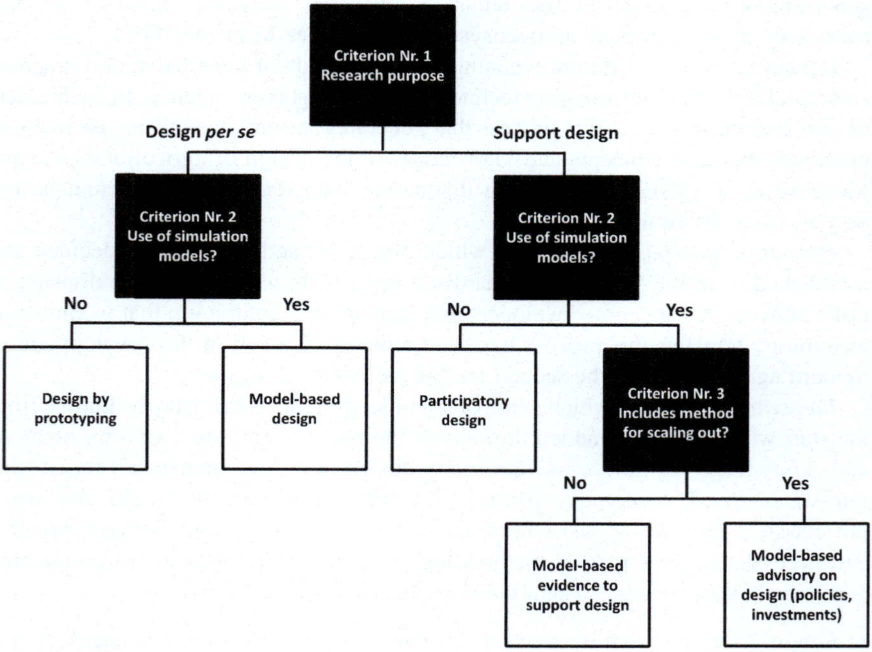

Fig. 8.1 A classification of methods used in agroecosystem design as reported in the scientific literature, based on research purpose, methods and the nature of research outcomes. (Adapted from Le Gal et al. 2010)

those that are based on actual prototypes. The second group includes methods that involve different stakeholders in a participatory process to support design, and methods that use models to the same end. A distinction is made between purely scientific exercises, on the one hand, and applications to inform actual design processes, including scaling-out mechanisms, on the other.

From a total 80 references reviewed by these authors, 59 of them corresponded to the categories model-based design (29) and model-based support (30), indicating that simulation modelling is the most widely spread method used in agroecosystem design following the scientific literature. In the light of the broad categories of innovative design proposed by Meynard et al. (2012), the use of models and the participatory methods are very effective ways of supporting *de Novo* design. Although the classification described above is a commendable, very informative piece of work, it is still the product of an intellectual process in which the actual boundaries between approaches and methods are very often hard to distinguish. For example, methods that combine some of these categories, such as participatory prototyping or participatory modelling, have also been reported in the design literature (e.g., Thornton and Herrero 2001; Tittonell et al. 2009; Dogliotti et al. 2014a, 2014b; etc.). Another limitation of this classification is the fact that it has been only restricted to the analysis of the scientific literature published in international journals, and not all processes of agroecosystem design that happen in reality are necessarily published or reported in a scientific article.

The process of agroecosystem design, irrespective of the specific method employed, is always described in terms of steps and very often as a cycle. If presented as a sequence of steps, the design process could generically consist of the following ones (*IDEAS*): Identify problems and objectives; Diagnose the current situation; Explore alternatives; Asses feasibility and potential impacts; Select and test best options (Fig. 8.2). In most of these steps there are instances of evaluation for which specific methods and indicators are necessary.

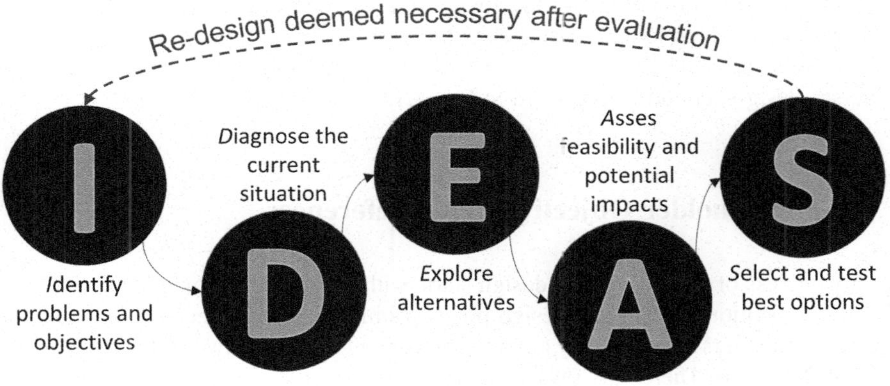

Fig. 8.2 A sequence of five proposed steps in the process of agroecosystem design that yield the acronym IDEAS. The sequence may become a cycle when redesign is deemed necessary after evaluation

The selected options (*S*) are then implemented, evaluated and monitored. While the steps *I* and *D* correspond to a phase of analysis, steps *A* and *S* correspond to the phase of synthesis (cf. Chap. 1, Fig. 1.3). Step *E* corresponds to both analysis and synthesis and implies also an evaluation of trade-offs between objectives (cf. Chap. 9). Models, field data, experiments, participatory processes, prototyping and diverse sources of knowledge can be used in combination throughout this process, challenging once again the general distinction of methods made in Fig. 8.1.

In depicting the design process as a cycle, many authors include also a diagnosis phase (e.g., Goewie 1997), while others:

(i) present 'design' as a step in a cycle (e.g. the DEED approach used in the NUANCES project – Giller et al. 2010);
(ii) present the entire cycle as an 'evaluation' process without explicitly mentioning the term design (e.g. the MESMIS sustainability evaluation framework – López-Ridaura et al. 2002); or
(iii) use terms such as 'intervention' or 'feasibility' studies to refer indirectly to design in a process of selection of alternatives (e.g. the land use cycle – Dent and Ridgeway 1986).

Irrespective of how the process is graphically represented, its cyclical nature implies that any new design requires progressive readjustments that may be detected either in ex-ante assessments or through ex-post monitoring and evaluation (Fig. 8.2). Evaluation may be seen thus as an intrinsic part of the design process. Ideally, after every evaluation cycle the system progresses towards a higher level in terms of achievement of the desired objectives or purpose (this could be represented more accurately as an ascending spiral rather than a cycle). But sometimes the objectives change along the way, especially in a situation of innovative design as described earlier in this chapter, often as a result of changes in the context in which the agroecosystem operates.

Through the rest of this chapter, we will first examine concepts and methods that pertain to the identification of objectives, criteria and indicators to guide the process of design, and then look into the specific methods of prototyping, participation and simulation modelling, without necessarily distinguishing between design *per se* versus design-support as proposed in Fig. 8.1.

8.3 Stakeholder Objectives and Preferences

Any process of agroecosystem design starts with the identification of objectives. As already mentioned, there are design processes in which the objectives are known *a priori* and others in which they emerge from the interaction between the stakeholders involved in it. There are several methods to identify stakeholders' objectives, priorities, preferences, criteria, etc. There are also methods to weigh the relative importance they ascribe to them. Such methods normally combine stakeholder participation through the development of lists or rankings with the use of some algorithm to assign numerical values that represent the preference stakeholders assign to each objective. Methods that fall under this category are known under the umbrella term of multi-criteria analysis (Romero and Rehman 2003).

The output of the participatory process is a preference matrix consisting of the objectives identified and the weights ascribed to them by different stakeholders (Table 8.1). Such a matrix can be an output in itself and used to inform decision-making, or used as input of a goal programming model, in which stakeholder goals are 'translated' into mathematically solvable objective functions subject to a number of constraints expressed too as mathematical functions (cf. Chap. 10). In the simple example presented in Table 8.1, two stakeholders evaluated two alternative management strategies using six indicators of economic and environmental performance. Management alternative 1 generates more income, soil erosion and nitrate leaching, uses more pesticides and leads to less species diversity and greater income variability than alternative 2. Whether alternative 1 is preferable to alternative 2 depends on the relative importance each stakeholder ascribes to the various objectives evaluated, and on the absolute differences between alternatives for each individual objective.

This is computed as follows: First, the value of each indicator is standardised with respect to the desirable target, which can be either the maximum or the minimum

Table 8.1 Example of a simple multi-criteria weighing of two alternative management strategies by two stakeholders considering six objectives (indicators)

Management strategies	Income	Soil erosion	Species diversity	Pesticide use	Nitrate leaching	Income variability	Totals (sum)
	($)	(kg ha^{-1})	(#)	(kg ha^{-1})	(kg ha^{-1})	(CV)	
Alternative 1	3000	1200	2	15	60	0.1	–
Alternative 2	1500	1000	5	3	23	0.2	–
Target (max/min)	3000	1000	5	3	23	0.1	–
Standardization							
Alternative 1	1.00	0.83	0.40	0.20	0.38	1.00	–
Alternative 2	0.50	1.00	1.00	1.00	1.00	0.50	–
Weights (preferences)							
Stakeholder 1	3	1	1	1	1	3	10
Stakeholder 2	1	2	2	2	2	1	10
Product S x W							
Alt 1/ stake-holder 1	3.0	0.8	0.4	0.2	0.4	3.0	7.8
Alt 1/ stake-holder 2	1.5	1.0	1.0	1.0	1.0	1.5	7.0
Alt 2/ stake-holder 1	1.5	1.0	1.0	1.0	1.0	1.5	7.0
Alt 2/ stake-holder 2	0.5	2.0	2.0	2.0	2.0	0.5	9.0

From: López-Ridaura et al. (2005)

value observed for each objective. For income, the most desirable value observed is \$3000; the standardised values for this objective are 1.0 for alternative 1 (3000/3000) and 0.5 for alternative 2 (1500/3000). In the case of nitrate leaching, alternative 1 (0.38 = 23/60) gets a lower value than alternative 2 (1.0 = 23/23) because a leaching rate of 23 kg ha^{-1} is preferable to one of 60 kg ha^{-1}. In allocating weights, stakeholders were given 10 points each that they had to distribute among the six objectives. The preference of each alternative for each stakeholder is calculated as the sum of the product of the standardised values for each objective times the weight assigned to them. In this example, alternative 1 is preferable over 2 (scores 7.8 vs. 7.0) for stakeholder 1, who assigned greater importance (weight) to income indicators, whereas alternative 2 is preferable over 1 (scores 9.0 vs. 7.0) for stakeholder 2, who showed preference for environmental performance indicators.

Translating stakeholders' goals and preferences into quantifiable and/or measurable indicators presents a number of challenges, not least when objectives are to be expressed as solvable equations. The identification of farmers' objectives and operational targets from general, sometimes abstract concepts such as that of sustainability has been undertaken in different ways. The bulk of the scientific literature dealing with this in different parts of the world appeared mostly in the 1990s and early 2000s. Pacini (2003) selected and classified seven sustainability evaluation frameworks in which stakeholders' goals were translated into formal indicators and/or objective functions (Table 8.2).

The frameworks presented in Table 8.2 cover hierarchical levels ranging from farm systems to landscapes, as well as different temporal dimensions of sustainability. Most of these approaches were based on systems theory. The framework proposed by Loomis (2000) found theoretical support in the capital theory as applied to sustainable development back in the 1990s (Stern 1997; Daly and Costanza 1992), whereas the one proposed by Van Mansvelt and Van der Lubbe (1999b) relied on Maslow's theory of human motivations. This interesting approach tries to (i) reflect the priorities and motivations that lead the actions of people that manage the agroecosystem (landscape) and (ii) act as a unifying concept to integrate natural, social and humanist disciplines. In agroecology we engage with stakeholders in agroecosystem redesign rather naturally, often without a specific method. Learning from the experiences – and often the failures – of the cases summarised in Table 8.2 may help us improve our approaches to stakeholder involvement.

8.4 Indicators

While stakeholders' goals are often expressed as rather abstract concepts, such as 'well-being' or 'sustainable growth', the indicators to represent these goals should comply with a number of requisites, such as being easy to determine or measure, intelligible and easy to communicate, reliable, informative, non-redundant, sensitive to the changes in the system and its driving variables, etc. (Fig. 8.3). An indicator is, per definition, a variable (cf. Chap. 2). Yet the difference between a variable and an

Table 8.2 Selected cases of sustainability evaluation frameworks that included identification and translation of stakeholders' goals into formal indicators and objective functions

Methodological framework	Theoretical background	Hierarchical level	Time scale	Geografical extent
1. Research Network on Integrated and Ecological Arable Farming Systems for EU and Associated Countries (Vereijken)	Systems theory	Farming systems	Short-term	Europe
2. Checklist for Sustainable Landscape Management (Van Mansvelt and Van der Lubbe)	Human motivations theory adapted to landscape management	Landscapes	Long-term	Europe
3. Framework for Assessing the Sustainability of Natural Resource Management Systems (MESMIS) (Lopez-Ridaura)	Systems theory	Peasant management systems	Long-term	Latin America
4. International Framework for Evaluating Sustainable Land Management (FESLM) (Smyth and Dumansky)	Systems theory	Landscapes	Short-term Medium-term Long-term	World
5. Problem-Solving Framework for Modelling Sustainability Issues (Weersink)	Systems theory	Farming systems	Long-term	Western countries
6. Indigenous Approaches to Holistic, Self-Determined Sustainable Development (Loomis)	Capital theory applied to sustainable development	Landscapes	Long-term	New Zealand
7. Environmental-Economic Framework to Support Multi-Objective Policy-Making (Pacini)	Systems theory	Farming systems	Short-term Long-term	Europe

Adapted from Pacini (2003)

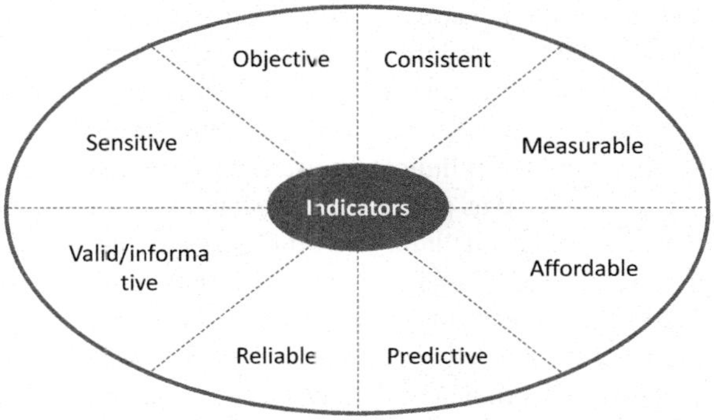

Fig. 8.3 Key characteristics of good indicators

indicator is that the latter contains reference values, which serve to transform raw data into information (Gallopin 1997). Reference values may take the form of thresholds, targets, benchmarks, etc. The following is a generic but not necessarily normative definition of reference values (Moldan and Dahl 2008):

- *Target*, or a desirable system state to be achieved;
- *Baseline*, or the current distance to a desirable state;
- *Threshold* value, or the distance to a collapse or tipping point;
- *Benchmark*, or the characteristics of another system considered as reference.

One of the most conspicuous recent examples of a reference value is the planetary boundaries proposed by Steffen et al. (2015) based on Earth-system processes. These boundaries are, by definition, thresholds that should not be crossed in order to remain within safe operating sustainability limits at planetary level. The boundary corresponding to CO_2 concentration in the atmosphere, for example, has been established at 350 $ppm_{(v)}$. Which means that to remain within safe operating limits in terms of global warming, this threshold should not be surpassed. Current levels (2023) are in the order of 419 $ppm_{(v)}$, beyond the proposed threshold and much higher than the pre-industrial value of 270 $ppm_{(v)}$.

An indicator is considered to be objective when it is factual, when it is not distorted by large subjectivity (feelings, opinions). In some cases, however, subjectivity may be unavoidable, and specific methods may be deployed to deal with that. An indicator is consistent when it can be unambiguously defined. A common case of lack of consistency emerges when several enumerators are collecting household or farm data through surveys. Each person may ascribe different meanings to a certain variable or use different ways to estimate it. The variable ‘total farm area’, for example, may in some cases correspond to the total area *owned* by the family, while in other cases it may correspond to the total area *farmed* by them. Indicators must be thus consistently defined to avoid skewedness and biases. An indicator is sensitive when it is able to reflect changes in the system in response to changes in the design or management of the system itself or in its broader context (scenarios, drivers). But sensitivity should not be too high as to exaggerate or amplify the real impact of such changes. Measuring and storing indicators must be possible and affordable with current technologies. Meta-analysis methods are available that produce reliable estimates of indicators, sometimes avoiding the need for sophisticated scientific methods to measure indicators.

The validity of an indicator can be defined in two ways. Firstly, by considering how informative the indicator is in relation to the process being monitored. One of the most common indicators used in the evaluation of agroecosystems is crop yield. Yield is defined as production of a harvestable product per unit area of land, and it is thus a measure of ‘land use efficiency’ or the ‘partial factor productivity’ of land. However, crop yield defined as such may not be very informative in cases in which land is not the most limiting factor. Labour productivity or productivity per unit energy may in such cases be more informative – and thus more valid – indicators of resource use efficiency than crop yield.

A second way to define validity refers specifically to the statistical validity of an indicator, which depends on its accuracy (distance between the estimate and the real value), precision (distance between estimates after repeated measurements) and repeatability (or statistical stability, when an experiment leads to significantly similar results when repeated). Finally, it is often desirable that indicators allow for some degree of predictability when extrapolated towards the future. Predictability does not necessarily always depend on the indicator itself, on its range of statistical validity, but also on the degree of certitude concerning the future. Extrapolations towards the short- to mid-term future are known as projections (cf. Chap. 2 – Fig. 2.5). Projections are applicable when it is possible to assume that the conditions defining the range of validity of an indicator will remain more or less unchanged within the forward period considered. Extrapolations towards more uncertain future scenarios are known as explorations.

8.5 Indicator Frameworks

An indicator framework is a system of multiple indicators used to assess multiple goals, criteria or performances in or of a system. In agriculture, most recent indicator frameworks have been developed as sustainability *evaluation* frameworks (cf. Table 8.2), and not explicitly to inform design. But, as already stated, the design process includes also the steps of diagnosis, ex-ante assessment and ex-post monitoring and evaluation, for which the indicators used in sustainability evaluation frameworks can be of great help. In policy-making, a distinction is made between hard targets such as political targets (e.g., to end hunger), and soft targets, such as those used as reference values in sustainability assessments. Designing agroecosystem or management strategies with soft targets in mind requires accurate information on reference values such as baselines (diagnosis), thresholds (e.g. minimum population of a certain species to ensure its reproduction; maximum level of a certain pollutant in a water body, etc.) and, when possible, benchmarks.

In many situations, targets, baselines and thresholds can be only provided as ranges, particularly in the case of soft targets. Such ranges may be associated with uncertainties, for example around the scientific underpinning of the processes involved, or with a diversity of preferences by key stakeholders who are unable to agree on a single reference value. Benchmarking by means of rankings of countries, regions or agricultural sectors can also be a useful way of providing reference values (e.g., income per capita, average yields, agricultural input sales, biodiversity indexes, etc.), provided that the ranking includes both 'good' and 'bad' examples. A system of pre-identified indicators, such as an indicator-based sustainability evaluation framework, is very useful in a process of rule-based design, in which objectives and targets are known from the start (e.g. Bohanec and Zupan 2004). However, in cases of innovative design, even the scale of analysis may not be known in advance. The objectives that are deemed important by different stakeholders may not be known, nor the best indicators to represent them or their reference values.

Indicators that are relevant to different stakeholders may also differ in terms of the scale at which they can be best assessed. Using generic system properties and attributes to derive indicators can be of great help in such cases. Instead of using a fixed set of indicators, the idea is to evaluate newly designed systems by examining key properties of interest (e.g., such as vulnerability), and then decide through a participatory process which criteria and indicators can best reflect such properties. The participation of stakeholders in the identification and definition of relevant criteria and indicators can contribute to their validity, intelligibility and meaning (cf. Fig. 8.3). The MESMIS framework (cf. Chap. 2) is one of the sustainability evaluation frameworks of this type most widely used in agroecology, beyond mere scientific applications (Masera et al. 1999; López-Ridaura et al. 2002; Astier et al. 2012). The framework considers generic agroecosystem properties such as those defined by Conway (1987) as productivity, stability, reliability, resilience and adaptability, and derives diagnostic criteria and indicators based on a structured baseline characterisation of the agroecosystem, as outlined in Fig. 8.4.

But before proceeding, lets first clarify what the terms properties, criteria and indicators mean through a simple example. If lack of stability (which is a property/attribute) in agricultural production is identified as a problem to be addressed through the re-design of an agroecosystem, then the possible causes underlying this are the ones postulated as diagnosis criteria. Causes of unstable productivity in the long term could be for example soil erosion, or soil compaction, or soil nutrient depletion. These are then postulated as diagnostic criteria; they can be

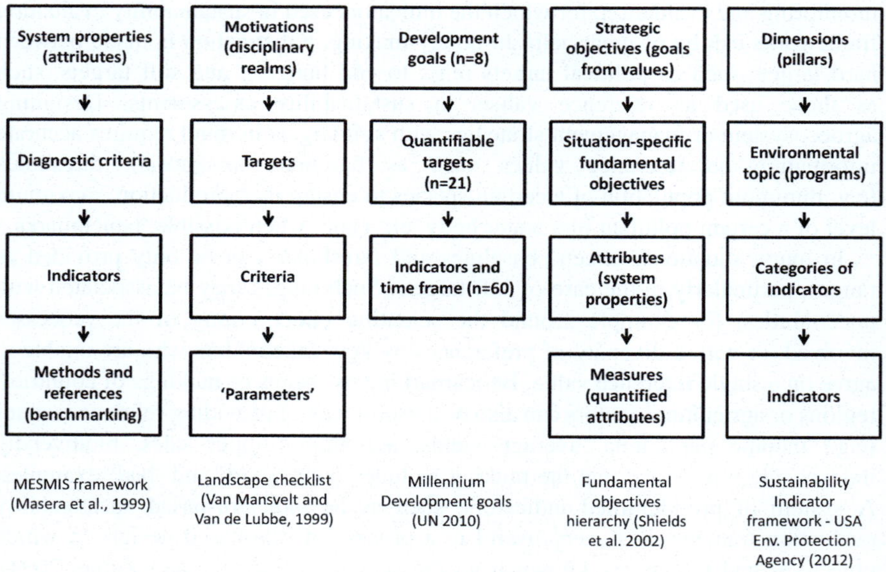

Fig. 8.4 A comparison of some selected indicator frameworks considering the hierarchy of entities that define each framework. In some frameworks the terms 'parameters' or 'measures' are used to refer to indicators

identified or corroborated through interaction with relevant stakeholders in a participatory process. Then, one or more indicators can be used to assess such criteria. For soil erosion, for example, an obvious indicator to be used would be 'tones of soil lost per hectare per year'. However, the amount of soil lost per unit area over a period of time can be estimated in different ways: it can be measured using sediment collection tanks or trenches, or estimated measuring changes in the thickness of the upper soil layer, by observing visual signs of soil erosion, through sediments collected in water bodies downstream, through remote sensing using some kind of spectral signature, through simulation modelling, etc. All these alternative methods to measure or estimate the indicator 'tones of soil lost per hectare per year' are likely to vary in their absolute and reference values. This means that an indicator should be specified not only with the proper reference values but also together with the method and units to be used in its assessment and quantification.

Using generic properties to guide design leads to having to consider dynamic system variables, for which reference values and indicators need to be defined not only at one point in time but also as a reference trajectory or as times series, with their inherent variability. The system of reference values most widely used in the applications of the MESMIS framework is benchmarking, by comparing current systems to 'improved' or alternative ones (Astier et al. 2012). The method by which to arrive at the reference values corresponding to the 'improved' system is not standardised, and not always explicit in the literature. López-Ridaura et al. (2005) introduced a number of modifications to the original framework to adapt it to multi-scale sustainability evaluations, using stakeholder preferences and knowledge to derive reference values and weigh objectives, and combined the framework with design-oriented modelling through goal programming.

Several other frameworks use a similar approach to arrive at indicators, although they may not explicitly use system properties (Fig. 8.4). For example, some approaches may use the concept of 'values', such as fairness or equity, and then define indicators that try to represent such values. At the beginning of this chapter we mentioned the possible use of 'principles' to guide design and evaluation. Other approaches use 'goals' or strategic objectives, and then derive indicators that measure the degree of fulfilment of such objectives. Both objectives and system properties are combined in the generic hierarchy of fundamental objectives proposed by Shields et al. (2002). The authors define two levels of objectives (Fig. 8.4). The strategic objectives that reflect goals derived from general values (such as 'equity', which could be a goal that is derived from an intrinsically human – or even animal – value such as 'fairness'). Then, the authors propose one or more levels of fundamental objectives, which are situation-specific, and attributes, which are used in a similar sense as system properties are used in other frameworks.[2] Finally, the authors

[2]The conceptual difference between properties and attributes in philosophy has been explained in Chap. 2: an attribute is considered to be ascribable while a property is something that can be possessed.

Table 8.3 Criteria used in the Landscape Checklist of Van Mansvelt and Van der Lubbe (1999a)

Quality of the biotic and abiotic environments		Quality of the social environment		Quality of the cultural environment	
Environment	Ecology	Economics	Sociology	Psychology	Physiognomy
1.1 Clean environment	2.1 Bio-diversity	3.1 Profitable farming	4.1 Local well being	5.1 Compliance to the natural environment	6.1 Diversity of landscape components
1.2 Sufficient and quality food and fibre	2.2 Ecological coherence	3.2 Greening the economy	4.2 Permanent education of farmers	5.2 Use of full landscape potential	6.2 Landscape internal coherence
1.3Regional carrying capacity	2.3 Eco-regulation	3.3 Regional autonomy	4.3 Access to participation	5.3 Presence of naturalness	6.3 Continuity of spatial arrangement
1.4 Economic and efficient resource use	2.4 Animal welfare		4.4 Landscape accessibility	5.4 Rich sensory qualities	
1.5 Adaptation and regional specificity				5.5 Experience of landscape unity	
				5.6 Experience of historicity	
				5.7 Cyclical developments	
				5.8 Careful management	

use the term measure – or quantifiable attribute – in the same sense that other frameworks use the term indicator.

The landscape checklist framework proposed by Van Mansvelt and van der Lubbe (1999b) more than 20 years ago, already presented in Table 8.2, defines human motivations at the highest hierarchical level, from which specific targets are then derived (Table 8.3). Human motivations are defined as pertaining to the three major realms of science, namely the natural (**β**), social (γ) and humanist (α) realms. The β-realm refers to the environment, resource conditions and ecological relationships; the γ-realm considers financial and service flows and the participation of stakeholders; the α-realm refers to subjective appreciations of the landscape and its cultural identity, considering both psychological and physiognomic aspects. Two groups of targets are defined per realm, loosely following disciplinary divides (e.g. environment, ecology, economics, sociology, psychology and cultural geography), and six checklists of criteria for sustainable landscape management are thus defined. Each criterion is quantified through a set of 'parameters', which is equivalent to the concept of indicators. The asymmetries between disciplinary realms in terms of the number of criteria proposed, and the possible measurable indicators that could be ultimately associated with them, is noteworthy and illustrative of the challenging disciplinary divides that still persist nowadays.

The sustainability indicators framework developed by the US Environmental Protection Agency (EPA) uses a taxonomy of indicators that starts off with the concept of sustainability pillars, or the environmental, economic and social dimensions of sustainability (Daly 1973). It considers the major environmental topics prioritised by this agency (air, ecological condition, human exposure and health, land, water), and four major categories of indicators: Adverse Outcome Indicators (destruction of economic or environmental value), Resource Flow Indicators (associated with the rate of consumption of resources), System Condition Indicators (state of the systems in question) and Value Creation Indicators (both economic and well-being). This is clearly a framework biased towards monitoring and evaluation. As many other frameworks of this kind, the focus is on assessing impacts and not necessarily on ways to reduce them.

Finally, the Millennium Development Goals was decidedly a target-oriented framework in which hard targets were set as reference values at global scale, using quantifiable criteria, short to mid-term time frames and indicators of progress towards such targets. For example, goal number 7 was "to ensure environmental sustainability", and was declined in four targets (A, B, C and D). Target 7C, for example, was stated as follows: "*To halve, by 2015, the proportion of the population without sustainable access to safe drinking water and basic sanitation*". The indicators corresponding to this target were two: (i) proportion of population using improved drinking water resources, and (ii) proportion of population using improved sanitation facilities. Eight global goals were defined in this way together with 21 quantifiable targets to be monitored by 60 pre-defined indicators (cf. Fig. 8.4). A similar approach has been followed in the definition of the 17 Sustainable Development Goals that replaced the Millennium Goals since 2015 (https://sdgs.un.org/goals).

Using a system of indicators is also a way of packing or reducing the information conveyed in primary or raw data. Indicators are analysed, often aggregated data. The highest possible level of aggregation can be obtained by computing indexes. There is a large number of indexes proposed in the literature to capture all aspects of agricultural sustainability in a single numerical value. They differ in the calculation method – although admittedly most of them are calculated as linear combinations of an array of (hopefully uncorrelated) indicators. In some cases, weighing coefficients are included in the equations designed to calculate such indexes. They are useful in regional assessments, which are often spatially explicit, in combination with geographical information systems. An example of such an index is the Agro-Eco-Index (Viglizzo 2003) used in the Mercosur to assess sustainability of agricultural systems. A limitation of the use of indexes is that possible informative interactions between the terms or individual indicators that make up the index tend to be masked. Likewise, counteracting effects of single terms in the index calculation model may be cancelled out, resulting in neutral net effects. Therefore, at the scales of analysis that concern the methods proposed in this book, the farm or the landscape, the use of such indexes is not recommended, as they tend to hide more often than reveal information.

8.6 Prototyping

The term prototyping has been used in a diversity of ways by different authors in the realm of agroecosystem design (e.g., Vereijken 1987; Sterk et al. 2007; Salembier et al. 2018; etc.). Fig. 8.5 is an attempt to represent the various steps of prototyping in a rather generic way, trying to capture the aforementioned diversity. Strictly speaking, the process of prototyping includes both a design and an evaluation phase. Strategic objectives, sometimes termed 'high level' objectives or goals, are first identified and categorised, or prioritised when necessary. At the same time, technical options to achieve such objectives plus the indicators to measure the degree of their achievement are identified and selected. Then a theoretical (sometimes termed hypothetical or 'virtual') farming system is designed by linking objectives to technical options and indicators. Such newly designed farming system (e.g. a crop rotation, a grazing regime, a poultry husbandry facility, an agroforestry system) is implemented in experimental and/or pilot farms, managed by researchers or by farmers, and assessed for a certain period on the basis of the indicators proposed. Finally, once the new system proves satisfactory to the various stakeholders involved in monitoring and evaluation, and based on the criteria set at the beginning, it is disseminated through the agricultural extension system or other scaling-out mechanisms such as demonstration units (farms, fields, etc.).

Approaches to prototyping tend to differ (i) in the process of objective definition, (ii) in the way in which technical options are identified, (iii) in the methods used for design and testing, and (iv) in the approach to dissemination of results. In some cases, authors refer exclusively to the first steps (the design phase) as prototyping. In other cases, the process starts from what innovative farmers are already doing, and their farms or management systems become prototypes or pilot cases. In yet other

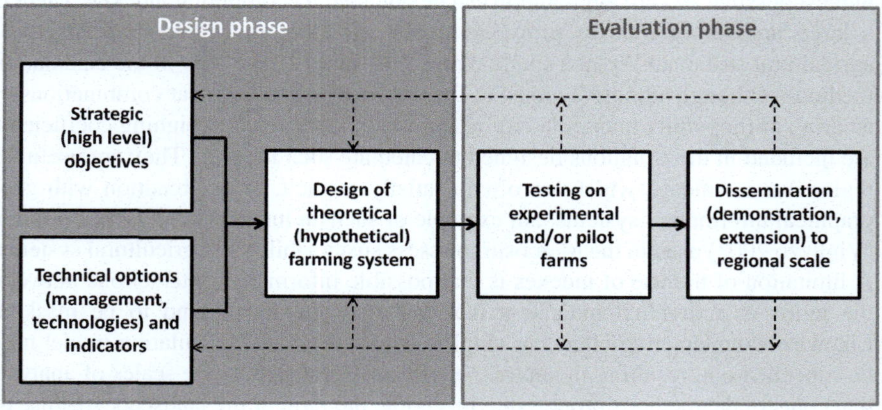

Fig. 8.5 A generic representation of the various steps involved in prototyping, denoting design and evaluation phases, as described in the classical literature on agricultural design. Dotted lines indicate possible feedback loops in the process

cases, farmers and other key stakeholders participate in the identification and categorisation of objectives, technical options and indicators, in the design of prototype farms and/or in the testing and adaptation phases (e.g. Waithaka et al. 2006; Chikowo et al. 2011). This is normally termed ‘participatory prototyping’. Simulation modelling can be used as part of the process to evaluate the viability of the hypothetical of virtual farm prototype (e.g. Tittonell et al. 2009; Dogliotti et al. 2014b). Nowadays we see an increasing number of approaches that use the principles of prototyping and that combine stakeholder participation and simulation modelling (e.g., Ryschawy et al. 2022). This is why it is so difficult to classify these as separate methods (and this is why the scheme proposed by Le Gal et al. 2010 has been so difficult to delineate, as the authors were ready to admit– cf. Fig. 8.1).

Prototyping in the context of smallholder agriculture in the tropics requires a number of adjustments to the generic methodology depicted in Fig. 8.5, which draws mostly from experiences in Europe. An important step in the context of smallholder agriculture is the diagnosis or characterisation of current diversity of farming systems and management practices (e.g. Box 8.1 – Hauswirth 2014). With some exceptions, and unlike farmers in more developed countries, smallholder farmers in the tropics seldom keep records of their activities for the tax officer or make explicit plans for the mid- to long term that concern for example crop rotations, renewal of machinery, sequential planting of fruit orchards to meet market demands, or animal replacement plans in productive livestock herds. This is not equivalent to say that there is no logic in the space-time organisation of smallholder landscapes. But the diversity of management practices is much wider than in commercial farms of temperate regions, where most farmers in a certain sector tend to plant their crops at the same time, use more or less the same cultivars, animal breeds and farming methods, or practice similar rotation schemes. For this reason, an initial step of sheer categorisation of the existing diversity is crucial to support prototyping in the context of smallholder agriculture. This initial characterisation and categorisation can also reveal the existence of interesting management systems. Such systems may become the new prototype for broader testing and dissemination.

Agricultural extension offices have typically used the last two steps depicted in Fig. 8.5 as part of their technology *transfer* process. Testing of theoretical prototypes can be done (i) on experimental farms, implemented and managed by researchers,[3] (ii) on pilot farms, in which the prototype is implemented and managed by researchers, by farmers or both, (iii) on pilot farms managed entirely by researchers, and (iv) on pilot farms managed entirely by farmers. Examples of pilot farms or prototype management systems are abundant worldwide and continue to be of great use. The ‘Russell Ranch’ of the University of California (UC) Davis (Fig. 8.6a) or the Farming Systems Experiment of Rodale Institute of Pennsylvania are good examples of this. An interesting case is that of whole-farm experiments or pilot

[3] A notable example of this is the case of agricultural research in Cuba, where researchers were made responsible of entire production units to make them not only productive but also economically viable, profitable (F. Funes-Monzote, pers. Comm.).

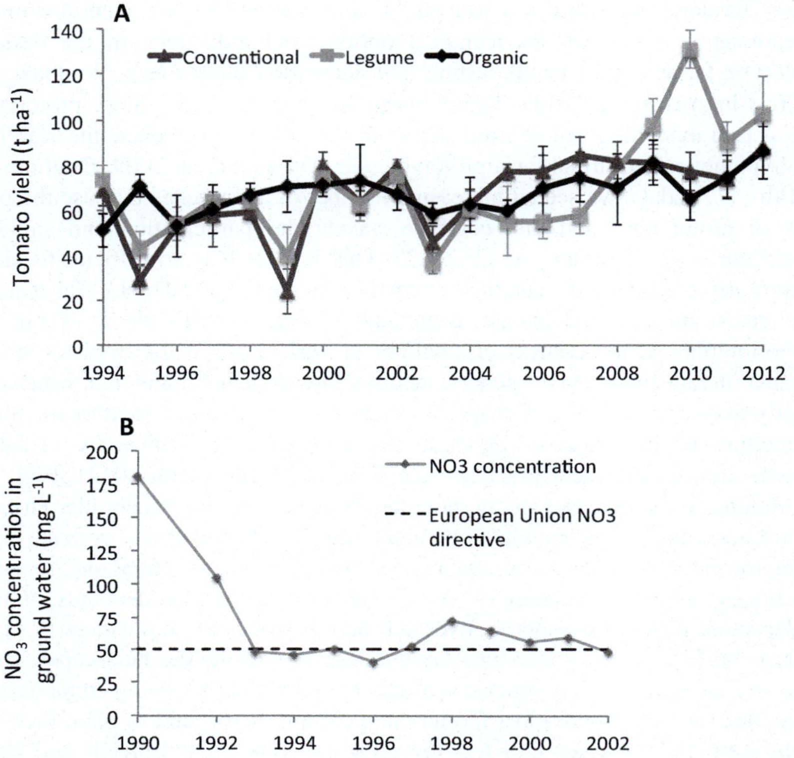

Fig. 8.6 Examples of long-term trends observed on pilot farms. (**a**) Comparative productivity of field grown tomato under three different crop rotation schemes and management systems (tomato-maize and tomato-maize-legume under conventional management, and tomato-maize under organic management) at the Russell Ranch of UC Davis, US (source: http://asi.ucdavis.edu/rr). (**b**) Nitrate concentration in ground water measured at the 'De Marke' pilot farm in The Netherlands, showing how the whole-farm management system introduced by researchers since 1990 drastically reduced ground water pollution with excess nitrate. (From Langeveld et al. 2007)

farms, such as the '*De Marke*' farm in The Netherlands, a dairy farm that has been entirely managed by researchers since 1992 (Fig. 8.6b). These pilot farms differ from other long-term agricultural research initiatives, such as the long-term soil management trials of Rothamsted in the UK or the GOK experiment[4] at FiBL, Switzerland. The latter are not strictly prototypes, but experiments designed with precise scientific objectives in mind to address a number of research questions. Yet they can certainly generate knowledge and information that is crucial for agricultural design in general and for prototyping in particular.

[4]The DOK trial compares biodynamic (D), organic (O) and conventional (K for German: "konventionell") production of arable crops such as wheat, potatoes, maize, soya and grass-clover leys since 1978, from where a number of scientific publications have been issued.

The concept of unidirectional technology transfer, from technology developers (researchers) to adopters (farmers), which dominated the agricultural extension paradigm during the last half century, was also the dominant one when most of the referred work on prototyping was published in the late 1990s. This approach is now being replaced by the concept of innovation systems and platforms, which envisage a more inclusive, participatory approach to technology development (e.g., Rossing et al. 2021). Yet forms of unidirectional technology transfer still persist, notably in the realm of plant or animal breeding or in the case of technologies developed by private companies. The feedback loops indicated in Fig. 8.5 are not necessarily always operational. When they are, underachievement of strategic objectives during the step of testing on experimental or pilot farms may lead to reviewing either the theoretical design, the technical options and indicators that were chosen, or even the strategic objectives. The same applies when undesirable outcomes emerge during the dissemination phase, such as lack of adoption by farmers or sub-optimal performances after broad scale adoption.

Box 8.1 Prototyping Hillside Cropping Systems in Northern Vietnam
The mountainous region of Northern Vietnam has seen a vertiginous expansion of rainfed maize cultivation on the hillsides of fertile valleys, fuelled by the creation of new feed markets for intensive livestock production. This resulted in deforestation and increased erosion of soils on steep slopes that are cultivated every season, often twice a year. Conservation agriculture, or the combination of minimum soil disturbance, permanent soil cover and crop diversification in space and time has been proposed as an option to stop soil erosion while maintaining crop productivity and resource use efficiency.

Rather than coming up with pre-designed cropping systems, Hauswirth (2014) started off with a characterisation of existing cropping systems in the region and an assessment of their productive, environmental and economic performance. Based on this, conservation agriculture prototypes were designed for the four main cropping systems identified (maize monoculture, maize-groundnut, maize-upland rice, maize-cassava) and tested in three pilot farms at *Suol Giang* (SG), *Phieng Luong* (PL) and *Chieng Hac* (CH). In the scheme below, prototypes are presented for two-year rotations, with main cash crops indicated in light grey, cover crops in black and mulches in stripped grey. Cover crops were established at the time of sowing the main crop (0), during its first weeding (1), during the second weeding of the first or second crop (2), or one month before harvesting the first or second crop (3). (T = tillage; NT = no tillage).

(continued)

Box 8.1 (continued)

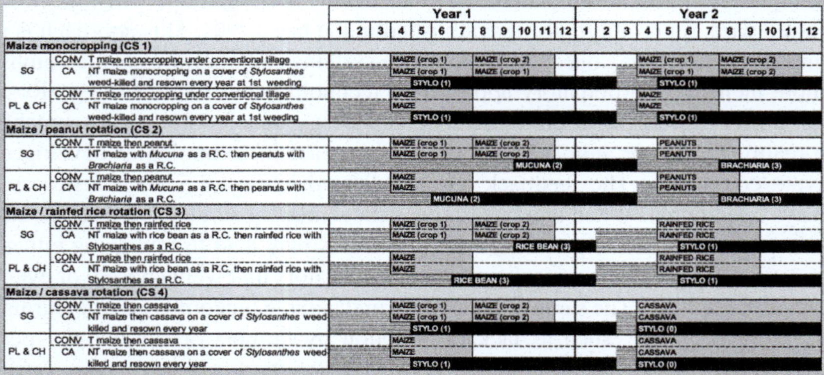

Hauswirth, D., 2014. Evaluation agro-économique ex-ante de systèmes de culture en agriculture familiale - Le cas de l'agriculture de conservation en zone tropicale humide de montagne (Nord Vietnam), PhD Thesis, Montpellier SupAgro, France, 150p.

8.7 Design Through Stakeholder Participation

Stakeholder participation is crucial in both gradual and *de Novo* design, and equally important in design *per se* (through prototyping) or in design-support processes (cf. Fig. 8.1). In the previous sections we saw that stakeholder participation can be engaged already from the identification of problems, strategic objectives and their prioritisation, during the choice of indicators, of alternative technical options and in their evaluation. In other words, stakeholders may participate in all the stapes of the IDEAS sequence (cf. Fig. 8.2). It is not the objective here to describe the participatory methods that are used in these instances. They are well described in the specialised literature on social science research and methods (e.g. http://www.researchtoaction.org). The aspect of stakeholder participation that is specific to the case of design is the way in which participants learn during the process and act upon this learning, with consequences for their management system.

Shared learning is central to the approach known as co-innovation, in which all stakeholders engaged learn through cycles of diagnosis and design (cf. Chap. 1, Fig. 1.5) in a process of gradual and continuous redesign of their agroecosystems (e.g. Dogliotti et al. 2014b; Douthwaite and Hoffecker 2017; Rossing et al. 2021). In the specific case of co-innovation, stakeholders are farmers as well as extension agents and other knowledge brokers such as technical advisors, researchers and, depending on the scale considered, also the representatives of local governments, of the private sector, agro-dealers, food chain agents, etc. A quite successful case of

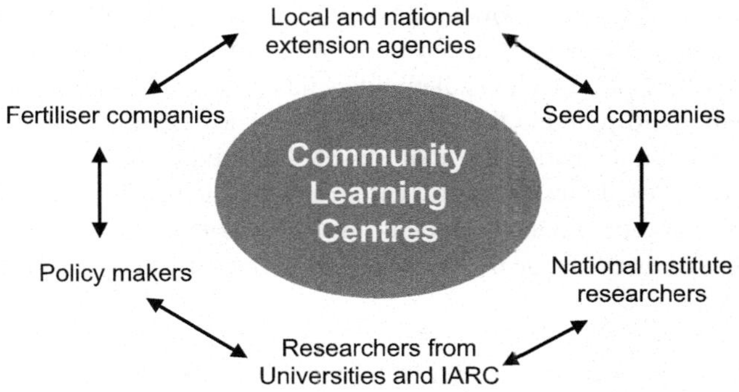

Fig. 8.7 Stakeholders that participate in community learning centres facilitated by the Soil Fertility Consortium for Southern Africa (SOFECSA). The community learning centre, constituted by local farmers, is the nucleus in an environment where information relay is non-linear. (From: Chikowo et al. (2011). IARC International Agricultural Research Centres)

multi-stakeholder platforms to facilitate co-learning and action research is the community learning centre, a concept developed and supported by the Soil Fertility Consortium for Southern Africa (SOFECSA) in several villages in Zimbabwe, Mozambique, Malawi and Zambia (Fig. 8.7).

These platforms are used to design, test, learn and disseminate technologies and practices for integrated soil fertility management (Chikowo et al. 2011). The level of engagement of different stakeholders in learning centres is not the same. Participation of policy makers, for example, is often much more sporadic than that of researchers from national institutes. In learning centres, farmers and other stakeholders go through full cycles of design and evaluation of management practices, starting from the collective identification of problems and objectives, the exploration of potential solutions and their implementation, either on farmers' individual land or on communal demonstration fields where researchers become involved in experimentation and management practices alongside the farmer.

The selection of stakeholders, for the learning centres and in general for any other multi-stakeholder process is an important step. Stakeholders can be classified according to (i) how relevant or important they are relative to the design process, and (ii) how influential they may be in shaping the process or imposing their views on others. If we take these two criteria, we can identify stakeholders that are:

- Important and influential
- Important and non-influential
- Non-important and influential
- Non-important and non-influential

The first two categories of stakeholders must be part of the design process, in one way of another. Participation of important and influential stakeholders is essential. Their participation can attract the participation of other stakeholders, as it often attaches relevance, commitment, legitimacy or 'seriousness' to the process. They are often also essential in the follow-up phases, in making things happen, in taking the first steps or in stimulating long-term continuity of design plans. Their degree of influence on other participating stakeholders has to be carefully managed; there are specific methods to regulate their influence down to operational levels. A typical example of important and influential stakeholders in the case of agroecosystem design is(are) the wealthiest farmer(s) within a community. They are often respected, and either admired or resented, by other members of the community. They may or may not be liked by everyone but their opinion is certainly heard and their behaviour tends to be imitated.

Participation by those that are important but non-influential, the second category of stakeholders, is most crucial. Often the actions that motivate the redesign process have these stakeholders as the key beneficiaries, and they may represent a majority within a rural community (e.g. the poorer households) or a minority (e.g. immigrant families, female-headed households, etc.). Their voices are not often heard and, due to local power relations or institutions, they are less able to exert influence on others. There are also participatory techniques designed to increase their influence and encourage their opinions during a design process. Sometimes these stakeholders are also less free to make management choices. Not just because of their limited resource endowment, but also because their decisions may have undesirable consequences for the *status quo* within the community, potentially affecting those that are more powerful.

A classical example of this can be seen in regions where communal grazing still persists. During the dry season, when natural grasslands are unproductive, the herd of the entire community is allowed to graze on standing crop residue biomass, even in the fields of poor farmers who do not own livestock (e.g. Rufino et al. 2011; Zingore et al. 2011; Diarisso et al. 2015; Dossouhoui et al. 2023). This contributes to organic matter and nutrient depletion in the fields of poorer smallholder farmers, and to their concentration in the fields of wealthier farmers who own the livestock and can use the manure to fertilise their fields. These practices are regulated by traditional by-laws and leave little room for change. If poorer farmers who do not own livestock decide, for example, to collect the crop residues to make compost to fertilise their own fields they will be met with severe opposition. Such behaviour would be deemed anti-social, with consequences for their position within the community.[5]

---

[5]For example, in times of land preparation at the beginning of the rainy season, non-livestock farmers may still have access to mechanized soil tillage through oxen-drawn plows facilitated by their wealthier neighbors, only when they entertain a good relationship with them.

8.8 Co-innovation and Model-Aided Design

As mentioned earlier, co-innovation is a process of gradual reflexion, learning and action by all the actors engaged in an innovation process. Learning cycles in collective, multi-stakeholder design process have been often equated to the concept of adult learning cycles and learning styles described in psychology by Kolb (1984). Kolb's cycle represents a learning process that starts from experiencing, or having a concrete experience, a feeling, then on to reflective observation through watching, reviewing and reflecting on the experience (Fig. 8.8). The cycle continues with an abstract conceptualisation, concluding, learning from the experience, and ends with the planning and trying out of what has been learnt through active experimentation. Kolb depicts this cycle using a four-quadrant diagram defined by two axes representing the processing continuum (from experimentation to observation) and the perception continuum (from concrete experiences to abstract concepts). On the basis of these, four learning styles are defined: diverging (feel and watch), assimilating (think and watch), converging (think and do) and accommodating (feel and do).

The concept of co-innovation is used in several disciplines. For example, it has been used to describe co-creation in the software development sectors, through collaboration between firms and individuals (Lee et al. 2012), where co-innovation is defined as the result of collective intelligence and consists of three pillars: (1) convergence of ideas; (2) collaborative arrangement and (3) co-creation of experiences. In agroecosystems design, Dogliotti et al. (2014a) adapted the concept of learning cycles and co-innovation to develop and implement a process of model-

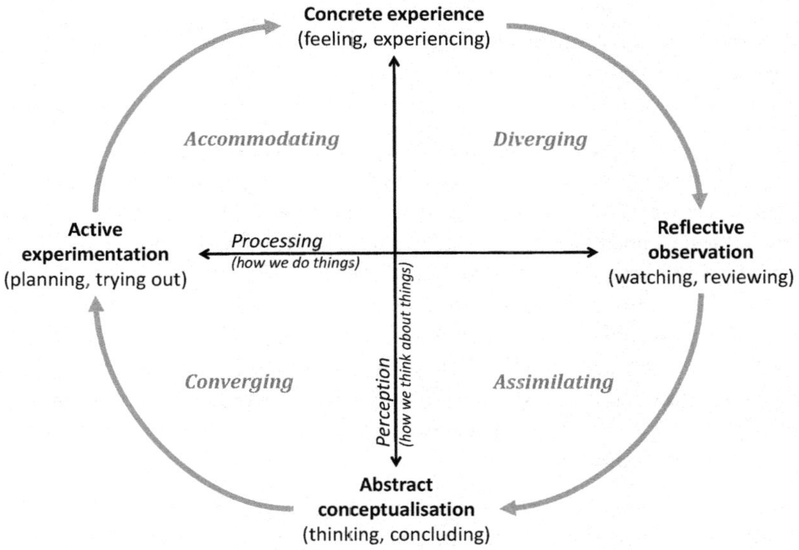

Fig. 8.8 Kolb's learning cycle and learning styles

aided redesign of farm management systems involving farmers and other stakeholders.

The adapted version Kolb's cycle to underpin agricultural systems innovations describes four steps (Rossing et al. 2021): Action, or implementing a 'bright' idea; Observation, or finding out about consequences; Analysis, or what are the implications? and Planning: which improvements are needed, how to make them happen? These steps in the learning cycle can be approximately assimilated with those of the DEED cycle referred to earlier: describe, explain, explore and design (Giller et al. 2010). Admittedly, all the work done under the DEED approach, of which I participated actively, has never reached the step of design. Similarly, Lefevre et al. (2014) describes the steps of reflexion, exchange, decision and action when reporting on their work on the co-design of soil improving cropping systems for organic farming in France. Looking back at the diagram that describes the process of prototyping in Fig. 8.5 it is possible to find multiple points of convergence with these various adaptations of Kolb's learning cycles.

An interesting aspect of the co-innovation approach proposed by Dogliotti et al. (2014b) is the use of linked field and farm scale simulation models to inform the design of farm planning and management strategies through the ex-ante evaluation of productive, economic and environmental indicators. For example, working with family vegetable farmers in southern Uruguay, this group of researchers explored, developed and implemented together with farmers alternative farm management plans to increase income and reduce soil erosion. The use of simulation models was essential, as planning of highly diversified and fast-rotating vegetable farms is normally a challenging task; let alone to try to explore the consequences of future scenarios. This is why the co-innovation approach is often described as resting on three pillars:

1. A complex systems approach;
2. A setting that enables social learning; and
3. A method to enable dynamic project monitoring and adaptive management.

Figure 8.9 shows the results obtained on two contrasting case study vegetable farms in Uruguay. The initial situation (i) of each farm in terms of average income and soil erosion levels was assessed at the beginning of the project, in a diagnosis phase that involved farmers and researchers. The situation after redesign (ii) was assessed at the end of the three-year period that the project lasted. The potential for improvement (iii) in terms of both indicators corresponds to explorations made with simulation models. It is clear that while the improvement brought about through redesign during the project was greater for farm B than A, in terms of reducing soil erosion and increasing family income. However, the potential for further improvement on these indicators with the management practices evaluated was greater for farm A than B, according to simulation modelling results.

This has been a successful project in Uruguay, from where the Ministry of Agriculture has drawn inspiration to design a new agricultural extension system. Applications of this approach to family livestock production systems in the hilly

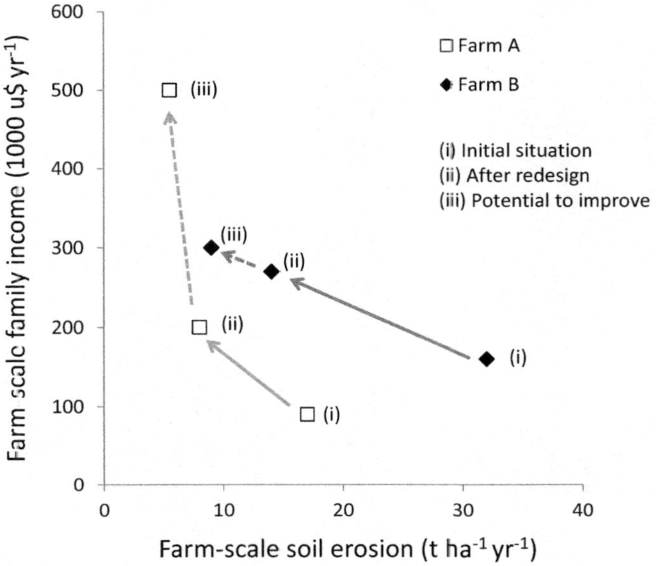

Fig. 8.9 Farm scale family income versus soil erosion on two small vegetable farms in southern Uruguay, modified from Dogliotti et al. (2014b)

region of eastern of Uruguay are yielding similarly promising results in terms of improving their economic viability, productivity and environmental sustainability (Ruggia et al. 2021). Another key element of this approach to co-innovation is dynamic project monitoring. During the implementation phase several regular meetings are celebrated with the different stakeholders to monitor and reflect upon the project progress. These meetings are known as PIPA (Participatory Impact Pathway Analysis) workshops (Douthwaite et al. 2003). Corrections or adjustments are discussed and implemented when problems or lack of effectiveness are detected. Additionally, a consultative group is appointed to observe and more critically evaluate and reflect on the entire process, and to feedback to the implementation teams once or twice a year as planned.

Other approaches that use models to aid decision-making and prototyping are often loosely termed 'participatory modelling'. There is no single set of steps or consolidated methodology and authors tend to report cases in which farmers participate from the onset of model building, such as the rule-based model GAMEDE developed by Vayssières et al. (2009) to analyse nutrient flows and balances at farm scale in La Réunion Island. In other cases, farmers participate in scenario analysis from where parameterisation sets are derived and used in simulation models. Model outcomes are then shared and discussed with farmers (e.g. Thornton and Herrero 2001). Other cases described in the context of smallholder agriculture include model-aided participatory prototyping. Farmers meet in groups to discuss and design what they considered to be the 'ideal' farm for their context (e.g., Waithaka et al.

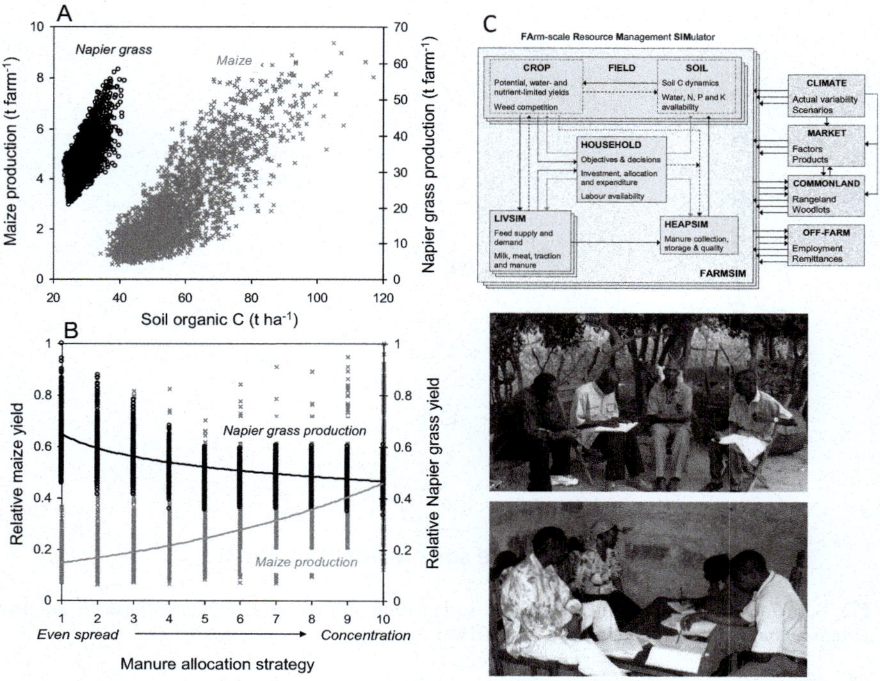

Fig. 8.10 An illustration of a model-based exploration of the biophysical feasibility of an ´ideal´ farm designed by farmers through participatory prototyping. Alternative scenarios of manure allocation to maize and Napier grass under variable climatic and market situations were simulated (**a** and **b**) using the model FARMSIM (Tittonell et al. 2009) represented in the diagram. (**c**) Pictures showing farmers working in small groups to design the ideal farm

2006). Then farm scale models are developed for a few representative case study farms in the community to explore strategies and pathways for the transition from the current to the 'ideal' farm situation, only when such ideal prototypes were actually feasible (Tittonell et al. 2009). Such exercises are very informative and an opportunity to learn for both researchers and farmers (Fig. 8.10).

At higher integration levels, such as at regional levels, model aided co-learning involving different stakeholders can also be very informative. Delmotte et al. (2017), working on land use and management options for the Camargue region of southern France, used a multi-scale agent-based system model connecting several computers on line and inviting local farmers to 'play' with a farm model built on real data from representative farms in the region. The Camargue is located in the Rhone delta and is an area of intensive rice cultivation, as a well as a classed *Ramsar* site and a nature reserve that host important diversity of migratory birds and other forms of biodiversity. Water pollution with excess nutrients and pesticides used in rice threatens the ecological integrity of these wetlands. The goal of the farm redesign exercise that farmers were asked to play was to reduce environmental pollution by, for example, converting to organic farming.

Each farmer 'played' on a single farm model consisting of a number of fields where he/she was able to simulate crop rotations with different management practices (irrigation, fertilisation, pesticide application, sowing dates, etc.). Simulations were run for periods of 10 years, to emulate a gradual transition phase. A central computer in the network conveyed the information from each individual farm model operated by the participating farmers. All this information was integrated and used in a regional scale model, which accounted for the relative distribution of farm types, soil types and irrigation districts within the region. The regional scale model outputs consisted of regional indicators such as pollution level in the major waterways and lagoons of the region, regional productivity and economic value of production, regional employment generation, etc. The participating farmers (which could be real farmers of other type of stakeholders) were able to see in real time the consequences of their management decisions had at the scale of the farm, the sub-regions and the entire region.

These examples illustrate the potential of using simulation models in combination with participatory research methodologies to guide co-innovation, co-learning and co-design. This combination is seldom used in agroecology and indeed one my goals when writing this book was to cast light on the possibilities this approach offers to inform agroecosystem design. Agroecologists can make good use of systems analysis tools such as simulation models, to better aid farmers and communities at exploring design options from multiple perspectives while integrating their knowledge, experience and aspirations. Agroecological farmers are also keen to know whether a new design they are about to implement will be economically viable, environmentally feasible, socially acceptable, and so on. The following chapters offer more examples of the use of models in agroecosystem design and evaluation.

8.9 Monitoring, Evaluation and Learning

Monitoring and evaluation (M&E) is an integral component of any project, program or policy, as it allows following up on the achievement of planned activities, expected results and intended impacts. As such, M&E can be a component of an agroecosystem design project, especially when agroecosystem are designed at the scale of a territory or a region. Monitoring, evaluation and learning (MEL) is a cyclical process through which individuals and organisations not only evaluate the impact of their actions (often organised around projects or programmes), but also assess processes, capacities and intentions through reflexive learning in order to propose adjustments. While traditional M&E primarily aims to generate data for accountability purposes or external evaluations, MEL takes a more proactive approach by engaging stakeholders (decision-makers) who are driven to enhance programs, integrating learning into program implementation. MEL ensures that knowledge gained through monitoring and evaluation activities is fed-back, processed and applied to enhance program delivery, leading to improved outcomes.

This approach emphasizes the importance of program adaptability, to be able to adjust assumptions and expectations through recurrent MEL iterations.

M&E systems, and now also MEL, rely on the same theoretical framework: the theory of change. A theory of change is a comprehensive description of why and how a desirable change is to come about within a particular context (Rogers 2014). It consists of explicating how the activities proposed through a plan or program are expected to lead to the achievement of the desired impact of the program. To do this, the theory of change is built on a chain of causalities that encompasses project or program inputs, activities, outputs, outcomes and impacts (Fig. 8.11). Inputs (material, intellectual, consumable, etc.) are used in a series of activities to deliver outputs, or immediate results. These outputs should ideally lead to outcomes, which are changes in the behaviour of relevant stakeholders (project beneficiaries, institutions, organisations) that may eventually lead to the expected impacts. Such impacts contribute to fulfilling the vision and mission of the project or program, or the organisations that implement them.

However, impacts are not within the realm of the intervention, be it a project, a program, or a policy. This is why in a theory of change we can distinguish a 'zone of control', which comprises the project inputs, activities and results. These are under the direct control of project managers and participants. The outcomes are within the 'zone of influence' of the action, which implies that the project or program will do the necessary to influence positive change in the behaviour of the relevant stakeholders. Such changes in the behaviour of stakeholders (e.g. adoption of practices,

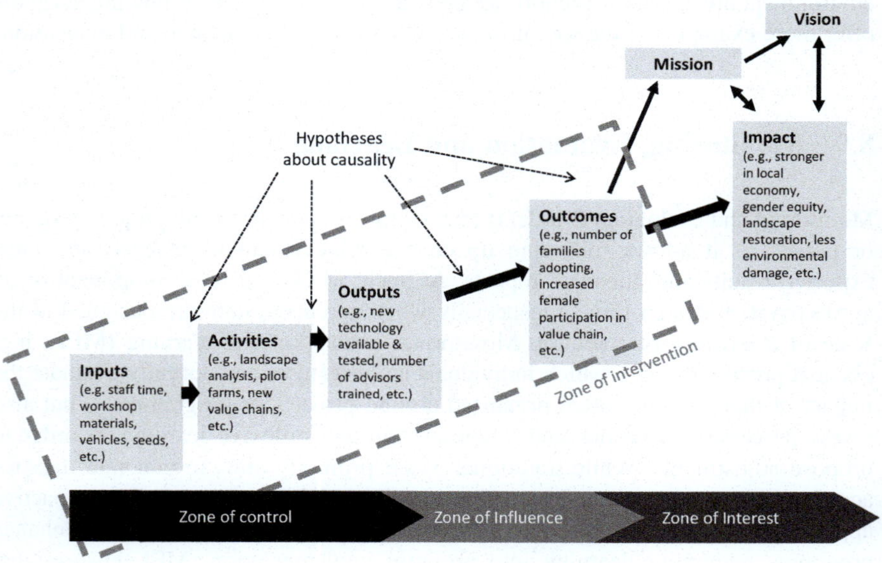

Fig. 8.11 A schematic representation of a theory of change depicting the logical steps and their interrelations (causalities) within the realm of project intervention, and the expected impact to fulfil the vision and mission of the project (in this example, a regenerative and circular farming project)

change of habits, etc.) will eventually lead to impacts, which are within the 'zone of interest' of the action. The entire chain of events, from inputs to impact, relies on a number of hypotheses about the causality at play between subsequent steps. The weakest hypotheses about causality are those linking outputs to outcomes, and outcomes to impact.

Let us use a practical example to illustrate this concept. Silo bags are long and narrow plastic bags used for storing grain after harvest for a variable period before selling. Although they were initially invented in Germany, the widespread adoption of this technology occurred in Argentina during the 1990s, after which it was exported to other grain-producing countries. Silo bags offer a convenient storage solution for farmers who lack on-farm or nearby storage facilities, eliminating the need to pay for storage services. The Instituto Nacional de Tecnología Agropecuaria (INTA) has extensively worked on further designing and adapting the silo bag technology, including the development of new plastic materials, grain feeders, drying systems, and more, to suit the local context. Using the theory of change framework, we can analyze the different stages of this example. The research and development efforts invested in designing the hardware components, such as machinery, plastic bags, and ventilation systems, can be considered inputs and activities (Fig. 8.11). The technology itself, the silo bag system, along with the knowledge surrounding its use and adaptation, represents the output. This stage is within the zone of control of INTA's research and development team. The outcome, which signifies a change in the behavior of relevant stakeholders, is reflected in the widespread adoption of this technology by large-scale farmers in Argentina and the increasing number of agricultural mechanization companies producing and selling it. Finally, the impact of this action can be seen in the additional million tons of grain that can be stored at the national level, its effect on commodity prices, and the cost reduction it provides for farmers, among other factors.

Explicitly or not, most research and development projects are designed using the theory of change. M&E fits well the logic of the theory of change, especially when it concerns the zone of control of a program, in the form of log frames, indicators and the like. MEL, on the other hand, fits less comfortably within the linear, unidirectional logic of the theory of change, and calls for adaptive project management through reflexive learning and permanent feedback between stakeholders and the teams in charge of implementing the MEL system. In transformative change processes such as agroecological transitions (cf. Chap. 10), a rigid theory of change and log frame can prevent reflexive MEL, which requires multiple iterations, readjustments and room to accommodate non-linear trajectories and emergent dynamics. Unfortunately, most donors and governmental organisations are not ready to abandon the rigid and control-oriented theory of change, preventing adaptive cycles and co-innovation.[6]

[6]For example, in a co-innovation project on climate change adaptation for rural household in Patagonia, we proposed to identify, implement and evaluate management solutions together with communities through reflexive learning and a dialogue of knowledge systems. The project funder,

8.10 Summary and Concluding Remarks

Agroecosystem design necessitates an evaluation method to assess whether the proposed new structures and functions result in the desired outcomes and contribute to the overall purpose and goals of the design exercise. In essence, it is practically impossible to design or redesign an agroecosystem without an evaluation framework comprising indicators and reference values. Farmers, in their decision-making process regarding design, intuitively employ a range of indicators to evaluate the impact of new structures and functions on various aspects of their agroecosystem such as production costs, labour requirements, productivity, aesthetics, cultural acceptance, personal fulfilment, and more. Farmers typically utilize a form of multi-criteria assessment, considering several indicators simultaneously. In systems analysis, an indicator is a variable that incorporates reference values, which can be thresholds, targets, baselines, or benchmarks. Indicators must possess certain properties to be useful, such as objectivity, consistency, measurability, affordability, predictability, reliability, validity or informativeness, and sensitivity. Several indicator systems have been proposed to assess agroecosystem sustainability, and less often so, to aid agroecosystem redesign. Most of them were introduced in the late 1990s and early 2000s, but they are still valid today. Numerous indicator systems have been proposed for assessing agroecosystem sustainability, and to a lesser extent, aiding agroecosystem redesign. These frameworks were significant breakthroughs as they successfully integrated multiple disciplines. In agroecology, evaluation frameworks that rely on generic system properties to derive measurable indicators are particularly relevant, such as the MESMIS framework. While the MESMIS framework does not explicitly mention design as an objective, it implicitly provides the necessary tools to support agroecosystem redesign. When selecting indicator frameworks, it is important to consider their adaptability for use in participatory design and evaluation settings, such as the co-innovation approach, which is highly favoured in agroecology. Simulation models can assist in design processes by allowing for ex-ante assessment of potential future scenarios. They can be combined with participatory prototyping to inform design decisions quantitatively. The several steps in the design and evaluation of agroecosystems can be described as a cyclical sequence that yields the acronym IDEAS:

---

however, wanted to know exactly how much money will these solutions cost before the project even started. They expected to see these activities already described in the log frame and costed in the budget. It was impossible to make them understand that in a co-innovation process we do not know these details a priori, because solutions emerge from the interaction with the community. Nevertheless, we had to guestimate the necessary budget to implement activities that we were not sure we would eventually implement. Such lack of flexibility prevents creative solutions to emerge trough co-innovation.

- *I*dentify problems and objectives
- *D*iagnose the current situation
- *E*xplore alternatives
- *A*ssess feasibility and impacts
- *S*elect and test best options

The selected options (*S*) are then implemented, evaluated and monitored, and eventually a new IDEAS cycle may start when readjustments are necessary. A combination of participatory methods and simulation modelling can be used in all these steps, as will be illustrated in the next two chapters.

References

Astier M, García-Barrios L, Galván Y, González-Esquivel C, Masera O (2012) Assessing the sustainability of small farmer natural resource management systems. A critical analysis of the MESMIS Program (1995–2010). Ecol Soc:17. https://doi.org/10.5751/ES-04910-170325

Bohanec M, Zupan B (2004) A function-decomposition method for development of hierarchical multi-attribute decision models. Decis Support Syst 36(3):215–233. https://doi.org/10.1016/S0167-9236(02)00148-3

Bos AP, Groot Koerkamp PWG (2008) Keeping of laying hens: development and co-design of new husbandry concepts for laying hens for sustainable egg production. Presentation at 8th European IFSA symposium (2008-07-09), Clermond-Ferrand, France

Conway GR (1987) The properties of agroecosystems. Agric Syst 24:95–117. https://doi.org/10.1016/0308-521X(87)90056-4

Daly HE (1973) Toward a steady-state economy. W. H. Freeman

Daly H, Costanza R (1992) Natural capital and sustainable capital. Conserv Biol 6:37–46

Delmotte S, Couderc V, Mouret J-C, Lopez-Ridaura S, Barbier J-M, Hossard L (2017) From stakeholders narratives to modelling plausible future agricultural systems. Integrated assessment of scenarios for Camargue, Southern France. Eur J Agron 82:292–307. https://doi.org/10.1016/j.eja.2016.09.009

Dent D, Ridgeway RB (1986) A land use planning handbook for Sri Lanka. FD 2, SRL79/058. Land use policy planning division, Ministry of Lands and Land. Development, Colombo

Diarisso T, Corbeels M, Andrieu N, Djamen P, Tittonell P (2015) Biomass transfers and nutrient budgets of the agro-pastoral systems in a village territory in South-Western Burkina Faso. Nutr Cycl Agroecosyst 101:295–315

Dogliotti S, Rodríguez D, López-Ridaura S, Tittonell PA, Rossing WAH (2014a) Designing sustainable agricultural production systems for a changing world: methods and applications. Agric Syst 126:1–2

Dogliotti S, García MC, Peluffo S, Dieste JP, Pedemonte AJ, Bacigalupe GF, Scarlato M, Alliaume F, Alvarez J, Chiappe M, Rossing WAH (2014b) Coinnovation of family farm systems: a systems approach to sustainable agriculture. Agric Syst 126:76–86. https://doi.org/10.1016/j.agsy.2013.02.009

Dossouhoui GI, Anita Yemadje PL, Cao DRV, Balarabe O, Tittonell P (2023) "Sedentarisation" of transhumant pastoralists results in privatization of resources and soil fertility decline in West Africa's cotton belt. Front Sustain Food Syst 7:1120315. https://doi.org/10.3389/fsufs.2023.1120315

Douthwaite B, Hoffecker E (2017) Towards a complexity-aware theory of change for participatory research programs working within agricultural innovation systems. Agric Syst 155:88–102. https://doi.org/10.1016/j.agsy.2017.04.002

Douthwaite B et al (2003) Impact pathway evaluation: an approach for achieving and attributing impact in complex systems. Agric Syst 78:243–265

Funes-Monzote FR (2008) Farming like we're here to stay: 1176 The mixed farming alternative for Cuba. PhD thesis, 1177 Wageningen University, Wageningen, The Netherlands

Gallopin GC (1997) Indicators and their use: information for decision-making. In: Moldan B, Billharz S (eds) Sustainability indicators. A scope, vol 58. Island Press, pp 13–27

Giller KE, Tittonell P, Rufino MC, van Wijk MT, Zingore S, Mapfumo P, Adjei-Nsiah S, Herrero M, Chikowo R, Corbeels M, Rowe EC, Baijukya F, Mwijage A, Smith J, Yeboah E, van de Burg WJ, Sanogo OM, Misiko M, de Ridder N, Karanja S, Kaizzi C, K'ungu J, Mwale M, Nwaga D, Pacini C, Vanlauwe B (2010) Communicating complexity: integrated assessment of trade-offs within African farming systems to support innovation and development. Agric Syst 104:191–203

Goewie EA (1997) Designing methodologies for prototyping ecological production systems. Course reader MSc. Ecological Agriculture (F800-204). Department of Ecological Agriculture, Wageningen Agricultural University, Wageningen

Hashimoto T, Stedinger J, Loucks P (1982) Reliability, resiliency, and vulnerability criteria for water resource system performance evaluation. Water Resour Res 18:14–20. https://doi.org/10.1029/WR018i001p00014

Hauswirth D (2014) Evaluation agro-économique ex-ante de systèmes de culture en agriculture familialeLe cas de l'agriculture de conservation en zone tropicale humide de montagne (Nord Vietnam), PhD Thesis, Montpellier SupAgro, France, 150p

Kolb DA (1984) Experiential learning: experience as the source of learning and development, vol 1. Prentice-Hall, Englewood Cliffs, NJ

Laborda L, Easdale M, Fallot A, Ocariz P, Tittonell P (2023) Rise from the ashes! Resilience patterns in Patagonia pastoralist communities. Sustain Dev. https://doi.org/10.1002/sd.2679

Langeveld H, Verhagen J, Neeteson J, Keulen H, Conijn J, Schils R, Oenema J (2007) Evaluating farm performance using agri-environmental indicators: recent experiences for nitrogen management in The Netherlands. J Environ Manag 82:363–376. https://doi.org/10.1016/j.jenvman.2005.11.021

Le Gal PY, Merot A, Moulin CH, Navarrete M, Wery J (2010) A modelling framework to support farmers in designing agricultural production systems. Environ Model Softw 25(2):258e268

Lee SM, Olson DL, Trimi S (2012) Co-innovation: convergenomics, collaboration, and co-creation for organizational values. Manag Decis 50(5):817–831

Lefevre V, Capitaine M, Peigne J, Roger-Estrade J (2014) Farmers and agronomists design new biological agricultural practices for organic cropping systems in France. Agron Sustain Dev 34(3):623–632. https://doi.org/10.1007/s13593-013-0177-2

Loomis TM (2000) Indigeneous populations and sustainable development: building on indigeneous approaches to holistic, self-determined development. World Dev 28(5):893–910

López-Ridaura S, Masera O, Astier M (2002) Evaluating the sustainability of complex socio-environmental systems. The MESMIS framework. Ecol Indic 2:135–148

López-Ridaura S, van Keulen H, van Ittersum MK, Leffelaar PA (2005) Multi-scale sustainability evaluation of natural resource management systems: quantifying indicators for different scales of analysis and their trade-offs using linear programming. Int J Sust Dev World 12:81–97

Masera O, Astier M, López-Ridaura S (1999) Marco para la evaluación de sistemas de manejo de recursos naturales incorporando indicadores de sustentabilidad MESMIS. Mundiprensa – GIRA – UNAM, México

Meynard JM, Aggeri F, Coulon JB, Habib R, Tillon JP (2006) Recherches sur la conception de systèmes agricoles innovants. Rapport à la direction de l'INRA, INRA, 55 p

Meynard JM, Dedieu B, Bos B (2012) Re-design and co-design of farming systems: an overview of methods and practices. In: Darnhofer I et al (eds) Farming systems research into the 21st century: the new dynamic. Springer, pp 407–431

Moldan HT, Dahl A (2008) Sustainability indicators: a scientific assessment. Bibliovault OAI Repository, the University of Chicago Press

Pacini C (2003) An environmental-economic framework to support multi-objective policy-making. A farming systems approach implemented for Tuscany. PhD thesis Wageningen University, Wageningen, The Netherlands, 173 p

Rogers P (2014) Theory of change, methodological briefs—impact evaluation no. 2, UNICEF Office of Research, Florence

Romero C, Rehman T (2003) Multiple criteria analysis for agricultural decisions, Developments in agricultural economics 11, 2nd edn. Elsevier Science B.V., Amsterdam

Rossing WAH, Albicette MM, Aguerre V, Leoni C, Ruggia A, Dogliotti S (2021) Crafting actionable knowledge on ecological intensification: lessons from co-innovation approaches in Uruguay and Europe. Agric Syst 190:1–16. https://doi.org/10.1016/j.agsy.2021.103103

Rufino MC, Dury J, Tittonell P, van Wijk MT, Herrero M, Zingore S, Mapfumo P, Giller KE (2011) Collective management of feed resources at village scale and the productivity of different farm types in a smallholder community of north East Zimbabwe. Agric Syst 104:175–190

Ruggia A, Dogliotti S, Aguerre V, Albicette MM, Albin A, Blumetto O, Cardozo G, Leoni C, Quintans G, Scarlato S, Tittonell P, Rossing WAH (2021) The application of ecologically intensive principles to the systemic redesign of livestock farms on native grasslands: a case of co-innovation in Rocha, Uruguay. Agric Syst 191:103148. https://doi.org/10.1016/j.agsy.2021.103148

Ryschawy J, Grillot M, Charmeau A, Pelletier A, Moraine M, Martin G (2022) A participatory approach based on the serious game Dynamix to co-design scenarios of crop-livestock integration among farms. Agric Syst 201:103414. https://doi.org/10.1016/j.agsy.2022.103414

Salembier C, Segrestin B, Berthet E, Weil B, Meynard J-M (2018) Genealogy of design reasoning in agronomy: lessons for supporting the design of agricultural systems. Agric Syst 164:277–290. https://doi.org/10.1016/j.agsy.2018.05.005

Shields DJ, Šolar SV, Martin WE (2002) The role of values and objectives in communicating indicators of sustainability. Ecol Indic 2(2002):149–160

Steffen W et al (2015) Planetary boundaries: guiding human development on a changing planet. Science 347:1259855. https://doi.org/10.1126/science.1259855

Sterk B, Ittersum MK, Leeuwis C, Wijnands FG (2007) Prototyping and farm system modelling—partners on the road towards more sustainable farm systems? Eur J Agron 26:401–409. https://doi.org/10.1016/j.eja.2006.12.006

Stern D (1997) The capital theory approach to sustainability: a critical appraisal. J Econ Issues 31(1):145–173

Thornton PK, Herrero M (2001) Integrated crop-livestock simulation models for scenario analysis and impact assessment. Agric Syst 70:581–602

Tittonell P (2014) Ecological intensification – sustainable by nature. Curr Opin Environ Sustain 8:53–61

Tittonell P, van Wijk MT, Herrero M, Rufino MC, de Ridder N, Giller KE (2009) Beyond resource constraints – exploring the physical feasibility of options for the intensification of smallholder crop-livestock systems in Vihiga district, Kenya. Agric Syst 101:1–19

Van Mansvelt JD, van der Lubbe MJ (1999a) Checklist for sustainable landscape management. Elsevier Science B.V, Amsterdam

van Mansvelt JD, van der Lubbe MJ (1999b) Checklist for sustainable landscape management: final report of the EU concerted action AIR3-CT93-1210: the landscape and nature production capacity of organic/sustainable types of agriculture. Elsevier Science Inc., 202 pp., hard cover

Van Noordwijk M, van Oel P, Muthuri C, Satnarain U, Sari R, Rosero P, Githinji M, Tanika L, Best L, Assogba G, Kimbowa G, Andreotti F, Lagneaux E, Wamucii CN, Hakim AL, Miccolis A, Abdurrahim A, Farida A, Speelman E, Hofstede GJ (2022) Mimicking nature to reduce agricultural impact on water cycles: a set of mimetrics. Outlook Agric 51:1–15. https://doi.org/10.1177/00307270211073813

Vayssières J, Guerrin F, Paillat J-M, Lecomte P (2009) GAMEDE: a global activity model for evaluating the sustainability of dairy enterprises part I – whole-farm dynamic model. Agric Syst 101:128–138. https://doi.org/10.1016/j.agsy.2009.05.001

Vereijken PH (1987) Van geintegreerde bestrijding naar geïntegreerde landbouw. Landbouwkundig Tijdschrift 99(5):19–21

Viglizzo E (2003) Manual AGRO-ECO-INDEX. Programa Nacional de Gestión Ambiental Agropecuaria-Proyecto de Eco-Certificación, Instituto Nacional de Tecnología Agropecuaria, Buenos Aires, p 2003

Vosin A (1993) Grass productivity. Island Press, Washington, D.C., 353 pages

Waithaka MM, Thornton PK, Herrero M, Shepherd KD (2006) Bio-economic evaluation of farmers' perceptions of viable farms in western Kenya. Agric Syst 90:243–271

Zingore S, Tittonell P, Corbeels M, Wijk MT, Giller KE (2011) Managing soil fertility diversity to enhance resource use efficiencies in smallholder farming systems: a case from Murewa District. Zimbabwe Nut Cycl Agroecosyst 90:87–103

Chikowo R, Mtambanengwe F, Mapfumo P (2011) Going through paces: empowering communities to use improved crop production practices to adapt to climate change and variability through field-based Learning Centres in Zimbabwe. A SOFECSA technical annual report for the sub-Saharan Africa Challenge Program (SSA-CP) prepared for the Forum for Agricultural Research in Africa (FARA). Soil Fertility Consortium for Southern Africa (SOFECSA), CIMMYT-Zimbabwe, Harare

Chapter 9
Trade-Offs Around Production and Livelihood Decisions

Abstract Trade-offs are inherent to decision-making about resource allocation to different activities in agroecosystems, particularly when resources are scarce or such activities are mutually exclusive. Making design and management decisions implies finding compromise solutions between objectives, or strive for synergies. Mapping out trade-offs and possible synergies is essential to inform such decisions. This chapter presents concepts and methods to analyse trade-offs and synergies between resource allocation objectives in agroecosystems. It focuses on decision making at the scales of (i) cropping systems and animal herds, (ii) farms, and (iii) landscapes/territories. The methods presented range from simple ad-hoc calculations, through participatory and qualitative methods, to quantitative trade-offs assessment using simulation models and scenario analysis. Common trade-offs in agriculture are illustrated with real life examples.

9.1 What Are Trade-Offs?

Trade-offs emerge when resources are limited, and when two or more competing or conflicting objectives must be met simultaneously. Trade-offs in resource allocation can arise at any scale or level of integration, from decisions about crop or animal husbandry to decisions on farm management, collective decisions by a community with regards to resource management in a territory, or policy decisions at regional or country level (Tittonell 2013). Farm-scale trade-offs, in particular, are central to agroecosystems analysis and design because individual decisions by farm managers may have implications for decisions and processes operating at both higher and lower scales. Decision-making at farm scale entails a number of choices that farmers need to decide upon, such as allocating land, labour or capital to competing land uses or farm activities. Farm-scale trade-offs have received much attention in agricultural systems analysis, they have been widely documented and several methods and tools have been developed for their assessment (e.g., Erenstein et al. 2015, Kanter et al. 2018, and references therein).

An example of farm-scale trade-offs typical for smallholder agriculture is the allocation of resources like crop harvest residues, often a scarce resource in these

P. Tittonell, *A Systems Approach to Agroecology*,
https://doi.org/10.1007/978-3-031-42939-2_9

systems, to different possible uses within the farm. Crop residues may be used to feed livestock, keep a mulch layer to cover the soil, make compost, use them as fuel or construction material (thatching, fencing), or sell them on the market as fodder, etc. Trade-offs around the use of crop residues have been largely documented in many different smallholder systems (e.g., Castellanos-Navarrete et al. 2015; Baudron et al. 2015; Tittonell et al. 2015), and several of these examples indicate that trade-offs at one scale may result in synergies at other scales (e.g., Andrieu et al. 2015). In several smallholder agroecosystems crop residues remain in the field during the dry season, traditionally, and the community's livestock feeds on them during this critical period (e.g., Diarisso et al. 2015). This system, known for example in France during the middle ages as the '*Vaine pâture*', is slowly disappearing nowadays as human and consequently livestock population densities grow in rural areas, intensifying competition for land and resources, and as farmers realise the importance of keeping their crop residues in their own fields to mulch or maintain their soils fertile.

Meeting a certain objective at farm scale may in some cases conflict with other objectives that are relevant at immediately lower or higher scales. For example, the objective of reducing labour needs at farm scale may conflict with the objective of generating rural employment to secure incomes in a region. In such cases, the decision-maker concerned with the objectives at stake will be a different one at each of these spatial scales, the farm or the region. The higher the scale or level of integration, the more complex decisions become, and the larger the number of stakeholders involved. Trade-offs may also emerge between objectives that are relevant at varying temporal scales when, for example, farmers need to trade off current high yields or attractive incomes in detriment of the longer-term sustainability of their natural resource base. There are cases in which trade-offs may not be there at present but they may emerge under certain scenarios determined by the combination of several possible contextual factors (e.g., climate, demography, policies, markets, etc.). Finally, synergies can be seen as the opposite to trade-offs, in the sense that more than on objective can be fulfilled at the same time.

9.2 Trade-Offs Analysis

When making decisions, two or more objectives may be in *conflict* when they are mutually exclusive, or in *competition* when the fulfilment of one of them results in a partial detriment for the other. In analysing trade-offs between two objectives, such an inverse relationship can be described in the simplest of cases as being proportional and constant. This is normally the case in agroecosystems when resources are finite and when no synergies between objectives are to be expected. In other words, when the fulfilment of one objective to the maximum level possible allowed by the availability of a given resource implies that the totality of that resource is used in the process, and therefore it is no longer available to fulfil another objective. For example, resources such as land, or labour or a given volume of irrigation water can be allocated in their totality to a certain activity, such vegetable production, in detriment of another activity, such as paddy rice cultivation. The substitution of one objective (activity, in this case) for another one can be described through a zero sum line (Fig. 9.1). The slope of the zero sum line indicates the elasticity of competition between the two objectives. This way of representing trade-offs implies that there is conflict between two partially mutually exclusive objectives. In some cases, trade-offs may not be represented as a continuum line or curve, as when two discrete activities or objectives are entirely mutually exclusive. For example, when using a given sum of money to buy either a tractor or a combine harvester; it is not possible to buy a *fraction* of any of them.

In most cases, however, trade-offs are more complex than in the cases described above, and the slope or elasticity is not necessarily constant over the entire range of possible levels of fulfilment of both objectives. Sometimes two or more objectives may compete with each other in a strongly substitutive regime, in such a way that fulfilling one objective makes it very hard to fulfil a second one. In such situations the absolute value of the elasticity of substitution is greater than one. In Fig. 9.1, increasing the level of fulfilment of objective A from A_0 to A_1' entails a reduction in objective B from B_0 to B_1' under the substitutive regime. Achieving a level of objective A as high as A_1'' would imply virtually resigning objective B, to a level as low as B_1''. There are cases however in which complementarities or compromise solutions are possible. If trade-offs are described by the complementarity regime drawn in Fig. 9.1, then achieving the level A_1'' of objective A can be traded off by a relatively small reduction in objective B from B_0 to B_1'. This is further illustrated

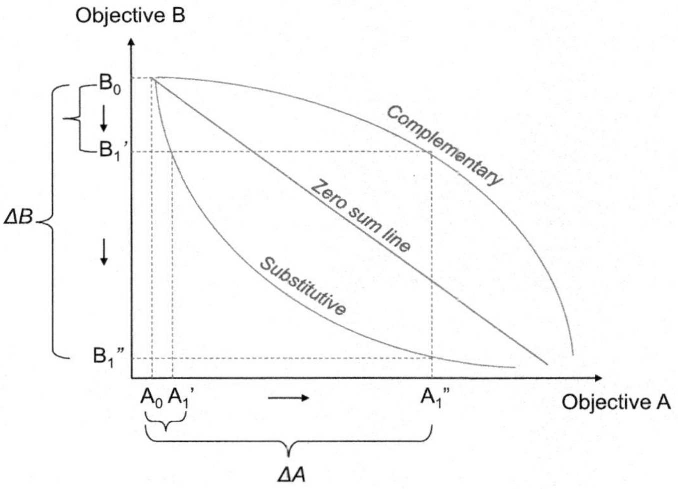

Fig. 9.1 Stylised representation of a trade-off map showing three possible trade-off regimes between two objectives

with the example of allocating crop residue biomass in smallholder agroecosystems in Box 9.1.

Box 9.1: Feeding the Cow or Feeding the Soil?*

In several smallholder agroecosystems farmers need to decide on the best use for their biomass resources, such as crop residues, for which there may be a local market as shown in Figure A. But keeping crop resides in the field is also important for the maintenance of soil fertility. These two competing objectives can be generically termed as 'utility of use as feed' (F) vs. 'utility of use as soil amendment' (S). In the simplified model represented in Figure B there is no other possible use for crop residues, so that the total biomass is partitioned between objectives F and S, without specifying units. The trade-off between these two competing objectives, which draw on mutually exclusive crop residue uses at a single point in time, may be best described by one of the three curves proposed in Fig. 9.1, termed here Regimes A, B and C. Regime A corresponds to a situation of strong competition between objectives F and S. Regime B corresponds to a situation of substitutability in which the rate of replacement or conversion from S to F or vice versa is inversely proportional. Regime C describes a situation in which complementarities are possible within a wide range of fulfilment of both F and S. Synergies between both objectives may also be possible when different time horizons are considered; for example, if soil amendments would allow for subsequent increases in feed productivity, then the effect of soil amendment on feed utility might be potentially positive in the longer term.

A

B

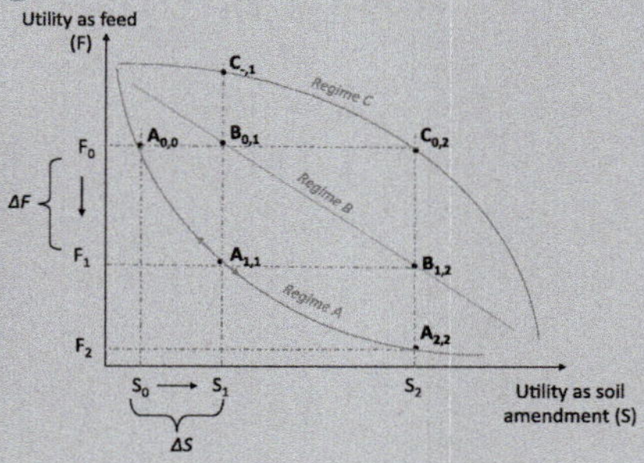

Let us first assume a competition scenario where the farmer uses most crop residues for F and little for S, and that the rate of substitution is described by Regime A. Increasing the allocation of crop residues to soil amendment (i.e. to increase the utility of soil amendment by an amount ΔS to a level S_1) will entail a strong reduction in the utility as feed (ΔF) down to a level F_1. The system experienced a shift from point $A_{0,0}$ to $A_{1,1}$. The utility S_1 could

(continued)

Box 9.1 (continued)
indicate, for example, a minimum target level of crop residue amendment necessary to maintain soil fertility in the medium term. The utility F_1 would then indicate the new level of livestock utility, that underwent a substantial reduction due to e.g. having reduced herd size and/or substantially lower herd productivity due to insufficient feeding. Within Regime A, soil conservationists may advocate S_1 to be insufficient for soil fertility maintenance in the long term and the corresponding need to further shift from point $A_{1,1}$ to $A_{2,2}$ to achieve a soil amendment target S_2 that is deemed preferable to target S_1 but further reducing utility of feed to F_2. The level F_2 could indicate, for example, a curtailed livestock utility because livestock productivity is now so low it barely provides any livestock functions. Regime A depicts a high degree of competition between the utility derived from feed and soil amendment and correspondingly severe trade-offs, whereas Regime B provides for substitutability and Regime C for complementarities – which imply increasingly favourable trade-off scenarios. The current (S_0) and minimum target (S_1) levels of allocation of crop residues to soil amendment would correspond with higher levels of feed utility when the trade-offs are described by Regimes B or C. The initial feed utility target level F_0 can be achieved with substantially higher utility levels within Regime B (point $B_{0,1}$, meeting the minimum soil fertility needs) and within Regime C (point $C_{0,2}$, meeting longer term soil fertility needs).

**This example is taken from Tittonell et al. (2015); the picture corresponds to a crop residue market in the Sahel and was taken by my colleague at CIRAD Dr. Rabah Lahmar; the title "Feeding the soil or feeding the cow" comes from a documentary movie on this issue produced by the African Conservation Tillage that is available here: https://www.youtube.com/watch?v=hAQbfXvE8Ec*

What can explain the differences between the substitutive and complementary regimes? In agroecosystems, different trade-off regimes may result from changes in the configuration of the system, in terms of structures and functions (cf. Chap. 2), such as changes in technological regimes or in crop or animal species, cultivars or landraces, in the organization of resource flows in the system, or changes in the boundary conditions within which the system operates, such as changes in policies, market prices, new opportunities, etc. Trade-offs may swap from a substitutive to a complementary regimen as the result of any such changes. Agroecological design aims at minimising trade-offs, maximising synergies, and identifying complementarities.

The first derivative of the trade-offs curve indicates the marginal rate of substitution between two objectives at any given level of the fulfilment of one of them. The average rate of substitution between objectives A and B in Fig. 9.1 can be calculated by relating ΔB and ΔA, as will be explained later on using more concrete examples.

Note, however, that whether a trade-off is substitutive or complementary depends also on the position in which the system is along the curves that describe such trade-off regimes. For example, the curve describing complementarities in Fig. 9.1 becomes highly substitutive towards high levels of fulfilment of objective A, when the absolute value of the elasticity of substitution becomes greater than one (i.e., towards the right hand side of the chart). This information is very valuable to assist decision-making on agroecosystem design and management, and justifies the need to map out trade-offs.

Let us examine now further cases in which changes in the configuration of the original system or in the contextual drivers that shape its behaviour allow shifts across system regimes to happen. On a mixed farm, for example, objectives A and B may represent respectively 'income from grain sales' and 'income from broiler meat sales'. The farmer needs to decide between selling his grain directly on the market or use it to feed broilers and then sell their meat. For simplicity, let us assume that the trade-offs between objectives A and B can be described through zero sum lines that represent two alternative regimes, I and II (Fig. 9.2). When the trade-offs are described by Regime I, moving from A_0 to A_1 implies a proportional decrease in objective B from B_0 to B_1. For example, increasing the amounts of grain sold directly on the market implies a reduction in the number of broilers that can be fed, or in their average weight gain, reducing the income from meat sales. By shifting from Regime I to Regime II the system experienced a possible synergy between objectives A and B. In the first place because in the new regime objective A can be fulfilled to a level A_0 without changes in B. Subsequently, if a decision maker is happy to relegate objective B to a level B_0, then the new regime allows a greater level of fulfilment of objective A, to the level A_2.

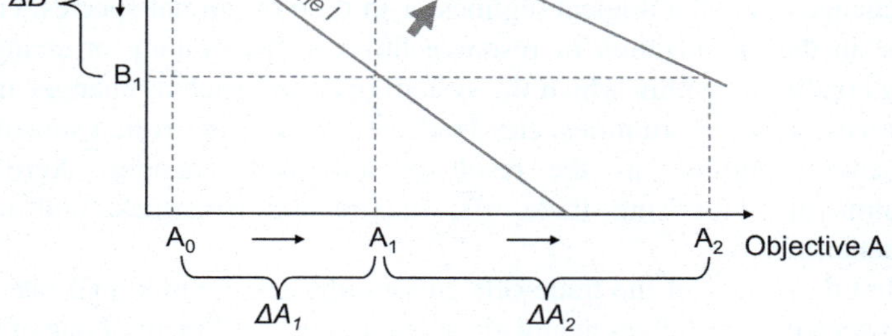

Fig. 9.2 Trade-offs between objectives A and B represented by two zero sum lines that correspond to two different system regimes

What can provoke such a regime shift? As indicated above, a regime shift can be the result of changes in either endogenous or exogenous, contextual factors, or both (simultaneously or sequentially). The former may include changes in the structure and function of the system that affect the nature or the parameters that describe its objective functions. For example, a new way of feeding or caring for animal housing or health, a different broiler breed, etc. could make the entire system more efficient at converting grain into meat. Likewise, exogenous factors such as changes in the meat-to-grain price ratio may allow greater gains to be made by allocating more of one resource to a single activity. Both examples, of endogenous and exogenous changes, may result in the ability of the farmer to maintain current income from meat sales levels while increasing their income from grain sales. Note that in this particular case the new regime implies a greater possible level of fulfilment for both objectives, and that the zero sum lines that describe regimes I and II are not parallel. This implies that the new system configuration under regime II allows improving both objectives, but predominantly objective A.

Once again, we see in the example represented by Fig. 9.2 that the shape, the origin, the slope and the range of the lines or curves that describe trade-offs have a well-defined meaning, and that it is important to have ways to not only quantify trade-offs as substitution rates at one single point, but to be able to map out trade-offs along a broad range of fulfilment of these objectives. The utility maximizing position for any regime and the associated trade-offs by moving along any regime are based on preferences, perceived benefits and risks, or sheer costs and constraints imposed by endogenous (e.g. resources) or exogenous factors (e.g. relative prices). New technologies, new agroecosystem designs, policies and/or development interventions may provoke (i) changes that result in system shifts within a certain regime, (ii) changes that allow system jumps from one regime to the next or (iii) changes that create new regimes. Based on this set of heuristics, it is possible to recognise cases in which the three regimes represented earlier in Fig. 9.1 may represent, either:

1. Three different socio-ecological contexts, being observed or proposed as scenarios;
2. Three different farm types within a certain territory (e.g. more or less resource endowed, market- vs. subsistence-oriented, etc.);
3. Three different scales of analysis of the same system; or
4. Three different system configurations alongside a trajectory of change or ecological intensification pathway.

In the latter case, when regimes are interpreted as different system configurations alongside a trajectory of change or ecological intensification pathway, elements from the ream of resilience thinking may help to conceptualise possible system shifts. Figure 9.3 illustrates how such a simple theoretical framework could be used to analyse system trajectories, using the example of crop residue biomass allocation to feeding animals versus feeding the soil (cf. Box 9.1). In Fig. 9.3a, moving from a current situation represented by point $A_{0,0}$ to a desirable situation represented by $C_{1,1}$ can be done following at least two alternative trajectories. On trajectory 1 (T_1), the shift is operated by first increasing crop biomass production and its allocation to soil

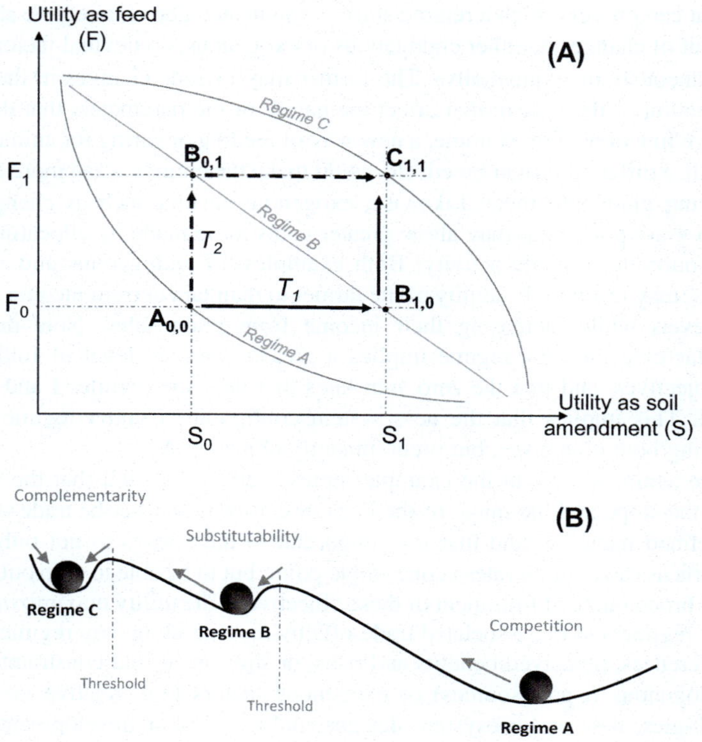

Fig. 9.3 Possible system trajectories and implications for the allocation of crop residue biomass to livestock feeding versus soil amendment. (**a**) A system that is currently in Regime A, whose relative allocation pattern described by point A0, may reach the target point C1 on regime C following two alternative trajectories, T1 or T2. (**b**) Moving across system regimes may imply that certain thresholds must be crossed, as represented by a stability landscape with three basins of attraction. (Adapted from: Tittonell et al. 2015)

(shift to $B_{1,0}$), so that greater soil fertility in the mid-term can lead in turn to greater biomass production and availability for livestock feeding to ultimately achieve a level F_1 in the future ($C_{1,1}$). On trajectory 2 (T_2), allocation to livestock feeding is first prioritised (shift to $B_{0,1}$), in order to increase livestock productivity (milk, traction, weight gain), generating greater incomes and/or animal manure to fertilise the soil in order to achieve the target fertility level S_1 in the future. Multiple combinations and intermediate trajectories between T_1 and T_2 are of course possible.

However, the assumption that income from livestock will allow investments in soil management, and *vice versa*, is certainly contestable. Higher incomes will not always necessarily translate into greater investments in soil fertility or in any other farming activity (cf. Tittonell 2014). For example, the diversity of trajectories that smallholder households may follow has been illustrated in a longitudinal study by Valbuena et al. (2015) in western Kenya, who show that better-to-do farmers invest increasingly in a portfolio of non- and off-farm activities. Back to Fig. 9.3, whether a trajectory T_1 is preferable to T_2 or not depends on the biophysical and socioeconomic contexts in which the system operates. It also depends in this case on the type of farm household concerned, in terms of resource endowment, of how risk averse the family is, on their immediate and long-term objectives, and on the current configuration of their farm system (e.g., surface area, current land use, herd size, feeding system, investment capacity, labour availability, etc.). Various studies on trade-offs of this nature in smallholder agroecosystems that were compiled in a special issue of the journal *Agricultural System* (Erenstein et al. 2015) seem to suggest that trajectory T_2 appears as a default strategy among poor farmers when they are able to own livestock. Farmers tend to value their livestock over their soils because they see their animals not just as productive assets but also as a means of saving.

Another noteworthy assumption in this simplified model is that the quality of crop residue biomass does not change across system regimes. In reality, however, inclusion of legume intercrops together with cereals may lead to greater biomass production and feed quality improvements (e.g., Naudin et al. 2014), thereby resulting in greater livestock productivity and potentially allowing regime shifts. If, instead, legumes are included in rotation with cereals this might eventually result in lower total annual biomass productivity and/or in faster decomposition of legume residues (e.g. Thierfelder et al. 2012), aggravating biomass trade-offs.

Shifting across system regimes requires investment, knowledge and support, and the trajectory may exhibit non-linearity, hysteresis and thresholds (cf. Tittonell 2014). Figure 9.3b illustrates shifts across regimes using the analogy of stability landscapes (Scheffer et al. 2006). The system is represented as a ball, and each regime is represented as a basin of attraction along an undulating landscape. When moving along the landscape, the ball tends to converge towards the bottom of each basin of attraction. A shift from regime A, characterised by strong competition between objectives S and F (cf. Fig. 9.1; Box 9.1), to regime B requires energy investments to 'push the ball upwards' and to overcome a threshold, beyond which the system would find its own new equilibrium at the bottom of basin B (substitutability regime). Figure 9.3b hypothesises that shifting from regime B to regime C requires less effort, less investment, albeit having to go through a new threshold.

This example, as the ones discussed so far in this chapter, indicates ways of conceptualising trade-offs between two objectives. In reality, however, farmers or communities have to trade off various objectives simultaneously. Such cases are more complex to analyse and represent graphically as done in Figs. 9.1, 9.2, and 9.3. Yet the analogy between trade-offs regimes and basins of attraction on stability landscapes deserves to be further explored, as it is a way of bridging the static analysis of trade-offs with that of agroecosystems dynamics (i.e., bridging synchronic and diachronic approaches – Chap. 2).

9.3 Methods for Trade-Offs Analysis

There are several ways of analysing trade-offs and yet not many attempts at categorising the various methods. A flexible classification of methods used in this emerging field in agricultural science is presented here just for didactic purposes. Methods most often used in trade-offs analysis, emerging from the scrutiny of a broad set of specialised literature include:

1. Ad-hoc analysis of data on (or observed) system responses
2. Quantitative analysis of elasticity using data on systems responses
3. Examining the results of experiments, surveys or model simulations
4. Participatory methods to formalise a problem and explore scenarios
5. Projecting changes using empirical statistical models (including GIS)
6. Multi-objective 'compromising' using optimisation routines in models
7. Agent based-systems representing collective decision-making

The list of methods may not be exhaustive. A major difference among these categories resides in their ability to explore compromise 'solutions' to the trade-offs. Categories 1, 2 and 3 are typical examples of *ex-post* trade-offs analyses. The first category groups a wide diversity of approaches – often not formalised – that have been used to analyse trade-offs. They do not have much in common, hence the term *Ad-hoc* used here to refer to them.

Trade-offs may be analysed quantitatively through the simple calculation of the elasticity of substitution between objectives (i.e., absolute or relative changes in A with respect to absolute or relative changes in B in Figs. 9.1, 9.2, and 9.3). Shadow prices, opportunity costs, payments for environmental services, etc., are all examples of such ways of quantifying elasticities grouped as Category 2 methods. Category 4 includes methods and support tools (e.g. fuzzy cognitive mapping, board games) to explore future scenarios involving key stakeholders. Categories 5–7 are those that propose different approaches to *ex-ante* trade-offs analysis. Category 5 includes the spatially explicit statistical projections done using land use maps, which most commonly rely on empirical models of varying complexity (from simple regression lines to artificial neural networks, Bayesian algorithms, etc.). Category 6 includes methods that compute the value of indicators that represent objectives, subject to a set of decision variables and constraints. Category 7 groups methods that use agent-based systems to represent negotiation and compromises between actors, typically used for the analysis of trade-offs in communal decision-making. The diversity of methods in these last three categories will be further classified in the section on quantitative approaches to trade-offs analysis using models, and will be discussed in more detail in Chap. 10.

9.4 Trade-Offs Analysis through Simple Calculations

The simplest way to approach the quantitative analysis of trade-offs is to calculate the slope and elasticity of lines or curves fitted through data describing tension between two objectives. Let us take again as an example the case of trading off between only two mutually exclusive objectives as represented in Figs. 9.1 and 9.2. The degree of change in objective B from B_0 to B_1, associated with a change in objective A from A_0 to A_1, expressed respectively as ΔB and ΔA, can be calculated simply as follows:

$$(B_1 - B_0)/(A_1 - A_0) = \Delta B/\Delta A \tag{9.1}$$

By looking at the substitutive and complementary regimes in Fig. 9.1, it is obvious that:

$$|B_1' - B_0|/(A_1' - A_0) > |B_1' - B_0|/(A_1'' - A_0) \tag{9.2}$$

Which means that although the absolute change in B, from B_0 to B_1' or $\Delta B'$, remains the same in both cases, the associated change in A was much greater in the case of complementarity (i.e., $A_1' - A_0 < A_1'' - A_0$). The change from B_0 to B_1' or $\Delta B'$ may represent the 'price' paid in terms of objective B units, or the acceptable loss of level in objective B, associated with a gain in objective A. Assuming that the level of fulfilment of objective A increase as we move to the right along the horizontal axis, and the level of fulfilment of B as we move upwards, then the actual change in B will be represented by a negative value.

Under the complementarity regime it is possible to obtain many more units of objective A (or $\Delta A''$) for the same amount of units of objective B 'lost' (or $\Delta B'$) than under the substitutive regime, where the gain in objective A corresponds to only $\Delta A'$. If the trade-off was described exclusively by the substitutive regime, so that the complementarity regime does not exist or is not possible, then the price to be paid for a $\Delta A''$ change in objective A will be equal to $B_1'' - B_0$, or $\Delta B''$. In other words, it will be much more 'expensive'. Since these are two objectives that are negatively associated, and we do not have enough information as to assert whether A is an independent variable to which B responds, the analysis made so far is also true *vice versa*. A simple example of trade-off calculations using this method is given on Box 9.2.

Box 9.2: The Cost of Soil Carbon

Let us assume a situation in which different natural resource management alternatives are being evaluated with respect to two main objectives: to maintain soil fertility in the long term and to increase income levels in the

(continued)

Box 9.2 (continued)
short term. These objectives are represented by the indicators soil organic carbon (t ha^{-1}) and the gross economic margin of the activity (\$ ha^{-1}). A particular aspect of this example is that both objectives operate within different temporal scales, represented by indicators that can be considered respectively slow and fast system variables.

Let us assume now that two alternative management systems are proposed, Alternatives I and II, which operate along the same trade-off regime as the current management system, described by a complementarity curve. Alternative I leads to doubling the gross margin with respect to its current level but with an associated 30% loss in soil organic carbon, whereas Alternative II would lead to 10% loss in soil organic carbon but with a 55% increase in gross margin. Which one is best?

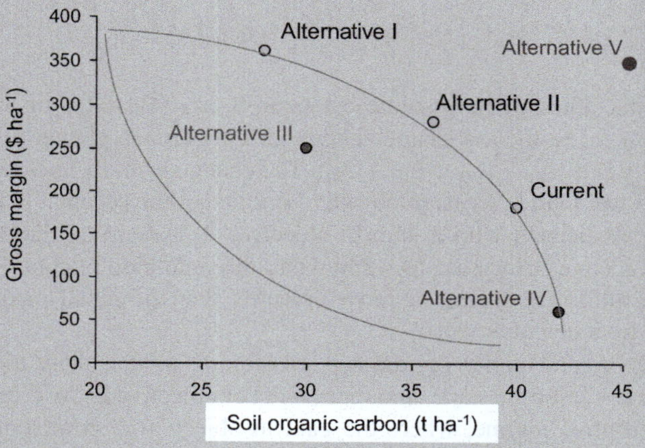

Objective	Indicator	Current	Alternative I	Alternative II
To maintain soil fertility	Soil organic matter (t ha^{-1})	40	28	36
To increase net income	Gross margin (\$ ha^{-1})	180	360	280
Cost of maintaining soil C:			**15 \$ t^{-1}**	**25 \$ t^{-1}**

If maintaining soil organic carbon is a priority and the decision is to continue with the current management system then it is in principle possible to estimate the costs of carbon in terms of how much income is being 'sacrificed'. By not implementing Alternative I, a total of 12 t ha^{-1} soil organic carbon are saved and 180 \$ ha^{-1} are sacrificed, at a cost of (180/12=) 15\$ per

(continued)

Box 9.2 (continued)
tonne of carbon. By not implementing Alternative II the loss of income comes to (100/4=) 25$ per tonne of soil organic carbon. Again, which of the two alternatives is best? This always depends on the objectives of the decision maker(s), and/or on the existence of meaningful thresholds for both objectives, but certainly Alternative II seems more attractive, because it pays better off for every unit of carbon lost and because the final soil organic carbon level is still 30% higher than under Alternative I. If there was a carbon market, how much should farmers get paid per ton of C to make sure they choose for carbon-saving practices?

Three other alternatives, III, IV and V are now also considered. Alternative III is an inferior one, as it leads to a net decrease in soil organic carbon and to a gain in gross margin that is below the levels already achieved by the previous two alternatives. Alternative IV favours the build up of soil carbon with a strong reduction in gross margin, which may likely be unacceptable to farmers. Alternative V represents a synergy, a simultaneous gain in both objectives, a shift towards a new trade-off regime that is more favourable than the original one. Alternative V illustrates the type of strategies at which agroecology aims: shifting to a new, synergistic regime.

When trade-off regimes are represented by zero sum lines, as in the two cases depicted in Fig. 9.2, then ΔB/ΔA represents the slope of the trade-off line and is constant throughout the full range of variation of objectives A and B considered. It is possible to assert in this case that:

$$|B_1 - B_0|/(A_1 - A_0) > |B_1 - B_0|/(A_2 - A_0) \tag{9.3}$$

in which the two slopes represent the two different regimes and can be expressed also as:

$$(\Delta B/\Delta A)_{Regime\ I} > (\Delta B/\Delta A)_{Regime\ II} \tag{9.4}$$

When trade-off regimes are represented by simple substitution or complementarity curves such as those represented in Fig. 9.1, then the slope is not constant. In absolute terms the slope calculated with Eq. 9.3 could be smaller, equal or greater than one depending on where we stand along these curves. This is dealt with by calculating the first derivative, or marginal change in one objective with respect to another, at any given point in time. But before moving to such calculations, note that the substitutive and complementary regimes in Fig. 9.1 are represented by stylised, monotonic curves that are more often the exception than the rule in reality. Trade-off curves may have both concave and convex phases, that is complementary and substitutive phases, alternating throughout the range of values of the indicators chosen to represent the objectives, as represented in Fig. 9.4. There may be

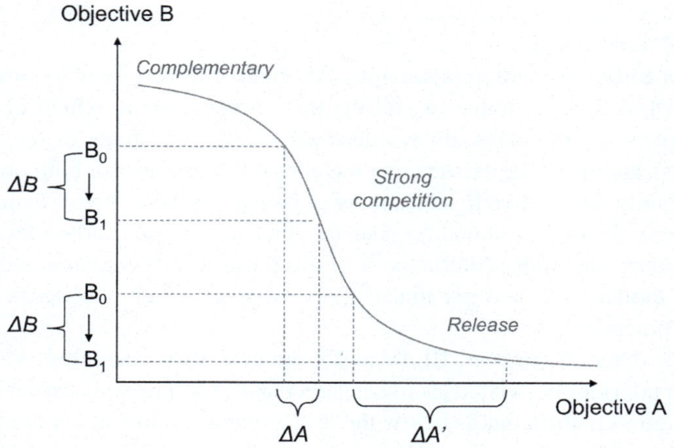

Fig. 9.4 A more complex trade-off regime between objectives A and B describing complementary and substitutive phases. Similar absolute changes in the value of an indicator representing objective B yield different degrees of change in the indicator corresponding to objective A depending on the position along the trade-offs curve

thresholds in the system, even within the same regime, that determine different phases as in Fig. 9.4. In such cases, the same level of absolute 'loss' in terms of objective B units (ΔB) will result in different levels of 'gain' in terms of objective A units (ΔA vs. ΔA′) at different positions along the curve representing a trade-off regime.

What could be the cause of different trade-off phases within the same system regime? The existence of thresholds that determine different phases within a same regime is often associated with the complexity of the system considered. Both, with the inherent complexity of the actual system (ontological) represented by multiple components, interrelations, scales and processes, and with the complexity that has been chosen to represent the system as a model (semiotic).

For example, a similar farm system may be analysed using a dynamic model that accounts for complex feedback between farm subsystems over time or with a simple spread sheet for the calculation of the economic gross margin of the farm by accounting for annual costs and revenues. Actual features that may lead to thresholds *within* a single regime (*NB*: for changes associated with shifting regimes see Fig. 9.3 and related text) include spatial heterogeneity, restrictions, synergies, cumulative effects over time and changes in scale. For example, if objective A in Fig. 9.4 represents crop productivity at farm scale and objective B livestock productivity, the gain ΔA may correspond to cases in which only marginal cropping land is allocated to crops, while the best cropping land is kept for livestock. This is not uncommon amongst livestock farmers who are transitioning towards crop production and yet show a tendency to favour livestock over crops (e.g. Cortez-Arriola et al. 2014). The gain ΔA′ in crop productivity will occur once the area of crops expands to a level at which the best cropping land starts being allocated to crops. An important threshold, often a cultural one, had to be crossed for this to happen.

The range of values within which the indicators chosen to represent objectives vary may differ substantially, as well as the importance allocated to each indicator/objective. This limits the use of Eqs. 9.1–9.4 to analyse trade-offs by simply calculating slopes. For example, back to Fig. 9.1, it is possible that a change in objective B from B_0 to B_1'' is deemed by a decision maker to be less relevant than a change in objective A from A_0 to A_1''. This could be the case when the relative intrinsic value assigned to both objectives differs. In a community, or in multiple stakeholder settings, different social actors may ascribe more or less importance to one or the other objective. Such differences are normally accounted for in quantitative trade-offs analysis by weighing the objectives using different methods, many of which include stakeholder participation. But the relative importance of both objectives may also differ when their original levels, A_0 and B_0, are closer to or farther from acceptable reference values. In other words, losing one dairy cow does not have the same economic impact for a large dairy operation milking hundreds of cows per year than for a smallholder farmer that typically owns two to three cows. The slopes of the trade-off curves may then be expressed in relative terms:

$$(|\Delta B|/B_0)/(\Delta A/A_0) \tag{9.5}$$

By using relative values the units disappear, which may be an advantage in some cases and a disadvantage in others. The use of relative rates of change was illustrated already in the example presented in Box 9.2. Equation 9.5 is also used when calculating the relative partial sensitivity of a simulation model, to quantify the relative changes in model outputs with respect to changes in model input values (cf. Tittonell et al. 2006). When such models are used in trade-offs analysis, the relative sensitivity of some of their outcome variables may already be considered as quantitative approximations to the actual trade-offs (Tittonell et al. 2007).

9.4.1 Crop-Livestock Integration Example

Mixed crop-livestock agroecosystems are very common worldwide and the co-existence of cropping and animal production activities normally leads to farmers having to make difficult decisions on the allocation of their resources. Figure 9.5 shows an example concerning land allocation to fodder production in intensive smallholder crop-livestock systems from the Kenya highlands. Assuming that the most important feed resource of the farm is Napier grass (*Pennisetum purpureum*), as is the case in most of the East African highlands, the maximum potential fodder production of the farm (p) is determined in principle by its total area suitable for cultivation (a). There may still be a feed gap at farm scale due to the small area available, or to the fact that livestock need more than just Napier grass for a complete diet. Current fodder production is represented by p' corresponding to an area a'. The

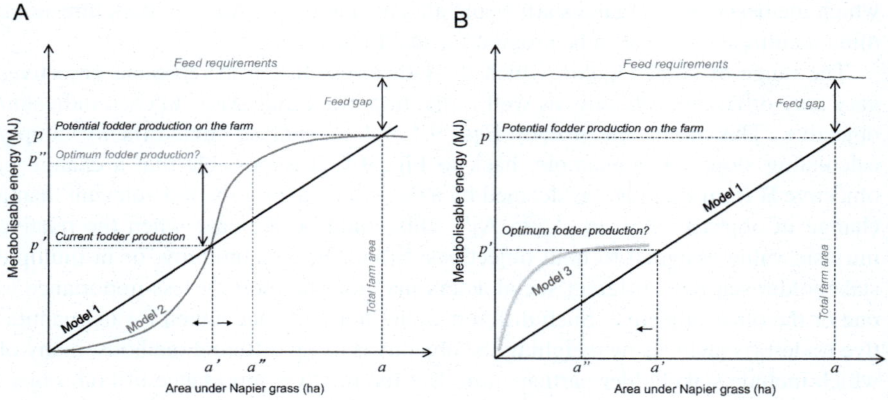

Fig. 9.5 Stylised representation of farm scale fodder productivity as a function of the area allocated to the cultivation of Napier grass, assuming different models to represent such relationship. Fodder production and feed requirements are expressed in terms of metabolizable energy, a unit that is commonly used in animal nutrition

simplest approach to the analysis of this farm will be to assume that fodder productivity is proportional to the area allocated to Napier grass, represented by Model 1 in Fig. 9.5a. Yet in reality we know that farms are heterogeneous in terms of soil quality (cf. Chap. 7) and that farmers in this region tend to allocate the land that is less suitable for cultivation to the production of Napier grass, as represented by Model 2, keeping the best land for the production of food and cash crops.

If Model 2 is right, then there seems to be also an optimal allocation of land to fodder production represented by the area a'' through which a fodder productivity p'' can be achieved (Fig. 9.5a). Beyond this point productivity gains would be less than proportional and it might be more profitable to purchase feed on the market, preserving land for other uses, rather than produce it on the farm. Thus, considering spatial heterogeneity through Model 2 gives us already a better insight on the actual trade-offs faced by the farmer when allocating land to fodder production. Moreover, if Napier grass productivity responds well to soil quality then perhaps the best strategy is to allocate only the more productive land to Napier grass, thereby reducing the area necessary to attain the same actual fodder productivity level, as represented by Model 3 (Fig. 9.5b).

Increasing the complexity with which we model the agroecosystem results in a greater complexity in terms of the results of our analysis, their interpretation, and the possible conclusions to be drawn from them. Moreover, in all the examples discussed so far trade-offs regimes were represented using continuous lines, assuming continuity and reversibility, and this assumption may also be questioned (cf. Tittonell 2014). This is why in trade-offs analysis we often speak of 'exploring' trade-offs rather than calculating them, making explicit reference to the assumptions made in building the models.

9.4.2 Agri-Environmental Trade-off Example

Let us now examine another example of using simple calculations to analyse trade-offs. Figure 9.6 shows relationships between average milk productivity and farm-scale nitrogen surpluses expressed per unit area (9.6A) or per kg of milk produced (9.6B) from small-scale dairy farms in Michoacán, México. These data, collected by

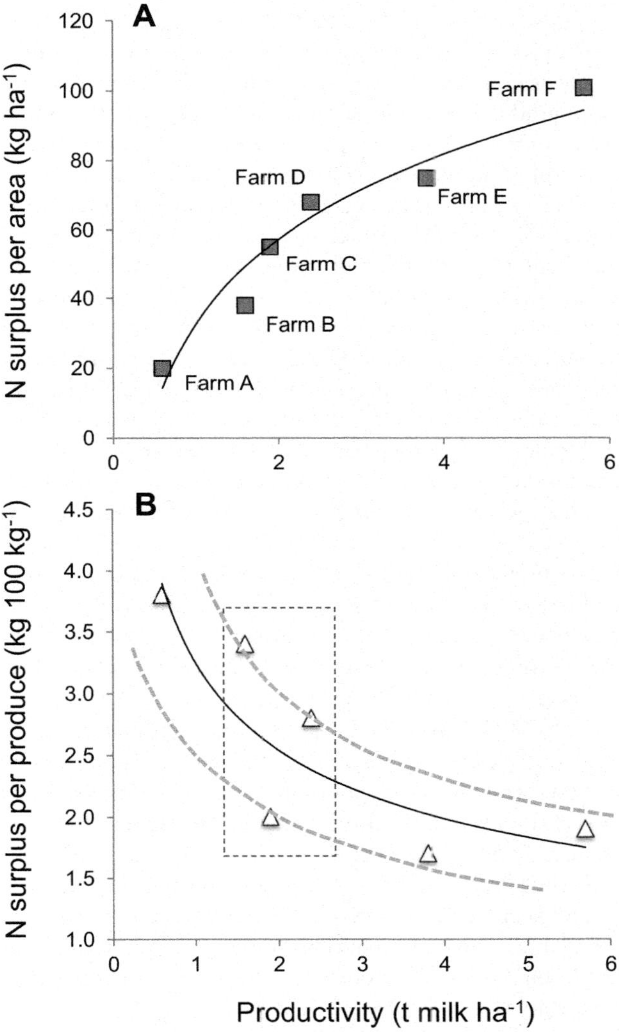

Fig. 9.6 Average milk productivity per farm versus nitrogen surpluses expressed per unit area (**a**) or per kg of milk produced (**b**) for six case study, small-scale dairy farms in Michoacán, Mexico. Solid black lines in both cases are curves fitted through the data just for illustrative purposes. The grey dotted lines represent other possible trade-offs regimes. (Data source: Cortez-Arriola 2016)

Cortez-Arriola (2016), correspond to six different farms that were followed for three consecutive years to analyse their efficiency and possible intensification pathways. Plotting the data and fitting a curve through them already provides an image of the nature of trade-offs that emerge when thinking about dairy intensification. From an environmental point of view, N surpluses are a problem, as excess N can lead to water pollution and modification of living communities in lakes and rivers, it is costly to produce N fertiliser using fossil energy, and it is anti-economic to bring in N into the system and use only a small portion of it for production. Minimising N surpluses is of great environmental and economic interest. Figure 9.6a shows that intensifying milk production from the level achieved by Farm A to that of Farm C (by 1.3 t ha^{-1}) leads to increasing N surpluses by 175% (from 20 to 55 kg ha^{-1}). But a similar gain in productivity from the level achieved by Farm D to that of Farm E (1.4 t ha^{-1}) costs only an extra N surplus of 7 kg ha^{-1}, a mere 10% increase.

Greater N surplus associated with higher productivity is the result of feed insufficiency. These small farms located in a region that faces a dry season every year are not self-sufficient in animal feed. They are not able to keep pastures productive year-round to meet the high requirements of genetically improved dairy cows. Farmers purchase concentrate feeds on the market, or produce maize for silage using N fertiliser. The N that enters the farm system in feed concentrates and in chemical fertiliser, and that is not used efficiently in the production process constitutes a surplus, and a potential environmental hazard. Under this logic of intensification, the more a farm is productive per unit area the greater its N surplus per unit area – a typical trade-off situation (Fig. 9.6a). Expressing N surpluses per 100 kg of milk produced instead of per unit area provides a different picture (Fig. 9.6b). It actually shows no apparent trade-off between production and the environment; but such is not the case.

Expressing N surpluses per kg of milk produced leads to the impression that the higher the productivity the lower the environmental impact. What generates impacts on the environment, however, is the total load of a pollutant, N in this case, irrespective of the productivity of the farm system from where that N originates. For example, Farm B has a surplus of 3.4 kg N per 100 kg milk produced, while Farm E has exactly half of this (Fig. 9.6b). But the annual surplus of N per unit area from Farm B, that is the total amount of N that can potentially leach to deeper soil water layers or run off with irrigation or rain water down to water bodies, is half of that from Farm E (Fig. 9.6a). If all the farms in the region of study had similar N emission potential as Farm E then the level of water pollution would be much higher than what it actually is at the moment (Farm B has 47 ha and 35 animals, while Farm E uses 133 ha and keeps 142 animals, so the total absolute N emission is also much greater from the latter). This information is highly valuable for decision makers and can improve the quality of the debate between encountered environmental and productive perspectives.

The real purpose of analysing trade-offs however is not just to point out problems, but to be able to inform decisions that can help finding the best compromises among the objectives at stake. For example, Fig. 9.6 shows that Farms B, C and D have comparable milk productivity levels (1.6–2.4 t ha^{-1}) but very different N surpluses

(38–68 kg ha^{-1}). Farm C in particular has relatively low N surpluses per produce (2 kg 100 kg^{-1} milk). This is perhaps the most important information from this study: it is possible, in the region of study, to attain an average milk productivity of about 2 t ha^{-1} with a relatively low N surplus. This particular farm owns 24 ha and keeps 31 animals, the highest animal density (1.3 animals ha^{-1}) of all six of them, and it is a good example of an intensive small-scale dairy system that may serve as a model for the intensification of other family farms in the region. One may assume that Farms B, C and D are fluctuating within different trade-offs regimes, as discussed already around Fig. 9.3, and indicated by the grey dotted lines in Fig. 9.6b.

Note also in Fig. 9.6a, b that the values expressed per 100 kg milk are not simply the result of dividing N surplus per area by milk productivity. The calculations are much more complex because the N surplus does not refer only to the dairy herd but to the entire farm system, including the production of maize, sorghum and pastures, plus the maintenance of animal categories that do not produce milk (e.g. calves, heifers, males, other species). Cortez Arriola et al. (2014) used a mathematical farm-scale model for these calculations, as will be discussed in Chap. 10.

9.5 Semi-Quantitative and Participatory Trade-Offs and Scenario Analysis

Trade-offs often imply normative choices between societal values, particularly when the level of integration and scale of analysis considered exceed the farm system or the agricultural landscape. There are situations in which trade-offs between two or more societal goals cannot be specified in quantitative units or mapped out using a pair of cardinal axes to represent objectives, as done so far in this chapter, or when trade-offs are expected to emerge in the future, under a given set of yet-to-occur circumstances. Then the best way to approach their assessment is through scenario analysis, and one may or may not count on quantifiable indicator variables for quantitative trade-offs analysis in future scenarios. Trade-offs scenarios can be analysed using qualitative methods, such as hand-drawn graphs, rich pictures or colour scores (e.g. traffic light colours) to indicate positive or negative relationships between various objectives. Semi-quantitative techniques allow generating an arbitrary numerical scale in which values are ascribed to 'levels' identified in a linguistic variable. For example, how good was the movie? The linguistic variable is the quality of the movie, and the levels are 1 = poor; 2 = acceptable; 3 = good.

Whenever societal choices are involved, stakeholder[1] participation is crucial to ensure legitimacy and meaning of the analysis. Stakeholder participation is also

[1]Stakeholder is a term first coined by R.E. Freeman in his book: 'Strategic Management: A Stakeholder Approach", to refer to those who can affect or be affected by the activities of a company. Other terms used in the literature to refer to stakeholders are actors, social actors or social subjects.

essential when assessing trade-offs associated with human agency and choices under alternative future scenarios. Stakeholder participation generally improves the trade-off analysis, as they bring in their perspectives, experience and knowledge about the systems being analysed. Their objectives and expectations may or may not be translated into quantifiable indicators. The formalisation of stakeholder perspectives in multicriteria decision-making is discussed and illustrated in Chap. 10. This section provides a few examples of methods that can be used to facilitate trade-offs analysis through stakeholder participation, considering also cases in which it is not possible to quantify trade-offs through the measurement or calculation of key indicator variables. Four methods are considered and briefly discussed using examples: (i) Fuzzy cognitive mapping; (ii) Fuzzy logic systems; (iii) Qualitative methods; and (iv) Games.

9.5.1 Fuzzy Cognitive Mapping

Fuzzy cognitive maps are semi-quantitative tools that allow 'translating' narrative stories into quantitative scenarios. Scenarios are nothing else than stories about the future, which can be expressed in words or in numbers. Scenarios are meant to offer a consistent and plausible explanation of how events may unfold over time (Gallopín 1997; Raskin et al. 2005). There are several methodologies to develop scenarios, and of particular interest in trade-offs analysis are those that allow involving stakeholder participation. Very often through participatory scenario development we obtain plausible images of the future in the form of narratives. Such narratives or story lines can be recorded, taped or written down, or they can be expressed using graphical representations, such as timelines, rich pictures, Venn diagrams, problem trees or flow charts. Flow charts are of particular interest as they allow representing cause-effect relationships, controls, drivers and influences.

Narrative storylines can be developed and linked to parameter sets in dynamic models using an iterative procedure, as was proposed by Alcamo (2001) and illustrated by Kok (2009). A panel of stakeholders develops stories, while a team of facilitators and modellers 'translate' these into quantitative scenarios, as described in Fig. 9.7. This approach is similar to the ones used in global exercises such as the Millennium Ecosystem Assessment (Carpenter et al. 2005) or the Global Environment Outlook (UNEP 2007), and several other studies spanning local to regional scales (e.g., EEA 2007; Kok et al. 2006; Kok and Van Delden 2009). Although this method allows structuring the process of scenario development and its quantitative assessment, the characterisation of the problem at stake may still remain descriptive, static or even superficial. This approach does not guarantee *per se* gaining any core understating of the underlying causes and dynamics of the trade-offs being assessed.

Fuzzy cognitive maps are built by relating causes and effects in a rather flexible way. Unlike proper flowcharts or graphical modelling languages, the boxes do not necessarily represent state variables but 'processes', 'events' or 'drivers'. The arrows that link them represent influences or causal links, but not necessarily flows of

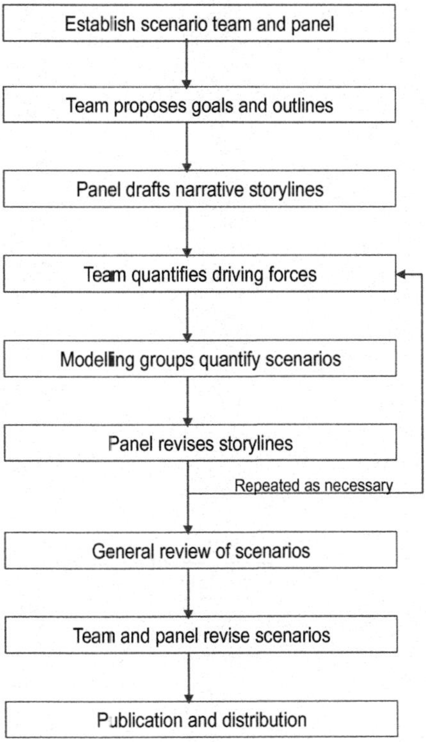

Fig. 9.7 A sequence of steps in a quantitative scenario analysis method known as 'story-and-simulation' approach, proposed by Alcamo (2001)

energy or matter. These are 'fuzzy' systems, with fuzzy components, fuzzy relationships and fuzzy boundaries. The arrowhead indicates the direction in which the causal effect operates. Next to the arrow, a positive or negative sign indicates the nature of the relationship. In Fig. 9.8a, a simple fuzzy cognitive diagram describes the consequences of the 'frequent flyer' policy that air companies implement nowadays. Flying frequently allows you to become a frequent flyer member and to accumulate free fly miles. Such free miles can be used to travel for free and in so doing they stimulate clients to fly more often. This is a positive feedback, and as such it leads to exponential, uncontrolled growth (in the number of actual miles actually flown per year in this case). Note that all the signs in the diagram are positive.

Figure 9.8b illustrates a case in which both positive and negative feedbacks coexist, and the final outcome, the dynamics of the system, is determined by the relative influence of one or the other process over time. In the example, a crop canopy is stimulated to grow as its photosynthetic surface increases, represented by the leaf area index. More leaf area per unit of soil implies a greater fraction of incoming solar radiation being intercepted by the crop plants and transformed into crop (leaf) biomass. A the same time, more crop biomass implies more energy needed for maintenance respiration, specially of plant organs that are not able to photosynthesise, such as roots or ligneous stems. This consumption of energy for

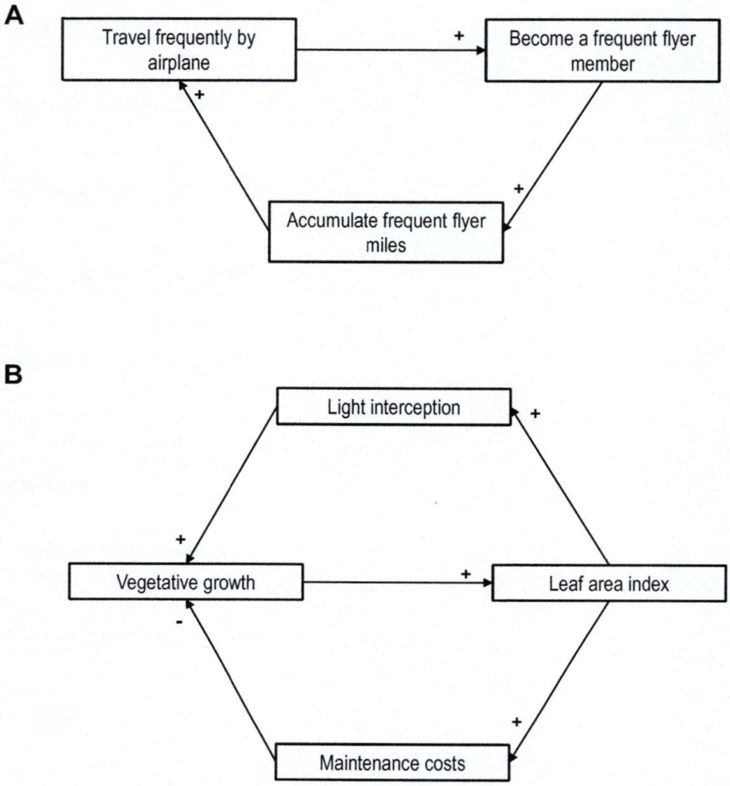

Fig. 9.8 Representation of feedback loops through a fuzzy cognitive mapping. (**a**) Illustration of a positive feedback using the example of the frequent flyer membership system that air companies propose to their clients. (**b**) Illustration of interlinked positive and negative feedbacks using the example of the vegetative growth of a crop canopy. The positive and negative signs indicate the nature of the influence between the processes represented

maintenance costs leads to negative net energy balances, especially by night, negatively affecting growth. This describes a regulatory mechanism that prevents crops from growing exponentially during the entire growing season. Regulation through negative feedbacks is a very conspicuous process in nature and generally leads to control and homeostatic equilibriums.

Fuzzy cognitive maps become semi-quantitative when a weighing factor is included next to the positive or negative signs ascribed to each arrow. Often values between −1 and + 1 are recommended for weighing relationships (cf. Kok 2009). The strength of the influence is determined by the distance to zero, and the direction by the + or – signs. This is illustrated in Fig. 9.9, which describes a process of deforestation through agricultural expansion, as it is commonly seen in South America. Some of the processes in Fig. 9.9 are represented as being self-enforced through the inclusion of positive feedback loops onto themselves. This is also known

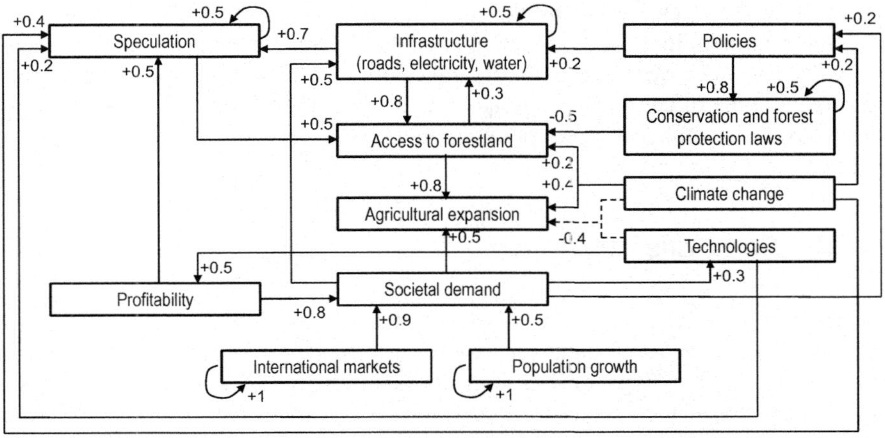

Fig. 9.9 Fuzzy cognitive mapping showing the relationship between drives of agricultural expansion and deforestation in the South American Chaco forest. The example is made up for illustrative purposes but the weights used are based on an actual participatory scenario exercise for the Brazilian Amazon forest documented by Kok (2009). The dotted arrows indicate two feedback loops that may or may not be identified during the participatory mapping exercise

as including a 'memory' in the system. For example, population growth in Fig. 9.9 is represented as a self-enforcing process, because a larger number of people leads to more births per year.

In participatory scenario development stakeholders define the processes or drivers to be included in the fuzzy cognitive map, as well as the relationships between them in terms of weights and signs. Some processes may be seen as either positive or negative loops, depending on the perception of the participants. In Fig. 9.9, this is exemplified by the relationship between climate change or technologies and agricultural expansion. In principle, technology development could increase agricultural yields per unit area thus reduce the need for agricultural expansion. In reality, however, increasing yield potentials lead to more profitability and to further agricultural expansion (a problem known as Jevons paradox). Negative feedbacks between climate change or technologies and agricultural expansion may still be included (the dotted lines) but using relatively small weights, smaller than those that represent the positive feedbacks between e.g. profitability and agricultural expansion.

These coefficients can be further used to analyse the resulting dynamics of the system through matrix calculations. The explanation of this procedure exceeds the scope of this book. Different groups of stakeholders will likely develop different fuzzy cognitive maps to represent the same problem (i.e., different semiotic systems to represent the same ontological system – Chap. 1). Then, the complexity of the fuzzy cognitive map developed by each group of stakeholders may help to reveal their perceptions and their understanding of the problems or trade-offs being analysed. This is illustrated by the work of Aravindakshan et al. (2021) with rice growers in Bangladesh, and by that of Alomia-Hinojosa et al. (2022) with crop-livestock farmers in Nepal. The complexity of the fuzzy cognitive maps in these examples was assessed through matrix indicators such as the number of components and the number of connections included in the map, the number of feedback loops, or the average number of flows per node in the fuzzy system.

9.5.2 *Fuzzy Logic Systems*

Sometimes processes cannot be quantified but experts can indicate the possible direction of change of an indicator variable or provide a well-informed estimate of the magnitude of an impact using scores or rankings. A common example of this is the assessment of the severity of weed infestation or of plant disease pressure in crops. Experts will use terms such as 'severe', 'moderate' or 'mild', or use a scoring scale such as 1 = very low; 2 = low; 3 = medium; 4 = high; 5 = very high. Another classic example is the livestock body condition score, which is a set of guidelines for visual assessment of the condition of domestic animals. Different countries and/or animal production systems have adopted different systems of scoring body condition, sometimes from 1 to 4, 1 to 9, etc. Table 9.1 presents an example of body condition scoring for beef cattle in use by the extension service of Virginia Tech University. In this case, the score comprises seven visual indicators of body condition, and all of them add the same weight to the total score. These scores could be linked through empirical evidence with, for example, a certain daily rate of live weight gain, a certain efficiency of feed conversion, etc.

Imagine a situation in which one aims to analyse future scenarios for which prospective simulation modelling is to be used. If an expert or visual scoring is the only information we can use in our analysis of a given process or indicator, and if we aim to include this process in our modelling exercise, then fuzzy logic systems may be of help at translating scores or 'opinions' into quantitative variables. In other words, translating linguistic variables into numerical ones. Fuzzy logic is a way of defining

Table 9.1 Reference table for body condition scores of beef cattle proposed by the Virginia Cooperative Extension system, Virginia Tech University, US

Reference point	Body condition scores								
	1	2	3	4	5	6	7	8	9
Physically weak	Yes	No	No	No	No	No	No	No	No
Muscle atrophy	Yes	Yes	Slight	No	No	No	No	No	No
Outline of spine visible	Yes	Yes	Yes	Slight	No	No	No	No	No
Outline of ribs visible	All	All	All	3–5	1–2	0	0	0	0
Outline of hip & pin bones visible	Yes	Yes	Yes	Yes	Yes	Yes	Slight	No	No
Fat in brisket and flanks	No	No	No	No	No	Some	Full	Full	Extreme
Fat udder & patchy fat around tail head	No	No	No	No	No	No	Slight	Yes	Extreme

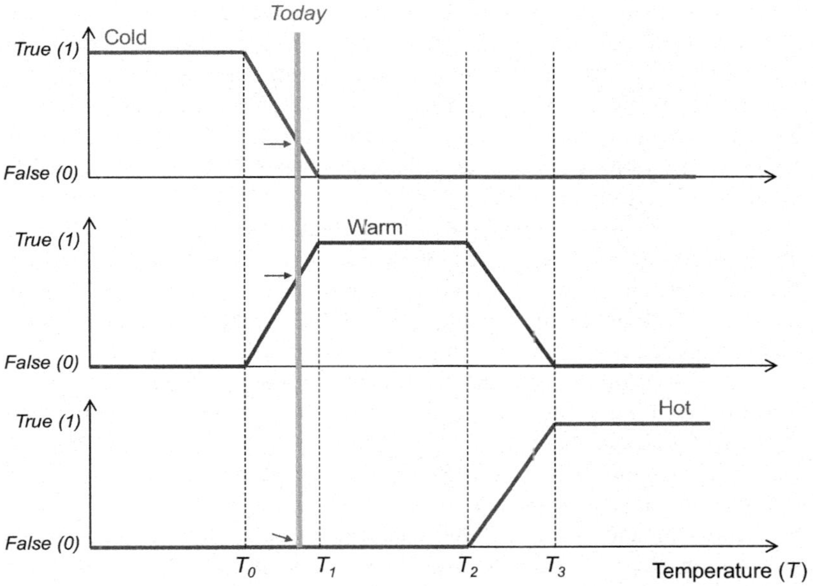

Fig. 9.10 An illustration of the concept of fuzzy logic sets using the example of the perception of the weather in a given day as a function of the daily mean air temperature. When the average daily temperature is $< T_0$ then the day is considered to be cold, the function 'cold' takes a value of $1 =$ True. Days are indisputably hot when $T > T_3$. The grey vertical line in the figure indicates today's temperature, which corresponds to certain membership values for the member functions 'cold' and 'warm', but to a zero value or False for the member function hot. There are 'fuzzy' areas or temperature ranges that describe the transition between a cold and a warm day, or between a warm and a hot day

degrees of truth concerning a process of interest, which will take real values that vary between 0 and 1 to reflect degrees between completely false to completely true. Figure 9.10 illustrates this with a simple example. Today's temperature, as indicated by the grey vertical line in Fig. 9.10 has a certain 'degree of truth' for the fuzzy function 'Cold', but a greater 'degree of truth' for the function 'Warm'. We could say that today's temperature is closer to warm than to cold. But it is certainly not a hot day, as indicated by the zero degree of truth corresponding to the function 'Hot'.

In the case of the weed infestation levels referred to earlier, different degrees of truth concerning weed infestation may be: 'no weeds', 'mild infestation', 'moderately infested', or 'severe weed infestation'. These degrees are fuzzy sets, and field observations can be assigned to these sets through a membership function; i.e. a function that defines to what degree an observation fits within a fuzzy set. The degree by which an object belongs to any of these fuzzy sets is denoted by a membership value. Expert knowledge is necessary to propose a way of translating visual indicators and thresholds into membership values. The term 'fuzzy' is used to indicate that such sets have no sharp boundaries, as they actually represent a continuum. Let us use an example from the literature to clarify all this.

Studying nutrient cycling efficiencies in smallholder crop-livestock systems of East Africa, Rufino et al. (2007) used a fuzzy logic system to assess nutrient retention efficiencies through manure collection and storage. Nutrients are lost from manure, from the moment faeces are excreted, through dung collection and storage and to the moment of manure application to crops. The overall efficiency of nutrient retention through these steps fluctuates typically between 0.1 and 0.3 in smallholder African systems. There is great interest in improving nutrient retention through proper manure handling to avoid nutrient losses and recycle them within smallholder crop-livestock systems. Manure can be stored up to one year in a heap or within a pit, where it is mixed with other materials such as crop residues and organic kitchen waste. Nutrient losses may take place through volatilisation or leaching, and these processes are affected by the type of storage system (heaps vs. pits), the period of storage (rainy vs. dry season), the duration of storage, the presence of a roof or any other form of cover, the nature of the floor where the pile rests, and by the quality of the stored material, such as its C to N ratio (e.g., Castellanos-Navarrete et al. 2015). Measuring actual losses from manure storage is actually very complicated and subject to substantial experimental error (cf. Tittonell et al. 2010).

To account for the effect of these management factors on nutrient losses from manure during storage in a dynamic farm system model, Rufino et al. (2007) proposed a fuzzy logic approach that considered the type of roofing, flooring and cover of the manure pile. Each criterion took three possible fuzzy categories. For roofing, the categories were 'unroofed', 'wooden roof' and 'iron sheets'; for pile coverage they were 'uncovered', 'branches' and 'polythene film'; and for the type of flooring they were 'sand', 'solid' and 'concrete'. Membership functions were defined for each of these fuzzy sets. Then, the three criteria were combined through a set of six rules to define a 'management factor', as follows:

1. If roof is *unroofed* and cover is *uncovered* and floor is *sand*, then management is *very poor*
2. If roof is *unroofed* and cover is *branches* and floor is *not sand*, then management is *poor*
3. If cover is *uncovered* and floor is *not sand*, then management is *poor*
4. If roof is *unroofed* and cover is *polythene film* and floor is not sand, then management is *good*
5. If roof is *not unroofed* and cover is *branches* and floor is *not sand*, then management is *good*
6. If roof is *not unroofed* and cover is *polythene film* and floor is *not sand*, then management is *excellent*

The new criterion in the model, the manure management factor, consists thus of these four fuzzy sets: 'very poor', 'poor', 'good' and 'excellent'. Then follows a step of *defuzzification* of these linguistic variables, as follows:

1. If management is *very poor*, then mass losses are *very large*
2. If management is *poor*, then mass losses are *large*
3. If management is *good*, then mass losses are *small*
4. If management is *excellent*, then mass losses are *minimal*

The resulting value of the management factor, which fluctuates between 0 and 1, is obtained by aggregating the membership values of the fuzzy sets of each of the six rules indicated above. If the reader is interested in the actual membership functions used in this study, they can be consulted in the original source (Rufino et al. 2007). For further details about these procedures see von Altrock (1995). Just as an illustration, the following scorings resulted from the aggregation of the three criteria in the East Africa study of Rufino et al. (2007): if roofing type = 0 (unroofed); cover type = 5 (uncovered) and flooring type = 3 (solid) the management is *poor*, and the resulting management factor is 0.49. If roofing type = 0 (unroofed); cover type = 10 (polythene film) and flooring type = 3 (solid) the management is *good* and the management factor gets a value of 0.80. The manure management factor (MF) thus calculated is used to compute the relative loss rate (RLR) of a give nutrient during manure storage, as follows:

$$RLR = MF \times (RLR_{BEST} - RLR_{WORST}) \tag{9.6}$$

Where, RLR_{BEST} and RLR_{WORST} correspond respectively to the minimum and maximum relative rates of nutrient loss during manure storage. These are parameter values that are obtained from experimental measurements. Whether fuzzy logic systems are accurate enough or not to simulate dynamic processes such as nutrient flows and cycling in agroecosystems depends on the objective of the study and on the amount of information available to create well-informed membership functions. In some cases, a good 'guestimate' may be preferable to a poor or unreliable measurement.

9.5.3 Qualitative Methods

Not being able to quantify indicators, or to use indicators that are quantifiable in their nature, or to use fuzzy logic systems to translate linguistic variables into numerical ones as just explained, does not imply that trade-offs cannot be analysed. There are different ways of representing and communicating trade-off situations in participatory settings, using visualisation methods such as rich pictures, analogies such as 'traffic light' scores (e.g., green for 'good', red for 'bad' and yellow for 'indifferent') or through semi-quantitative methods that assign more or less 'tokens' to statements or objectives depending on how much one agrees or disagrees with them (Box 9.3). Figure 9.11a shows an example of the qualitative assessment of trade-offs between five objectives associated with three crop management systems. The darker the colour the greater the score pertaining to each objective. It is important in this case to indicate the direction of desirable change for each indicator. With an indicator such as 'yield' the desirable direction of change is clear. But when it comes to assessing costs, for example, it is important to indicate whether a darker colour implies high costs or low costs. For this reason it is preferable to express the objective as 'cost reduction' or 'lower costs' as done in the example of Fig. 9.11a, and thus again the darker the better.

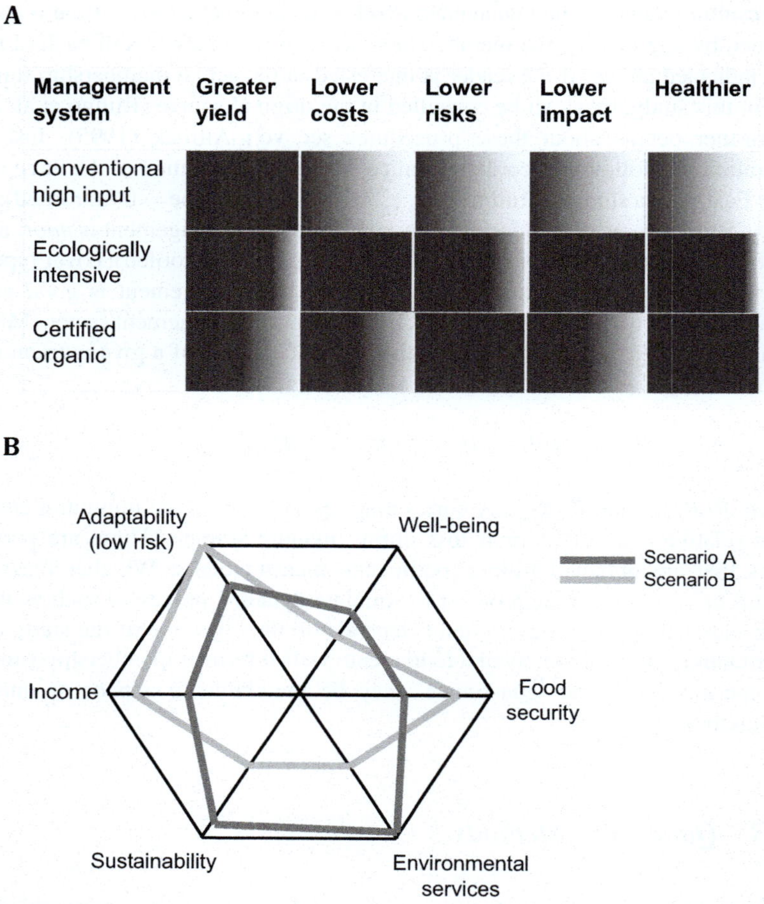

Fig. 9.11 Examples of visual assessment of trade-offs. (**A**) Trade-offs between five objectives comparing three crop management systems. Full grey indicates the highest score for the objective considered, and the degree of greyness indicates the relative distance to the maximum score. (**B**) Trade-offs between six objectives comparing two alternative future scenarios. Each objective takes a relative value from lowest to highest moving outwards from respect to the centre

Figure 9.11b shows an example of qualitative assessment of six objectives to compare two possible future scenarios. There seems to be a trade-off between objectives such as income and environmental services, or between these and food security. Sustainability in this case is presented as an objective, under the more narrow, ecological definition of the term: the ability of a system to maintain its structure and function over time. There is a trade-off between this objective and adaptability, since adaptation may sometimes require profound transformation of the original system. Broader definitions of sustainability actually consider all these objectives simultaneously, and thus sustainable systems are those that are productive, adaptable, that lead to well-being and to the delivery of ecosystem services, etc. This type of multi-criteria representation, known as radar, amoeba or spider-web graph, is also used in quantitative trade-offs analysis, for which the values of the indicators have to be standardised. In quantitative analysis, each objective must be represented by one or more quantifiable indicators, as will be explained in Chap. 10.

Box 9.3: Semi-quantitative Participatory Scoring

In her PhD thesis on nutrient management options for smallholder farm households in the mid hills of Nepal, Victoria Alomia Hinojosa used a graphical board method to understand farmers' decision-making through their perceptions, preferences and priorities. The board depicted the various fields of a farm and/or the possible land use and management options, which were 'scored' by farmers by means of tokens. Different tokens represented cash income, labour, yields, costs, etc. In the picture below we see the results of comparing different ways of growing a maize crops, either sole or intercropped with different legume species, planted in rows or broadcasted, etc. These scores can be recorded and used in semi-quantitative assessments of trade-offs associated with the adoption of these technologies.

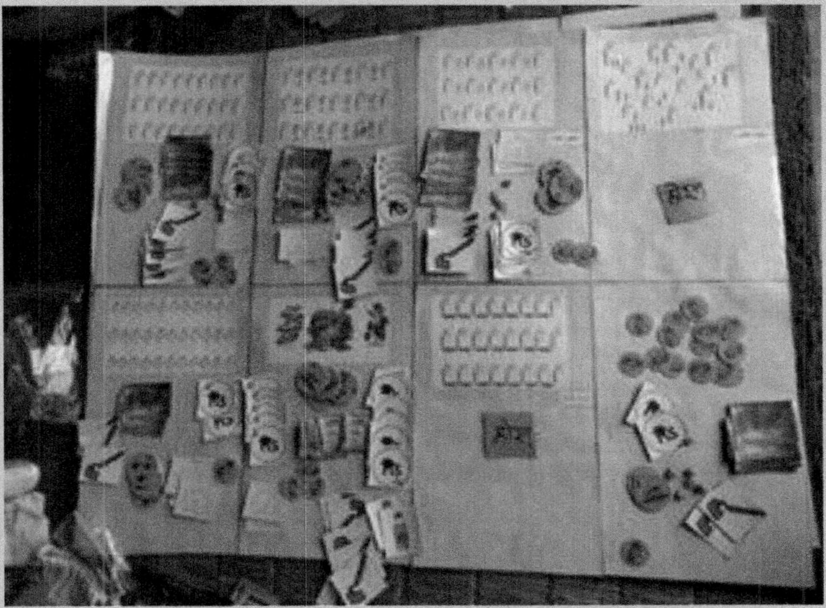

Source: *Alomia Hinojosa, M.V., Sustainable intensification of cereal-based agro-ecosystems through crop-livestock integration and diversification in the Terai and Mid-hills of Nepal, on going PhD Thesis, 2018, Wageningen University, The Netherlands.*

9.5.4 Games

Another way of approaching participatory, semi-quantitative analysis of trade-offs in complex systems involving stakeholders is the use of adult board games. These are particularly suited to allow participants to safely enact and explore the benefits and challenges of complex collective land use decision-making during the learning process (cf. García-Barrios et al. 2011). Games can be used as a tool to reflect upon alternative land use decisions by a community, or by researchers, to learn more about the way a community, a household or an individual farmer, make decisions (Michalscheck et al. 2020). Such games were first developed to facilitate learning in business education (Duke 1974), and since then they have been used in a variety of settings and for different goals, always involving stakeholder participation (Wilkinson 2016). Games, in particular role-playing ones, have been used in combination with agent-based models (cf. Speelman et al. 2014) to address problems regarding collective decision processes, common property, or co-ordinated actions among actors, etc., often crossing disciplinary boundaries and embracing system complexity (e.g., Etienne 2014).

In the field of agriculture and natural resource management, games have been used as discussion and decision support tools (e.g. Barreteau 2003; Etienne et al. 2015). Such games are commonly developed in such a way that goals and rules have many degrees of freedom and therefore the solution space of the game is mostly unknown. These are known as open-ended games. Bousquet et al. (2001), founding members of the community of practice in game development known as COMMOD (which stands for Companion Modelling: http://www.commod.org), argue that games with an unknown solution space are difficult to reproduce, and therefore the systematic comparison of their results is unpractical. Closed games, on the other hand, are those in which the goals and rules define a large but countable set of solutions which can be revealed through analytical and simulation methods. These are generally simpler and more stylized games. They can be used in experimental set ups designed in a way that enables the replication of results with various groups of participants (Falk and Heckman 2009). Closed games allow testing specific experimental hypotheses about the relation between game outcomes and the attributes and behaviours of players (García-Barrios et al. 2011).

Speelman et al. (2014) used this approach to unravel collective decision-making on land use and natural resource management in a community located next to a Man and Biosphere Reserve in Chiapas, Mexico. A role-playing board game was specifically developed to actively involve smallholders with conflicting interests and activities in the process of designing more sustainable agricultural landscapes. The game, known as RESORTES (literally coil-springs in Spanish), which is the Spanish acronym for Social Networks and Sustainable Land use Planning (Speelman and García-Barrios 2010), depicts an agricultural landscape and tries to capture the major challenges faced by the local community. These are, on the one hand, risk and uncertainty in monetary returns associated with different land use types. On the other, the coordination of land use activities among farmers, in order to (i) obtain financial benefits from payment schemes that reward the provisioning of ecosystem services from non-fragmented forest areas, or (ii) scale-related additional returns or reduced costs for open field land use types. When playing the game, participants make land use decisions and through them they can gain 'payment for ecosystem service' points and/or 'economy of scale' points. The rationale of the game is briefly explained in Box 9.4.

Box 9.4: The RESORTES Board Game (Speelman and García-Barrios 2010)

RESORTES (*Redes Sociales y Ordenamiento Territorial Sustentable*) is a board game developed to explore, discuss and evaluate the land use planning process. The goal of the participant is to win the game by accumulating the largest number of points. The target audience are smallholder farmers and/or communities that are planning their agricultural landscape; researchers; NGO offices; students that work on land use cooperation issues. The game requires six participants, one facilitator and one assistant, and it requires 15 min for preparation, 30–60 min for playing and 45–90 min for debriefing.

Technical specifications

Cooperative; non-zero-sum; goal-seeking; common access game (See classification of Klabbers 2009 p. 42). Non-electronic intellectual skill game - Board game (See classification of Ellington et al. 1982 adapted by Klabbers 2009 p. 37). Game board consisting of 37 connected hexagons divided in four equally-sized quadrants; Field cards – 24; Land use cards – 4 sets of 24 cards; Dice – 2 with one small range and one large range of numbers but with same average; Scoreboard; Monopoly money; Computer facilitated version of the game available in Netlogo 4.2 (Wilensky 1999) by contacting the corresponding author (erika.speelman@wur.nl).

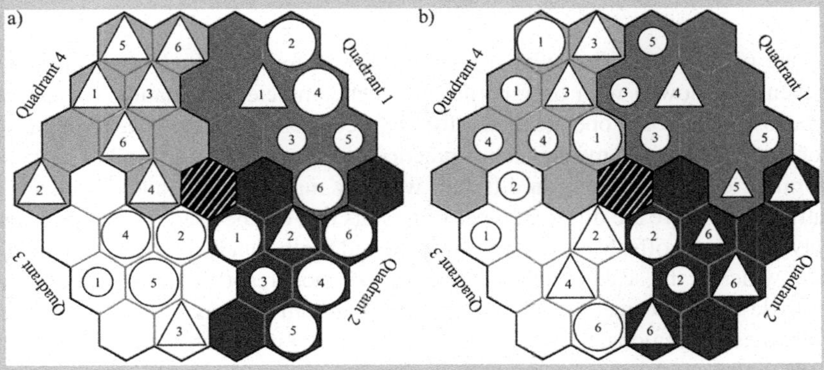

(continued)

Box 9.4 (continued)

Mechanics and rules

A typical game session starts off with extensive game explanation through trial rounds. Once the facilitator has assured him or herself that all players understand the game, field allocation starts. Players take turns and select one field location per round. When all fields have been selected land uses are added to the fields, one per field. When all players are satisfied with their selected land uses, both dices are thrown and the facilitator and the players jointly check if any additional points through the planning schemes are earned. Then, that round's points are calculated and all players receive their points. Then, the facilitator highlights the current state for obtaining additional points through one or two of the planning schemes. The game continues as long as the players want. After the game, the game debriefed in the form of group discussion. Two incentive schemes for landscape planning can lead to additional points: (1) Payment of Environmental Services (PES) and (2) Economies of Scale (EoS). The PES scheme requires eight fields per quadrant with forest-cover (virgin or forest-based land use) and rewards all those who hold a field in the respective quadrant five additional points per round. The EoS scheme requires a minimum of ten of the same cleared-field land uses on the whole board and rewards every cleared-field land use four additional points per round. At the end of every round, players receive that round's points. Points result from standard points dependent on risk-level of current land use choices, and additional points from one or both planning schemes.

9.6 Quantitative Trade-Offs Analysis Using Simulation Models

In spite of having to deal with so many different forms of trade-offs, farmers are frequently endowed with only partial, often incomplete information on how their agroecosystem would operate under different scenarios, or in the long term, or when adopting alternative management practices. Here is where models can help informing decision-making in different ways by mapping out trade-offs in a quantitative way. There are different types of models that can be used for the quantitative analysis of trade-offs. Most of them are designed for the analysis of trade-offs at farm scale, which is the scale at which farmers make most resource allocation decisions. Yet trade-offs also emerge at field or cropping system level as well as at the level of landscapes and regions, for which models can be very informative tools (Kanter et al. 2018). Some models are specifically designed to analyse trade-offs, whereas others can be used to quantify objectives and map out trade-offs in a more ad-hoc fashion (cf. Sect. 9.4). The major difference between the former and the latter resides in the use or not of optimization functions in the model, and on whether the dynamic feedbacks in the system are captured or not within the model. Based on these criteria, models used for trade-offs analysis can be roughly grouped as follows:

A. Models that optimize or approximate objective functions
   - Static, single-time-step models with point solutions
   - Static, single-time-step models with solution spaces and/or frontiers
   - Dynamic models coupled to global search algorithms

B. Models that do not optimize or approximate any objective function
   - Static, single-time-step models (quali- or quantitative)
   - Statistical models (including spatially explicit land use analysis models)
   - Dynamic simulation models (e.g. of cropping, livestock or farm systems)

The list may not be exhaustive, and the differences may not be always sharp between categories, but this rough grouping of models allows comparing approaches in terms of the advantages and disadvantages of each type, data requirements, assumptions and type of results to be expected. The next chapter will present and discuss examples of models used for trade-off analysis at farm scale, following this rough classification, but also a few examples of models used for trade-offs analysis at lower (field) and higher (landscape) integration levels.

The various types of models for trade-off analysis at farm scale can be very useful at informing *discussions* with farmers. In such cases, it is rather more useful to present information on both optimal and sub-optimal solutions, in the form of trade-off maps instead of point solutions. Only in a few exceptional cases these models could be used to support *decisions*. They are not necessarily meant to 'predict' outcomes, as would a crop simulation model for example, but rather to 'explore' possibilities (cf. van Ittersum and Rabbinge 1997). The first three categories of models described here are also suited for trade-off analysis at higher scales, such as the landscape or regional scales. They can be combined with approaches that consider local actors explicitly such as agent-based systems (e.g. Baudron et al. 2015). A limitation of farm scale models in general is that they are often parameterized for a few representative farms – mostly due to intensive data requirements. The analysis generally starts with a categorization of farm diversity by means of surveys and farm typologies (cf. Chap. 4), and by the selection of one or a few representative farms per farm type to be used in the modelling study. Capturing the probability distribution of the model outcomes of farm types in a territory would require modelling techniques that simplify data requirements at farm scale (e.g., minimum data approaches – Antle et al. 2016).

Although models may be powerful tools, as they allow finding the best compromises between more than two objectives simultaneously, their weaknesses reside in the identification of relevant indicators that represent farmers' objectives and rationale, and in the risk of over-interpretation of the results obtained. Identifying indicators – and their meaningful thresholds – is an important step in agroecosystem design (cf. Chap. 8). Such indicators may be expressed as objectives in multiple criteria and multi-scale analyses, representing the goals of different stakeholders.

The definition of meaningful indicators and thresholds also allows the negotiation and participatory prototyping of desired agroecosystem functions. An example from western Kenya shows how the feasibility and the transition towards prototype farms designed by local farmers could be evaluated by means of a simulation model (cf. Tittonell et al. 2009 – presented in Chap. 8). Several approaches using models as *discussion* support tools to enhance participatory learning and action were developed worldwide, using different type of models. The Companion Modelling approach referred to above is one of them (Simon and Etienne 2010), which combines participatory tools such as role-playing games with the simulation of the interaction between actors in a community using agent-based modelling techniques.

9.7 Summary

Trade-offs emerge when resources are scarce and imply the conciliating two or more competing objectives. Trade-offs are common in agriculture and natural resource management. Stakeholder participation in trade-offs analysis is crucial when societal values are involved. The concept of alternative trade-offs regimes, and the assumption of thresholds within and between regimes, allows linking up trade-offs analysis with resilience thinking theory of multiple stable states. There are several methods to assess trade-offs, from the calculation of simple elasticity indicators when examining the substitutability between pairs of objectives to the use of semi-quantitative methods that allow stakeholder participation and simulation modelling coupled with scenario analysis. In this chapter we looked at examples of quantitative, semi-quantitative and qualitative trade-offs analysis methods. Several examples considered trade-offs in mixed crop-livestock systems, where the allocation of resources between cropping and livestock activities often presents competing objectives. The use of simulation models for trade-offs analysis will be examined in more depth in the following chapter.

References

Alcamo J (2001) Scenarios as tools for international environmental assessments. In: Environmental issue report 24. European Environment Agency, Copenhagen

Alomia-Hinojosa V, Groot JCJ, Andersson JA, Speelman EN, McDonald AJ, Tittonell P (2022) Assessing farmer perceptions on livestock intensification and associated trade-offs using fuzzy cognitive maps; a study in mixed farming systems in the mid-hills of Nepal. Systems Research and Behavioral Science 40:146–158. https://doi.org/10.1002/sres.2830

Andrieu N, Vayssieres J, Corbeels M, Blanchard M, Vall E, Tittonell P (2015) From farm scale synergies to village scale trade-offs: cereal crop residues use in an agro-pastoral system of the Sudanian zone of Burkina Faso. Agric Syst 134:84–96

Antle JM, Basso B, Conant RT, Godfray HCJ, Jones JW, Herrero M, Howitt RE, Keating BA, Munoz-Carpena R, Rosenzweig C, Tittonell P, Wheeler TR (2016) Towards a new generation

of agricultural system data, models and knowledge products: design and improvement. Agri Syst. https://doi.org/10.1016/j.agsy.2016.10.002
Aravindakshan S, Krupnik TJ, Shahrin S, Tittonell P, Siddique KHM, Ditzler L, Groot JCJ (2021) Socio-cognitive constraints and opportunities for sustainable intensification in South Asia: insights from fuzzy cognitive mapping in coastal Bangladesh. Environ Dev Sustain 23: 16588–16616
Barreteau O (2003) The joint use of role-playing games and models regarding negotiation processes: characterization of associations. J Artif Soc Soc Simul 6
Baudron F, Delmotte S, Corbeels M, Herrera JM, Tittonell P (2015) Multi-scale trade-off analysis of cereal residue use for livestock feeding vs. soil mulching in the Mid-Zambezi Valley, Zimbabwe. Agric Syst 134:97–106
Bousquet F, Lifran R, Tidball M, Thoyer S, Antona M (2001) Agent-based modelling, game theory and natural resource management issues. J Artif Soc Soc Simul 4:1–5
Carpenter SR, Pingali PL, Bennett EM, Zurek MB (eds) (2005) Ecosystems and human well-being (Volume 2): scenarios. Findings of the Scenarios Working Group of the Millennium Ecosystem Assessment. Island Press, Washington, DC
Castellanos-Navarrete A, Tittonell PA, Rufino MC, Giller KE (2015) Feeding, crop residue and manure management for integrated soil fertility management – a case study from Kenya. Agric Syst 134:24–35
Cortez Arriola J, Groot JCJ, Amendola Massiotti RD, Scholberg JMS, Mariscal Aguayo DV, Tittonell PA, Rossing WAH (2014) Resource use efficiency and farm productivity gaps of smallholder dairy farming in North-West Michoacán, Mexico. Agric Syst 126:15–24
Cortez-Arriola J (2016) Exploring opportunities for on-farm innovations in smallholder dairy systems of Michoacán, Mexico. Wageningen University, Wageningen
Diarisso T, Corbeels M, Andrieu N, Djamen P, Tittonell P (2015) Biomass transfers and nutrient budgets of the agro-pastoral systems in a village territory in South-Western Burkina Faso. Nutr Cycl Agroecosyst 101:295–315
Duke RD (1974) Gaming, the future's language. Wiley, New York
EEA (2007) Land-use scenarios for Europe: qualitative and quantitative analysis on a European scale, EEA Technical Report No. 9. EEA, Copenhagen
Erenstein O, Gérard B, Tittonell P (2015) Biomass use trade-offs in cereal cropping systems in the developing world: overview. Agric Syst 134:1–5
Etienne M (2014) Companion modelling: a participatory approach to support sustainable development. Springer, Berlín
Etienne M, Le Page C, Bousquet F (2015) Le collectif ComMod : modélisation et accompagnement
Falk A, Heckman JJ (2009) Lab experiments are a major source of knowledge in the social sciences. CESifo Working Paper Series No. 2894. Available at SSRN: https://ssrn.com/abstract=1536400 or https://doi.org/10.2139/ssrn.1536400
Gallopín GC (1997) Branching futures and energy projections Renew Energy Dev 10(3):4–5
García-Barrios L, García-Barrios R, Waterman A, Cruz-Morales J (2011) Social dilemmas and individual/group coordination strategies in a complex rural land-use game [Dilemas sociales y estrategias de coordinación individuales/grupales en un juego complejo del uso del suelo rural]. Int J Commons 5(2):364–387. https://doi.org/10.18352/ijc.289
Kanter DR, Musumba M, Wood SLR, Palm C, Antle J, Balvanera P, Dale VH, Havlik P, Kline KL, Scholes RJ, Thornton P, Tittonell P, Andelman S (2018) Evaluating agricultural trade-offs in the age of sustainable development. Agri Syst. https://doi.org/10.1016/j.agsy.2016.09.010
Kok K (2009) The potential of fuzzy cognitive maps for semi-quantitative scenario development, with an example from Brazil. Global Environ Change 19:122–133. https://doi.org/10.1016/j.gloenvcha.2008.08.003
Kok K, Van Delden H (2009) Combining two approaches of integrated scenario development to combat desertification in the Guadalentín watershed, Spain. Environ Plan B 36(1):49–66

Kok K, Patel M, Rothman DS, Quaranta G (2006) Multi-scale narratives from an IA perspective. Part II. Participatory local scenario development. Futures 38(3):285–311

Michalscheck M, Groot JCJ, Fischer G, Tittonell P (2020) Land use decisions: by whom and to whose benefit? A serious game to uncover dynamics in farm land allocation at household level in Northern Ghana. Land Use Policy. https://doi.org/10.1016/j.landusepol.2019.104325

Naudin K, Bruelle G, Salgado P, Penot E, Scopel E, Lubbers M, Ridder N, Giller K (2014) Trade-offs around the use of biomass for livestock feed and soil cover in dairy farms in the Alaotra lake region of Madagascar. Agri Syst 134:36–47. https://doi.org/10.1016/j.agsy.2014.03.003

Raskin P, Monks F, Ribeiro T, Vuuren D, Alonso A, Field C (2005) Global scenarios in historical perspective. Ecosystems and Human Well-being 2

Rufino MC, Tittonell P, van Wijk MT, Castellanos-Navarrete A, de Ridder N, Giller KE (2007) Manure as a key resource to sustainability of smallholder farming systems: analysing farm-scale nutrient cycling efficiencies within the NUANCES framework. Livest Sci 112:273–287

Scheffer M, Van Geest GJ, Zimmer K, Jeppesen E, Søndergaard M, Butler MG, Hanson MA, Declerck S, De Meester L (2006) Small habitat size and isolation can promote species richness: second-order effects on biodiversity in shallow lakes and ponds. Oikos 112:227–231. https://doi.org/10.1111/j.0030-1299.2006.14145.x

Simon C, Etienne M (2010) A companion modelling approach applied to forest management planning. Environ Modell Softw 25(11):1371–1384

Speelman EN, García-Barrios LE (2010) The RESORTES (Redes Sociales y Ordenamiento Territorial Sustentable), unpublished boardgame

Speelman EN, García-Barrios LE, Groot JCJ, Tittonell P (2014) Gaming for smallholder participation in the design of more sustainable agricultural landscapes. Agric Syst 126:62–75

Thierfelder C, Cheesman S, Rusinamhodzi L (2012) A comparative analysis of conservation agriculture systems: benefits and challenges of rotations and intercropping in Zimbabwe. Field Crop Res 137:237–250. https://doi.org/10.1016/j.fcr.2012.08.017

Tittonell P (2013) Towards ecologically intensive smallholder farming systems. Design, scales and tradeoffs evaluation. In: Vanlauwe et al (eds) Agro-ecological intensification of agricultural systems in the African Highlands. Earthscan/Rutledge, Oxon, pp 132–144

Tittonell P (2014) Livelihood strategies, resilience and transformability in African agroecosystems. Agric Syst 126:3–14

Tittonell P, Leffelaar PA, Vanlauwe B, van Wijk MT, Giller KE (2006) Exploring diversity of crop and soil management within smallholder African farms: a dynamic model for simulation of N balances and use efficiencies at field scale. Agric Syst 91:71–101

Tittonell P, van Wijk MT, Rufino MC, Vrugt JA, Giller KE (2007) Analysing trade-offs in resource and labour allocation by smallholder farmers using inverse modelling techniques: a case-study from Kakamega district, Western Kenya. Agric Syst 95:76–95

Tittonell P, van Wijk MT, Herrero M, Rufino MC, de Ridder N, Giller KE (2009) Beyond resource constraints – exploring the physical feasibility of options for the intensification of smallholder crop-livestock systems in Vihiga district, Kenya. Agric Syst 101:1–19

Tittonell P, Corbeels M, van Wijk MT, Giller KE (2010) FIELD – a summary simulation model of the soil-crop system to analyse long-term resource interactions and use efficiencies at farm scale. Eur J Agron 32:10–21

Tittonell P, Gérard B, Erenstein O (2015) Tradeoffs around crop residue biomass in smallholder crop-livestock systems – what's next? Agric Syst 134:119–128

UNEP (2007) Global Environment Outlook 4. Environment for Development Assessment Report. Malta by Progress Press Ltd., Malta

Valbuena D, Groot JCJ, Mukalama J, Gérard B, Tittonell P (2015) Improving rural livelihoods as a "moving target": trajectories of change in smallholder farming systems of Western Kenya. Reg Environ Chang. https://doi.org/10.1007/s10113-014-0702-0

Van Ittersum MK, Rabbinge R (1997) Concepts in production ecology for analysis and quantification of agricultural input-output combinations. Field Crops Res 52:197–208 von Altrock C (1995) Fuzzy logic and neurofuzzy applications explained. Prentice Hall PTR, New Jersey

Wilkinson P (2016) A brief history of serious games. In: Dörner R, Göbel S, Kickmeier-Rust M, Masuch M, Zweig K (eds) Entertainment computing and serious games, Lecture notes in computer science, vol 9970. Springer, Cham. https://doi.org/10.1007/978-3-319-46152-6_2

Chapter 10
Pathways for Agroecological Transitions

Abstract Agroecological transitions entail a reconfiguration of the social and ecological dimensions of agroecosystems to produce and sustain rural livelihoods and ecosystem services following the principles of agroecology. Agroecological transitions are designed by and with the stakeholders involved. The practice of co-creating, implementing and evaluating knowledge among different actors has been defined as co-innovation in Chap. 8. There are different methods and tools to support co-innovation in practical terms, which draw on participatory approaches. Although this is not a text book about agroecological design, some of these methods will be briefly presented and discussed in this chapter. At the field, farm or landscape level, a complementary methodology to support reflection, discussion and decisions in a co-innovation process is using mathematical simulation models. Models are helpful when assessing (ex-ante) the likelihood of a particular impact, be it positive or negative, associated with a new design. But also, to explore scenarios and trade-offs together with stakeholders, or to generate 'images' of the future. Any mathematical model can be used to support co-innovation. Here, however, I will focus on those that can be used to inform farm-level design decisions and compare transition pathways, by explicitly addressing trade-offs and synergies. Finally, this chapter discusses the necessary changes at broader scales, such as value chains and food systems, and proposes a series of steps to conceptualise and shape agroecological transition pathways and the corresponding domains of innovation alongside the transition.

10.1 Understanding Agroecological Transitions

An agroecological transition can be defined as the gradual (social, ecological, organizational and technical) reconfiguration of an agroecosystem over time necessary to produce and deliver food, fibre and other substances, plus ecosystem services, following agroecological principles (cf. Chap. 1). The nature and extent of such reconfiguration depend strongly on the starting point of the transition, and on the scale at which the transition is to take place. Agroecosystems may transition from a degraded estate, from a simplified, industrial state, from an already agrobiodiverse

P. Tittonell, *A Systems Approach to Agroecology*,
https://doi.org/10.1007/978-3-031-42939-2_10

state, etc., which represent clearly distinct starting points (cf. Tittonell 2020 and Chap. 1). Transitions may take place at farm scale, sometimes even within a portion of a farm, or at landscape (community) level, at territorial, regional, or even at national level. Transitions at higher levels of integration may require profound reconfigurations of the legal, commercial and institutional environments through policy innovation. In other words, agroecological transitions at territorial to national level need alignment of the socio-economic, ecological, institutional and political dimensions of agroecology.

Further, the notion of agroecological transitions varies across schools of thoughts, organisations and individuals, which is partly due to the multiple definitions of agroecology that coexist. Well-known pieces of literature have defined agroecology as 'science, practice and movement', that is, a scientific discipline, a set of farming practices and a global social movement (Sevilla Guzmán and Woodgate 2013; Altieri and Nicholls 2017; Wezel et al. 2018; etc.). However, encountered visions still persist that emphasise any one of these dimensions over the other two, sometimes to the point of finding, suggesting or instilling potential conflict between them (e.g. Toledo and Barrera-Bassols 2017; Giraldo and Rosset 2018). In this book, I adhere to the spirit of the science, practice and movement definition (or rather agroecology as *the* science of practice and movement – Chap. 1). However, this definition conveys a strong narrative but is rather impractical when it comes to operationalising such concepts in the form of policies, development programs, technological innovations or monitoring and evaluation systems to support transitions.

Here, I propose to assess the complexity of agroecological transitions from multiple perspectives, by focusing on:

- the knowledge systems on which they rest;
- the set of practices, norms, organisations and discourses that stakeholders deploy to generate, share or implement such knowledge; and
- the facilitating actions that pave the way for the transition, which include political and institutional change.

Science, practice and movement cut across the three dimensions proposed above, so there is no intrinsic contradiction between both definitions. And, since these three dimensions (in short: knowledge, innovation and policies/institutions) imply changes at different levels and scales, we should not refer to an agroecological transition but to several simultaneous, interacting agroecological *transitions* (Tittonell 2019). Agroecological transitions may imply different levels of complexity, types of stakeholders and spatial-temporal scales (Duru et al. 2015; Anderson et al. 2019; Tittonell 2019). This represents the *breadth* of the transition process. Agroecological transitions may sometimes imply relatively small and discrete technological changes (e.g., farmers starting to use biofertilizers or reduced tillage), to thorough reconfiguration and redesign of production systems and landscapes, or even whole food system transitions including changes in trading, manufacturing, distribution and consumption patterns. These are two contrasting examples in terms of the *depth* of the transition process. Therefore, agroecological transitions can be categorised with respect to their breath and their depth. Examples of very broad and deep transitions are not yet abundant unfortunately.

Agroecological transitions are designed by and with the stakeholders involved in them, be them farmers, civil servants, researchers, extension agents, traders, processors, retailers or consumers. The interplay between knowledge implementation, practices and reflective learning is also often termed 'co-innovation system' (Dogliotti et al. 2014; Rossing et al. 2021). Knowledge and practices represent two key dimensions of change in agroecological transitions. They may be indeed mobilised, created or fostered by different actors of the innovation system, such as the research, extension and training sectors, the private sector and the proximate institutional environment. Social movements are essential to facilitate the creation of enabling environments for the transition to agroecology at local, regional or national levels (or even international, through green or fair trading). Thus, the science, practice and movement dimensions of agroecology operate at different levels. Let us use a conceptual model to represent their roles.

10.1.1 A Three-Gear Engine

Operationally, I propose to consider three *levels* to categorise complexity in agroecological transitions: (i) the farming system,[1] with its own set of 'intrinsic' knowledge and practices, (ii) the innovation support system, characterised by local organisations and institutions that foster new knowledge and practices, and (iii) the enabling conditions that result from the broader social-ecological and political context. These three levels, represented as gears in Fig. 10.1, influence one another. Local innovation support mechanisms that are conspicuous in agroecology are the multi-actor innovation platforms (cf. examples in Tittonell et al. 2016), but also the classical agricultural extension systems where they still exist. The gear model suggests that small – slow – changes in the broader environment or enabling conditions, sometimes just signals, may translate into fast changes in the two lower levels, particularly at the farming system level. This is however a hypothesis that needs to be tested in different contexts by examining current dynamics in agroecological transitions and their drivers.

Evidence in favour of this mechanism may be found in the example of countries like Brazil, where public policies have positively influenced the transition towards agroecology of thousands of family farmers over the last two decades (the upper gear in Fig. 10.1). Such a transition has been supported by a large number of institutions from NGO's, farmer unions, markets, universities and even local churches (e.g., Teixeira et al. 2018), that constitute the innovation support agents (the middle gear in Fig. 10.1), and that existed before the enabling conditions were established. This

[1] Here, I refer to farming system as an abstract concept, broader than the agroecosystem, which is concrete and delimited in space.

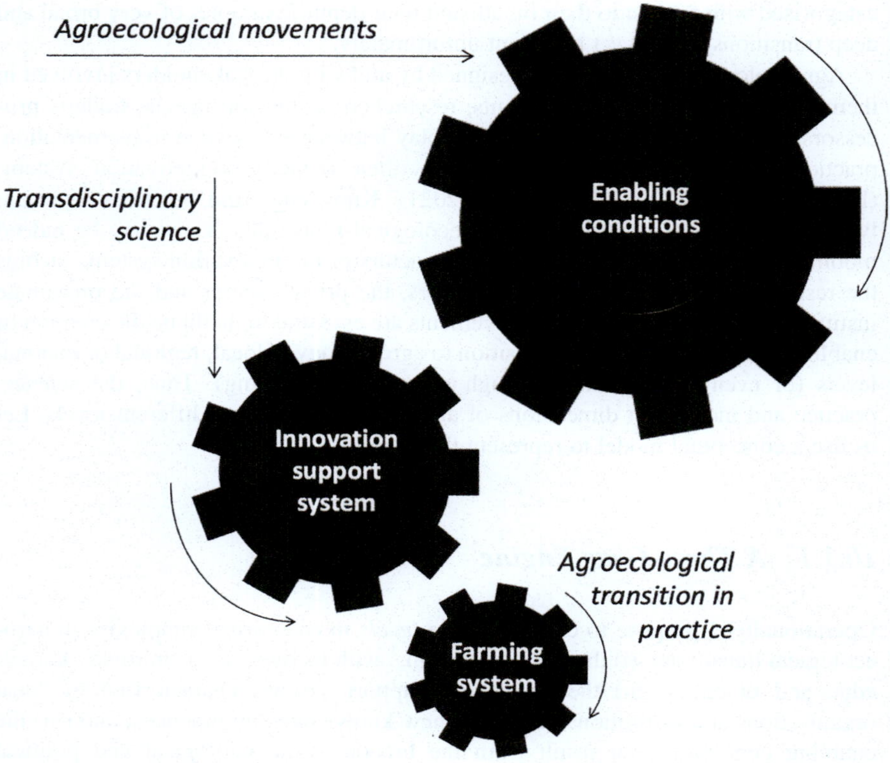

Fig. 10.1 Gear model to represent multiple levels in agroecological transitions: Can slow pace changes in the enabling conditions translate in fast changes in both innovation support and farming systems?

resulted in a broad agroecological transition in Brazil. Yet, if such a mechanism was everywhere true and generalisable, then agroecological farmers would not be special or exceptional in a given region or territory, as they tend to be today, even in Brazil (e.g., in Minas Gerais, self-defined agroecological farmers represent ca. one third of the farmers in a territory – Teixeira et al. 2018). Moreover, given the current emergence of policies in favour of agroecology across the world, we should be experiencing a smooth and broad transition to agroecology across different farming system types and regions, which is mostly not the case. This suggests that there are still elements that block the turning of the smaller gears. So, what else is necessary to make sure that these gears turn smoothly, transmitting their power (dynamics) across levels?

To begin, if the middle gear, which represents innovation support systems, is absent, the forces that favour the turning of the larger gear (enabling conditions) would not result in the movement of the smaller gear (agroecological transition). Furthermore, for smooth turning, gears require regular lubrication. In this case,

I refer to three elements that act as the oil between the gears, which I call "catalysers": knowledge, innovations, and policies/institutions. Without these catalysers, changes in enabling conditions would lead to heterogeneous and scattered responses in farming and innovation support systems. These responses may as usual exhibit pioneering behaviours among farming families and institutions, but would not result in smooth, broad, and deep agroecological transitions. For instance, let's consider the well-known case of agroecological transition in Minas Gerais, Brazil (Cardoso et al. 2001; Teixeira et al. 2021), which I frequently reference in this book. The local NGO *Centro de Tecnologias Alternativas da Zona da Mata* (CTA) has played an essential role as a middle gear over the past 25 years. The presence of CTA was a prerequisite for national policies such as the *Fome Zero* Program or the National Agroecology Law to succeed in this territory, stimulating successful transitions by providing support and guidance to local initiatives.

In other cases, the innovation support system may be present, in the form of national research and extension systems, or even the political environment be enabling to agroecology, but the necessary agroecological innovations may not be available. This is a situation faced currently in countries in which their governments support agroecology (Mexico, France, Argentina, etc.) but where the classical agricultural research and extension systems are not yet prepared[2] to support agroecological transitions. This is a preoccupation often shared by colleagues working in such organisations, as they are eager to support agroecological transitions in practice but lack the locally relevant innovations to do that, or the knowledge available worldwide may not always be applicable in their context, or the social movements that may help develop the local innovation support system may be absent. Table 10.1 summarises the interplay between the levels at which agroecological transitions take place, those represented by the gears in Fig. 10.1, and the catalysers of the transition, those that represent the oil in between the gears: knowledge, innovation and policies/institutions.

Finally, there are the drivers, the ones that make the entire gear system move (represented as arrows in Fig. 10.1). Socio-cultural changes, new consumption patterns, political movements, environmental concerns, media trends, new technologies, etc., they may all influence the turning of the gears, particularly the upper one representing the enabling conditions. Social movements have a very active role to play at provoking shifts in the enabling conditions, by e.g. pushing for national laws and policies that promote agroecology, advocating new forms of production and consumption through communication campaigns and awareness raising, promoting alternative markets and economies, creating networks that connect knowledge and innovation systems, or lobbying for institutional change. Transdisciplinary science, through the co-construction or knowledge and co-design of practices together with different stakeholders is essential to support innovations, both technical and organisational.

---

[2]Prepared in the sense of 'willing to', but also in the sense of having the necessary knowledge, budget, technologies and personnel to support agroecological transitions.

Table 10.1 The interplay between the interrelated levels at which agroecological transitions take place (the gears in Fig. 10.1) and the catalysers necessary for each dimension of the transition: knowledge, innovation and policies/institutions (the oil between the gears)

	Catalysers ('the oil in between the gears')		
Levels ('the gears')	Knowledge	Innovations (technical and organisational)	Policies and institutions
Farming system[a] (agroecosystem)	Local (traditional and new) agroecological knowledge National and global knowledge systems (scientific literature, knowledge repositories, databases, case studies, observatories, etc.)	Redesign of production processes, inputs and outputs Restoration, improvement and maintenance of the natural resource base (incl. agrobiodiversity) Redesign of value chains and market channels	Favourable regulations (environmental, economic, sanitary) Incentives for agroecological transitions (credit, subsidies, tax) Agroecological markets and value chains (bio-economy)
Innovation support system	Applied knowledge stored, actualised and delivered by research and extension services Know-how from value chain and collective (associative, cooperative) organizations Knowledge created, stored and delivered by NGO's (local and global) Limitation of the negative effects of secrecy, patents and privatisation of knowledge	Co-innovation among stakeholders or value chain actors (farmers, traders, processors, retailers, consumers) Knowledge brokers and innovation champions (from public and private academic, research and extension organisations) Financial incentives to sustain co-innovation systems Organisational engineering to support shared learning	Innovators (influencers, quarter masters, knowledge brokers) within governmental structures and among policy makers Fluid communication channels between research/innovation and policy making Cooperation with social and political movements Role of national and international influence groups (WWF, IUCN, IPBES, IPCC)
Enabling environment	Mechanisms for co-creating and sharing (from distal to peer-to-peer) knowledge Available (proven, accessible, actionable, affordable) management techniques, inputs, tools and technology packages Fundamental and applied (transdisciplinary) agroecological science	(Local) governance of co-innovation systems Social norms (e.g. gender) and value systems that promote equitable participation in co-innovation processes Professional networks and scientific societies (e.g., SOCLA, ENAF) Freedom of choice and elimination of coercive unsustainable practices (e.g. compulsory fumigations)	Policy action in support of agroecology Existing agroecology policies (e.g. Brazil, France, Mexico) Facilitated and equitable access to natural resources Fair distribution of benefits across value chains Green, fair, responsible markets Financial incentives to absorb transaction and transition costs and risks

[a]Contrary to what was expressed in Chap. 2, I prefer to use here the term farming system instead of agroecosystem, as transitions may imply changes in a single farm but also over an entire farming sector

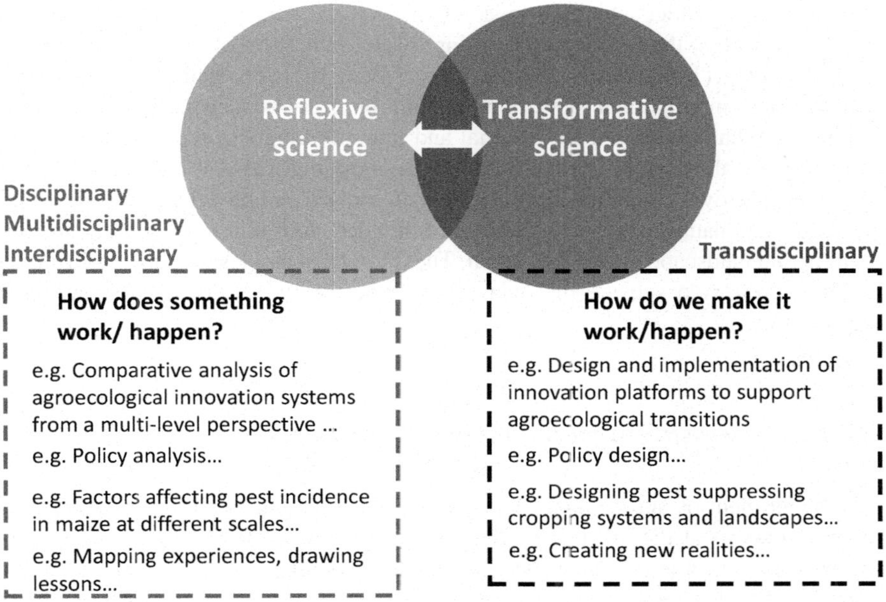

Fig. 10.2 A representation of the differences between reflexive science (a step forward from reflective science) and transformative science in support of agroecological transitions, indicating the most common approaches to knowledge generation and learning (disciplinary, multi-, inter- and transdisciplinary) in agroecology

There are different instances of knowledge generation at which different approaches may be necessary. For example, in the realm of reflexive science, disciplinary, multi- and interdisciplinary approaches may be needed at different stages to understand how something works or why and how something happens (Fig. 10.2). Transformative science, at least in agroecology, requires transdisciplinary approaches to co-innovate among stakeholders to be able to transform realities (cf. Chap. 1). Both phases, reflexive and transformative, are necessary and complementary. They interact to continuously support innovations. It is often assumed among agroecology activists and enthusiasts that transformative, transdisciplinary science is the only approach valid in agroecology. This is a mistake. It is the interplay between both types of science, the smooth flow of information, hypotheses and questions between these two phases what makes agroecological knowledge progress. Understanding and then acting, but also understanding through acting, or even while acting.

Note also the small but important difference between reflexive and 'reflective' thinking. Reflective implies pondering about what we have learned, while reflexive implies considering also how such learnings affect the broader context of our actions. Relevant questions to be addressed in order to draw generalisable lessons from the study of successful (or unsuccessful) transitions to agroecology include, among others: Under which socioeconomic, cultural, legal, political or ecological conditions did(is) the transition take(−ing) place? What sort of mechanisms were in place to support producers in their transition? Were these enabled by governmental, commercial, international development, aid or grassroot institutions? Local or regional? What were the main external and internal drivers of the transition? Were they ecological, cultural, legal, political, or economic drivers? What was the role of the market? How was the innovation system organised and how did it emerge? What were the key management practices or redesign decisions that fostered the transition at farm and landscape scale? And so on. These are all examples of questions from the realm of reflexive science that can inform transformative science and action in agroecology.

10.1.2 Assessing Agroecological Transitions

The assessment of agroecological transitions requires understanding the necessary transformations at the level of production systems and their impact on its various performances, but at the same time the enabling conditions and innovation support mechanisms that foster such transitions (cf. Fig. 10.1). Let us examine some theoretical frameworks (reflexive approaches) proposed to assess agroecological transitions at multiple levels. Transitions have been studied by several authors and from different perspectives, in fields as diverse as technology, policy, consumption, energy or sustainability transitions. Prof. Steve Gliessman from the US described agroecological transitions as a sequence of three phases: efficiency gains, input substitution and system redesign. Although this has been broadly used in the agroecology literature with reference to Gliessman (2006), it is actually an adaptation of the ESR model (efficiency-substitution-redesign) proposed earlier by Prof. Stuart Hill from Australia (Hill and MacRae 1996). Building on these ideas, Fig. 10.3 provides a selection of frameworks from my own past work at conceptualising agroecological transitions. These are based on approaches of increasing complexity, which were borrowed from different disciplinary domains and adapted to the study of agroecological transitions.

Starting with the simplest, the single-scale approach in Fig. 10.3a uses the concept of state and transition models from ecology to depict transitional dynamics at the level of agroecosystems (without focusing on enabling conditions or innovation systems) and their services, as their ecological structure changes (degrades or aggrades) (Tittonell 2018). This concept has been recently expanded to consider multiple alternative states when assessing the contribution of agroecology to building resilience and adaptability of agroecosystems (Tittonell 2020). Figure 10.3b represents the various phases in the transition towards agroecology, more or less in the same way as originally put forward by Hill and MacRae (1996), and later Gliessman (2006), but as influenced by both technical and institutional innovation (cf. Tittonell 2014a). Figure 10.3c represents multiple-level transitions taking place simultaneously at technical, social-ecological and institutional scales through four

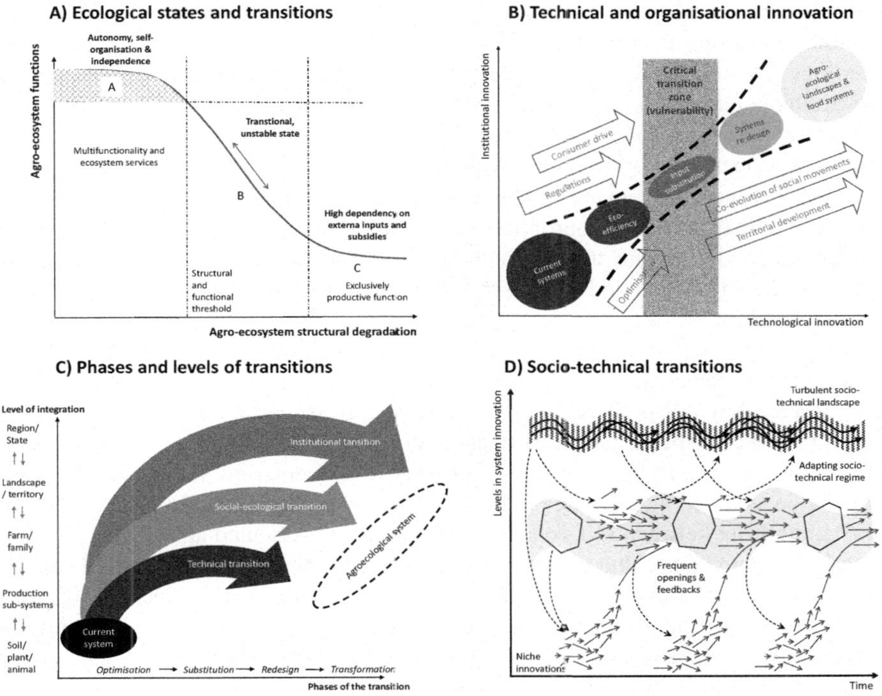

Fig. 10.3 Four increasingly complex perspectives that can be used to assess agroecological transitions as proposed in my previous work. (**a**) Agroecosystem dynamics and functions using state and transition models (Tittonell 2020); (**b**) Agroecological transitions as the result of technical and institutional innovation (Tittonell 2014a); (**c**) Multiple-level transitions from system optimisation to transformation (Tittonell 2019); (**d**) Agroecology as niche innovation using Geel and Schot's multi-level perspective (Tittonell et al. 2016). See text for explanation

phases: optimisation, substitution, redesign and transformation (Tittonell 2019). Finally, Fig. 10.3d uses the multi-level model proposed by Geels and Schot (2007) to illustrate how a 'turbulent' socio-technical landscape is nowadays generating frequent openings in the sociotechnical regime that characterizes agriculture and food production for agroecological niche innovations to emerge and become part of the regime (cf. Tittonell et al. 2016).

In spite of their increasing complexity, none of these approaches indicates how the various levels influence one another. In other words, the mechanisms by which the turning of a gear communicates movement to the next level are not explicit (cf. Fig. 10.1). The sociotechnical regime in Fig. 10.3d comprises technologies, institutions and, in the particular case of agriculture, natural resources (their status and quality). Using the same approach, Anderson et al. (2019) proposed six domains of transformation to define the interphase between niches and regimes: 1. Access to natural ecosystems; 2. Knowledge and culture; 3. Systems of exchange; 4. Networks; 5. Equity; 6. Discourse. These domains of transformation determine agroecological practices and are influenced by and influence governance systems. Within each domain, there are factors, dynamics, structures, and processes that constrain agroecology, and those that enable it. However, while these authors identify enabling or disabling conditions within each domain, I refer to enabling conditions in Fig. 10.1 more broadly, as those that shape the regime itself and therefore generate the necessary openings for niche innovations, thereby promoting transitions. In other words, beyond focusing on enabling or disabling innovations around particular agroecological practices, I propose to take a whole-system perspective in which enabling conditions are a combination of socio-technical landscape and regime (cf. Fig. 10.3d) factors and processes that allow, promote or prevent agroecological transitions, not just innovations.

Enabling factors or levers to promote agroecological transitions may be technical (e.g., new management practices), institutional (e.g., market shifts, innovative R&D support systems) or political (e.g., 'sticks and carrots'). The relative importance of each of these factors depends on contextual, cultural, demographic and sometimes historical conditions that affect the propensity of farmers, rural families and firms to transition towards agroecology. The relative importance of such factors will also certainly differ between self-consumption and market-oriented agriculture. In major market-oriented agricultural sectors, such as horticulture, intensive animal production and export-oriented commodity and specialty crops, the propensity of value chain actors for the transition to agroecology is variously affected by policies and regulations, accessibility, markets, technologies, production costs and natural resources (Table 10.2). For example, while horticultural farmers are largely influenced by market demands, so that the transition to agroecology may be strongly and directly steered by consumers, the transition to agroecology in animal production is much more affected by the availability, status and quality of natural resources, notably access to space, water, supportive soils and quality grasslands. In both cases, however, beliefs, discourses, networks and value systems will still primarily influence the propensity of actors to transition to agroecology.

Table 10.2 Some factors influencing the propensity of actors in different value chains to transition to agroecology

Factors	Horticulture (e.g. fresh vegetables)	Animal production (e.g., beef, milk, wool)	Export commodities (e.g., maize, soybean, oil palm)	Export specialties (e.g. coffee, tropical fruits, cacao)
Policies & regulations	++	+	+	++
Accessibility	++	–	–	–
Markets	+++	++	++	+++
Technology	+	–	+++	+++
Production costs	–	+	++	+
Natural resources	–	+++	+	+

(–) no or minimal influence; (+) moderate influence; (+++) strong influence

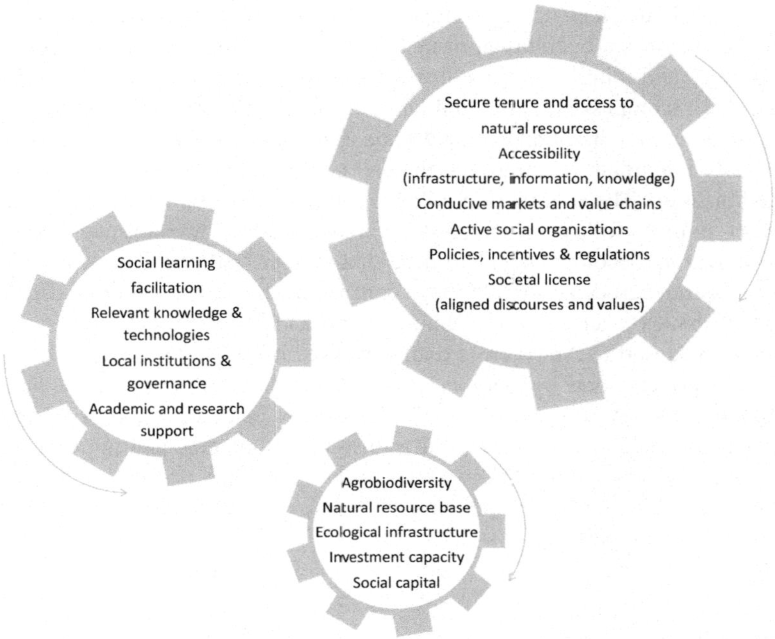

Fig. 10.4 A representation of the three levels relevant to assessing agroecological transitions and examples of what to look for at each level

The factors enumerated in Table 10.2 are thus not exclusive nor exhaustive, they involve more than one of the six domains proposed by Anderson et al. (2019), and they operate at the three levels proposed in Fig. 10.1, the farming system, the innovation support system, and the enabling conditions. These factors and their relative importance will vary across types of farming systems (self-consumption- vs. market-oriented) and world regions. Agroecological transitions entail much more than changes in practices or production technologies: transitions may often require thorough redesign of agroecosystems and value chains. To operationalise the concept of multi-gear transitions, Fig. 10.4 lists sets of enabling and disabling factors at each level that may influence the propensity of farmers, communities and firms to transition to agroecology.

10.2 Designing the Transition with Stakeholders

A crucial initial step in the process of agroecological transition with stakeholders is to identify the desired end goals of the transition. This involves envisioning the future configuration of the agroecosystem in terms of its structures and functions. Identifying this future picture of the agroecosystem is, in itself, an exercise in design.

As discussed in Chap. 2, it is important to differentiate between management practices (which are operational or tactical in nature) and design principles (which are strategic and long-term in nature) when addressing agroecosystem design. The design can encompass an individual farm, a specific component or process within a farm, or an entire community or landscape unit comprising multiple farms.

Equally important to identifying the end goal is characterizing the starting point of the transition. The starting point of an agroecosystem significantly influences the pathway it must follow, whether it be transitional or transformative. Does the transition to agroecology begin from a smallholder traditional system or a large-scale industrial agroecosystem? Does it start from a degraded soil or landscape? Is it within a context rich in (agro-)biodiversity and local knowledge or within a "biocultural desert"? Are there existing market incentives or not? What are the prevailing policies and regulations? Is the socio-cultural environment favourable or unfavourable? Answering these questions helps shape the appropriate trajectory for the agroecological transition and informs the strategies and approaches to be employed. Understanding the starting point is crucial for developing tailored interventions that consider the specific context and challenges faced by the agroecosystem in question.

10.2.1 Prototyping and Back-Casting

The process of agroecosystem design comprises two major phases, known as prototyping and back-casting. During these phases, the team developing the design plan (ideally a diverse team balancing gender, age, educational background and field experience) needs to consider four major aspects that will shape their space for manoeuvring in terms of what is possible, what is feasible, and what is viable:

- The context
- The resources
- The livelihood system
- The objectives

These four aspects are central to delineating a realistic design and they need to be well characterised through a baseline study. Based on these aspects, prototyping consists of coming up, in a participatory way, with a vision for an agroecosystem that is able to fulfil all of the objectives that drive the new design, be them economic, social or environmental. During prototyping, and to be able to think creatively and freely, participants must assume that financial investments and future costs or labour requirements are not an obstacle to achieving the new design. This is why, when doing this exercise with rural communities, prototyping is described as delineating the 'ideal farm', in which all the necessary resources are assumed to be available. Fig. 10.5 below shows an example of participatory prototyping during a field workshop in Kenya.

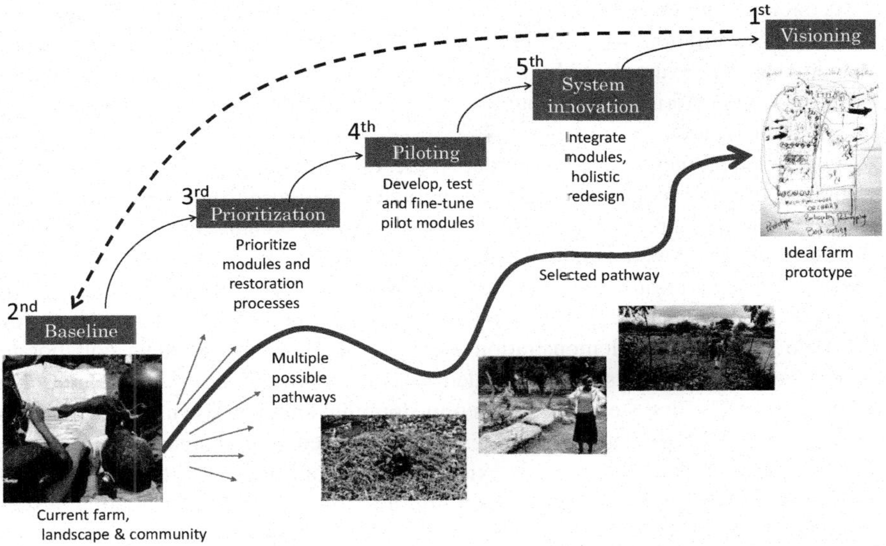

Fig. 10.5 A scheme representing a participatory prototyping and back-casting exercise with a smallholder farming community in western Kenya. The exercise starts by visioning, delineating an 'ideal farm' prototype. From such vision, and using the baseline information in terms of context, resources, livelihoods and objectives, transition pathways are selected that undergo iterative phases of prioritization and system innovation. In this example, the design was organised in 'modules', such as a livestock module, a food crops module, etc. in which farmers piloted and tried out management practices. The most promising pilots from the modules were finally integrated into a holistic design (whole system innovation)

The back-casting phase consists of delineating a realistic implementation plan for the redesign, from the current farm (baseline) to the ideal farm (vision). There may be a wide diversity of possible pathways in this transition but they may not be all implemented simultaneously (Fig. 10.5). This is why it is important to undergo a prioritization phase to identify which practices will be first implemented, bearing in mind that implementing new practices may often come with 'yield penalties' or initial productivity loses while the new practice is adjusted and/or properly mastered. To arrive at the ideal farm prototype, only the options that are biophysically possible, economically viable, and socially acceptable (feasible) are selected and prioritised, and then tested through establishing pilots. In an earlier exercise in Vihiga, also in western Kenya, Tittonell et al. (2009) observed that when delineating the ideal farm participants tended to overestimate for example the flows of nutrients that can be achieved within an average-sized farm. For example, the amount of manure that can be produced to fertilise soils or the amount of crop residues that will be available to feed to livestock (cf. Chap. 9). This is where calculations, using back of the envelope mathematics or complex simulation models, are of key importance to inform agroecosystem design and prioritization. Priorities may also be defined according to economic pressures or market/value chain opportunities, or according to the urgency of restoration measures in farms or landscapes where resources are severely degraded. In the example of Fig. 10.5, the community chose as guiding principles only those trajectories that favoured soil regeneration, recycling (circularity) and social inclusiveness.

One practical way to start an agroecological transition pathway on the ground is to define different modules within the desirable, ideal farm. For example, a poultry module, a nutrient recycling module, a multipurpose orchard module, etc.. These modules are developed gradually, and subsequently connected at farm scale. The redesign of the various modules can be done by installing pilots, demonstration units, or modules. These can be implemented on all participating farms, on a few example farms, or on demonstration areas in which all participants collaborate. When the transition starts from degraded soils and landscapes, restorative measures take priority. The first step consists then in stopping, slowing down and/or reducing the impact of the degradative factors (e.g. reduce soil erosion, increase water capture, balance nutrient flows, reduce soil oxidative conditions, etc.). Once the first restorative practices are in place and functioning, it is time to start incorporating system innovations that may require greater investments in terms of money, knowledge, time and/or labour. Here is where most practices under circular agriculture start to make sense, by connecting the various modules of the farm to achieve a holistic farm management. These are but a few ideas and concepts on agroecosystem restoration and redesign. This is a book about systems analysis, not about design; a forthcoming book on agroecological design will elaborate on these steps and methods further.

10.2.2 Transition Pathways

Once the starting point is characterised, the end goal defined, and the design principles identified, it is time to choose a pathway. There are several practices and design principles that are proposed as alternatives to conventional or industrial agriculture, using ecological principles to variable extents. They can be classified based on two criteria (Fig. 10.6). In the first place, on their reliance on ecological replacement, that is, to what extent they replace external inputs with ecological functions. Then, on the scale or level of integration at which they operate or are designed and implemented. There are many exceptions to the broad categories represented in Fig. 10.6, and fuzzy rather than sharp boundaries between them. If we define industrial agriculture as the result of classical agronomy (Chap. 2: Autoecology), which reasons at field scale and relies mostly on external inputs, then its reliance on ecological replacement can be depicted as nil or minimal, as done in Fig. 10.6 as 'current agriculture'.

Conservation agriculture or agroforestry are also designed at field scale, but they exhibit greater reliance on ecological replacement (Fig. 10.6). Agroforestry, however, is a wide categorisation and we may find a diversity of approaches under this umbrella term. Integrated pest or nutrient management combine agrochemicals with

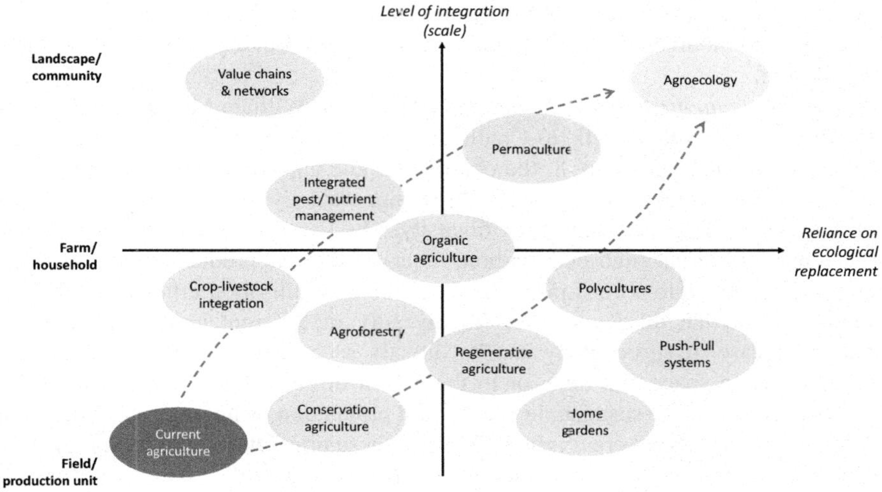

Fig. 10.6 Alternative approaches to industrial agriculture that can be engaged in a transition to agroecology, classified according to their reliance on ecological replacement and the level of integration at which they operate. Organic agriculture is placed arbitrarily at the centre, as it normally entails a combination of farm-sale design, input substitution and ecological replacement (e.g. pest biocontrol). Dashed arrows illustrate two alternative pathways towards agroecological landscapes. Modified from Tittonell (2018)

natural regulation or organic inputs, and because of this, they tend to consider levels of integration that are higher than the single field. Polycultures and push-pull cropping systems are designed within a single farm, but they often imply changes at scales broader than the individual field. Organic and low input agriculture tend to integrate crop and livestock production and recycling, which imply farm-level design. These systems do not necessarily use agrochemical inputs, but they may use other forms of inputs. For example, in low input smallholder agro-pastoralism, free grazing livestock transfer nutrients from communal grazing areas to agricultural fields; such transfers are inputs to the farm system.

The various approaches represented in Fig. 10.6 may be seen as alternative pathways towards agroecology. Agroecology is depicted as the approach that exhibits the greatest reliance on ecological replacement, and operates at landscape rather than farm level. It must be said, however, that actual agroecological farms are often 'islands' of diversity operating within 'seas' of industrial agriculture, especially in large scale systems (cf. Tittonell et al. 2020). In most parts of the world I have worked, the idea of a complete agroecological landscape is far from being realised. There are of course notable examples of highly diverse landscapes with low or no chemical input use in South America, such as in native grassland-based livestock systems in Argentina and Uruguay (Modernel et al. 2018), or in the example of agroecological transitions in Mina Gerais, Brazil where up to one third of the farms on the landscape are agroecological ones (cf. Teixeira et al. 2018). Landscapes that host smallholder farming systems in e.g. several parts of Africa may also function as agroecological landscapes 'by default'. That is, because inputs are not used and biodiversity is still high, but not as a result of a purposive agroecological design.

"*Design is the first sign of human intention*", said William Macdonough, one of the proponents of the Cradle-to-Cradle approach to industrial design and architecture. As explained in Chap. 8, the cradle-to-cradle approach relies on three major principles that are also largely applicable in agroecology: (i) waste is food, (ii) use current solar income, (iii) celebrate diversity. The first principle refers to recycling and reusing materials (nutrients, carbon, water) in different production processes, the second one points to a maximisation of capture and utilisation efficiencies of solar radiation, and the third one refers to diversity in different ways, which in the particular case of agroecology can be directly associated with the idea of agro-biodiversity in space and time, or to the notion of combining diverse knowledge systems. Many of the sustainable agricultural production technologies and practices proposed in Fig. 10.6 were originally built on these principles, namely on recycling, efficiency and diversity, which are the key principles behind ecological intensification (Tittonell 2014a). A strong implication of these principles is that they respond to the need for a gradual decoupling of agriculture from the mining and petrochemical sectors and/or from any other form of exploitation of non-renewable resources. Indeed, when it comes to promoting an ecological intensification of agriculture through agroecology there are four major design principles to be observed (Tittonell 2018):

1. Decouple agricultural production from non-renewable resources
2. Rehabilitate degraded soils
3. Create synergies through crop and livestock integration
4. Manage landscape-level processes

These are not the only design principles necessary in an agroecological transition, but they represent an absolute minimum requirement. Transitions entail much more complex socio-ecological processes as depicted in different ways in Figs. 10.1, 10.3 and 10.4. But when it comes to concrete actions to be taken on the ground, targeting the biophysical dimension of agroecosystems, these four grounding principles may be of help as a starting point (examples of these are provided in Tittonell 2018).

10.3 Supporting Redesign: Model Simulation of Farm Scale Trade-Offs

In this section, I will provide examples on how simulation models can be used to support decisions on agroecosystem design and transition pathways, focusing on a common use of simulation models: the quantitative analysis of trade-offs and synergies at farm scale. These are, in my view, the most relevant case of model-based studies that can support agroecological design and transitions and hence the only type of models I will describe in some detail here.[3] This section builds on the definition of mathematical simulation models presented in Chap. 2, and on the classification of models suitable for trade-offs analysis proposed in Chap. 9. If the reader is not familiar with models, it is recommended to read the corresponding sections of Chapters 2 and 9 before proceeding with the rest of this section.

Trade-offs are central to rational decision making when resources are scarce (Tittonell 2013). Model-based explorations may contribute to capturing, at least in part, the complexity of the interactions between the socio-economic and bio-physical dimensions of agroecosystems. Trade-offs analysis using models consists of assessing *ex-ante* the consequences of operational (management) or strategic (design) decisions for the fulfilment of two or more farm-level objectives, under different scenarios. The models used for this vary in their complexity (Chap. 9), and their implementation requires an initial identification of trade-offs formulated as hypotheses. Sometimes farmers or researchers are not aware of trade-offs, and their identification requires methodologies that generally rely on participatory approaches as explained in Chapters 8 and 9. Let us recap the six main types of models that can be used for trade-off analysis that were proposed in Chap. 9:

A. Models that optimize or approximate objective functions:

   A.1 Static, single-time-step models with point solutions: These models provide a single solution that optimizes an objective function (or multiple ones) at a specific point in time. They do not consider variability or dynamic changes over time.
   A.2 Static, single-time-step models with solution spaces and/or frontiers: These models consider multiple solutions that form solution spaces or frontiers. They provide a range of options that optimize one or more objective functions at a specific point in time, allowing for decision-making based on trade-offs and preferences.
   A.3 Dynamic models coupled to global search algorithms: These models incorporate dynamic processes and are coupled with global search algorithms to explore and optimize objective functions over time. They can account for changes, uncertainties, and interactions within the system and identify optimal solutions based on the desired objectives.

B. Models that do not optimize or approximate any objective function:

   B.1 Static, single-time-step models (qualitative or quantitative): These models provide qualitative or quantitative assessments of the system without explicitly optimizing any specific objective function. They can be used for

[3] Modelling at lower or higher integration levels, such as crop growth modelling or land use change modelling, can provide valuable insights to scientists and contribute to reflexive science and understanding (as depicted in Figure 2). However, these types of models may not directly yield practical recommendations for farmers regarding agroecosystem redesign, unlike farm-scale modelling.

descriptive analysis, scenario comparisons, or identifying patterns within the system.

B.2 Statistical models (including spatially explicit land use analysis models): These models utilize statistical techniques to analyse patterns, relationships, and trends within the system. They do not aim to optimize specific objective functions but provide insights into statistical associations, which may be hypothesised as cause-effect relationships.

B.3 Dynamic simulation models (e.g., cropping, livestock, or farm systems): These models simulate the dynamic behaviour of the system over time and capture the interactions and feedback loops within the system. They do not focus on optimizing objective functions but allow for exploring different scenarios and understanding the system's behaviour under various conditions.

These different types of models serve various purposes in understanding, analysing, and making decisions within agroecosystems. They can provide valuable insights into system dynamics, trade-offs, and potential outcomes, depending on the specific goals and objectives of the analysis.

10.3.1 Static, Single-Time-Step Models with Point Solutions

Trade-offs at farm scale are inherent to farmers' decisions. Trade-offs may also emerge when objectives are prioritized by different stakeholders, considering different spatial and temporal scales, and responding to diverging concerns. For example, trade-offs are made between today's productivity and long-term sustainability, between farm economic efficiency and regional employment, between food security and negative externalities in the form of environmental impact, etc. Figure 10.7 shows an example of trade-off curves obtained through multiple-goal linear programming (MGLP – Type A.1), between the objectives of achieving food self-sufficiency at the farm scale and maximising the value of agricultural production sold on the market at regional scale (Lopez-Ridaura 2005). The curves in Fig. 10.7 were obtained through multiple model runs or calculations (one per point), modifying the mathematical constrains in the model at each step. In this particular example the optimisations where done at farm scale and at regional scale, resulting in two different trade-off curves. Thus, strictly speaking, this is not just a farm system model but a multi-scale optimisation model that considers the integration levels of farm, village and region. The lack of a line joining the various points along the curve is not trivial: each point is the result of an independent simulation, not a continuum of points in a solution space.

Often the solution landscape around each solution point in Fig. 10.7 is not known, and this is perhaps the major disadvantage associated with goal programming models. An optimum solution may be just what is known as a local optimum. Which is equivalent, in climbing terms, to reaching the top of a boulder on a misty

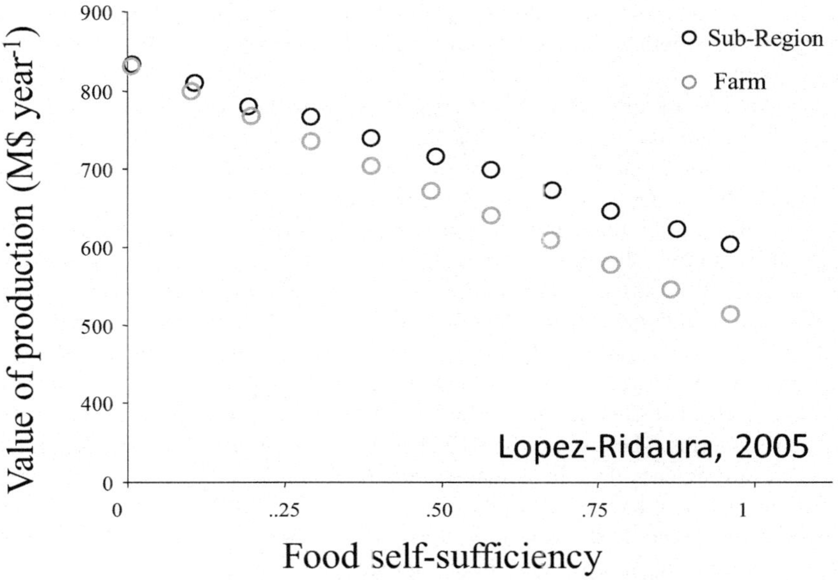

Fig. 10.7 Trade-offs analysis through multiple-goal linear programming considering two scales of analysis, the farm and the (sub-) region, and two objectives: farm scale food self-sufficiency and regional value of agricultural production. The trade-off between these two objectives is stronger when the analysis (optimisation) is done at farm scale than when it is done at the scale of the region, as indicated by the steeper slope in the former case

morning and thinking that one is already at the top of the mountain, when the top is not yet visible due to the mist. The top of the mountain represents the absolute optimum, and what we just reached is a local optimum, but we are not aware of it due to the mist that surrounds us. A way to unravel the solution landscape around an optimum is to use the second category of models (Type A.2), those that are able to generate explicit solution spaces.

10.3.2 Static, Single-Time-Step Models with Solution Spaces or Frontiers

This category includes models that use Pareto algorithms to map out efficiency frontiers. Fig. 10.8a provides an example of a trade-off map that results from alternative model solutions evaluated with respect to two objectives, in this case the organic matter balance at farm scale and the farm's operating profit (Groot et al. 2016). The red dot in the graph represents the current farm system. Each green or blue dot is a different solution, a different farm configuration and/or management strategy, and the external dashed line represents the Pareto efficiency frontier, or the most favourable trade-off regime between both objectives in this case. The blue dots represent the solutions that improve on both objectives with respect to the current farm system.

Unlike the case of goal programming, the solution landscape is known from this type of simulations, which provide the option of not only choosing optimal solutions but also sub-optimal ones. Fig. 10.8b shows simulations using the same model (FramDESIGN – Groot et al. 2016) for two smallholder farms in Malawi (Timler et al. 2020). The solutions indicate that increasing soil C inputs requires more labour (e.g., for manure collection, composting and application, or to grow cover crops, etc.) and the most favourable trade-off regimes are those in which soil C inputs are maximised while minimising labour needs. The solution space differs greatly between these two case study farms, one of them (Farm A) better endowed in resources than the other. The size and shape of the solution space can also be of interest to characterise 'windows of opportunity' for different types of farms under various scenarios. The distance between the current position of the farm (red dot) and the Pareto efficiency frontier, or most favourable trade-off regime, represents the efficiency gap that has to be fulfilled. In the case of Farm B, the efficiency frontier lies relatively closer to the current farm situation than in the case of Farm A, which has more room for improvement in both objectives.

Although only two objectives or indicators are shown in both graphs in Fig. 10.8a, b, the solutions were obtained by optimizing several objectives simultaneously (seven in the first case, five in the second). For any particular point of interest within the solution space, it is possible to trace back the combination of parameters (i.e. management or land use decisions) that led to that solution, and this can be very informative in discussing with farmers and other stakeholders. The advantage of this method is that no specific choice is made concerning optimal solutions. The entire solution space is presented to the user, allowing a greater transparency in model outcomes. Another advantage is that the objectives can be evaluated in their original units, without having to express them in monetary units.

10.3.3 Dynamic Models Coupled to Global Search Algorithms

The next category (Type A.3) of models used in trade-offs analysis at farm scale includes a very heterogeneous group of tools that can be built by linking dynamic simulation models to global search algorithms. Such procedures are sometimes termed 'inverse modelling techniques', because the desirable outputs are known or defined a priori, and the model is 'asked' to find all the possible combination of parameters (e.g., farmer management decisions) that lead to such results or set of desirable solutions. In essence, the approach works using the same principles of Pareto optimization described above, and mathematically speaking it operates in the same way as when the parameters of a dynamic model are calibrated through

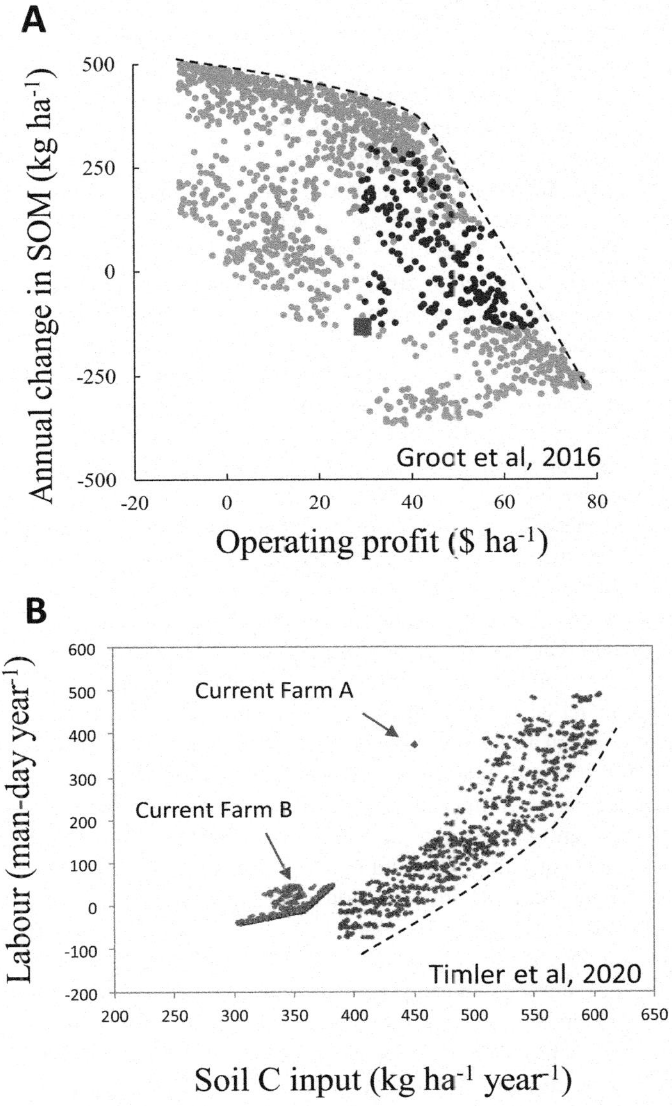

Fig. 10.8 Examples of farm scale analysis of trade-offs using Pareto algorithms to map out efficiency frontiers. (**a**) Relationship between profit and soil organic matter (SOM) balance at farm scale in a dairy system in western Europe; (**b**) Relationship between carbon inputs to soil and labour needs (family + hired labour) in two smallholder farms in Malawi; negative values for the labour balance means that the family's labour force is underutilised on the farm. In both graphs the red dots indicate the current position of the farm system with respect to these indicators, and the dashed lines are hand-drawn to indicate the most favourable trade-off regime

multiple runs to minimize the difference between observed and predicted output variables. In the case of inverse modelling, objective functions are defined that maximize or minimize the value of a certain objective indicator, which is also an output variable of the simulation model.

Figure 10.9 shows an example of a farm-scale trade-off analysis between nitrogen losses and farm productivity that was done using inverse modelling techniques, by linking a dynamic, daily time step crop-soil simulation model (DYNBAL – Tittonell et al. 2006) and a Metropolis-type global search algorithm (MOSCEM) (Tittonell et al. 2007). The example refers to a smallholder farm in Kenya (2.2 ha) that has a number of fields under different crops, a certain level of resource endowment, such as family labour or animal traction, and a certain initial capital to invest at the start of the growing season, either on hiring labour, purchasing fertilisers, etc. The tool was set up to explore multiple combinations of 12 parameters that included, among others, the amount of labour invested in land preparation or in weeding, the partitioning of organic manures between the various fields of the farm (preferential allocation to certain fields), the timing of activities, etc. The objectives to be optimized were to increase farm food production, to reduce soil losses through erosion, and to increase farm-scale nitrogen use efficiency (to minimize N losses). The dynamic crop-soil model was parameterized for each field and crop on the farm, and a central farm decision-making module allocated land, labour and inputs to the various fields of the farm. This central module also integrated the outcomes from simulations of the various fields and crops, calculated the degree of satisfaction the objective functions at farm level, and ranked the results on the basis of this.

When the model is run for a sufficiently large number of iterations the 'clouds' of possible solutions will also evolve towards trade-off frontiers, as in the case of the models presented in Fig. 10.8 and related text. The clouds of solutions in Fig. 10.9a correspond to three different scenarios of financial liquidity at the beginning of the growing season: the current one (light grey), a two-fold increase (dark grey) and a five-fold increase (black dots) in investment capacity (expressed in Kenya Shillings, KSh). They are plotted with respect to two of the three objective functions, and show a trade-off frontier that indicates that total N losses can be kept at 80 kg farm^{-1} year^{-1} when total farm grain production is below 6000 kg; beyond that productivity level there is a clear trade-off between productivity and N use efficiency. The advantage of this method – which may be computationally demanding – is that it does not use fixed input-output relationships as the previous two approaches. The feedback mechanisms that are simulated with the dynamic model are also still captured in the farm scale analysis.

10.3.4 Models that Do Not Optimize or Approximate any Objective Function

Among the models that do not optimize or approximate an objective function, the first two types (B.1 and B.2) are rather straightforward and present no particular features that would be worth discussing here. In essence, any statistical model

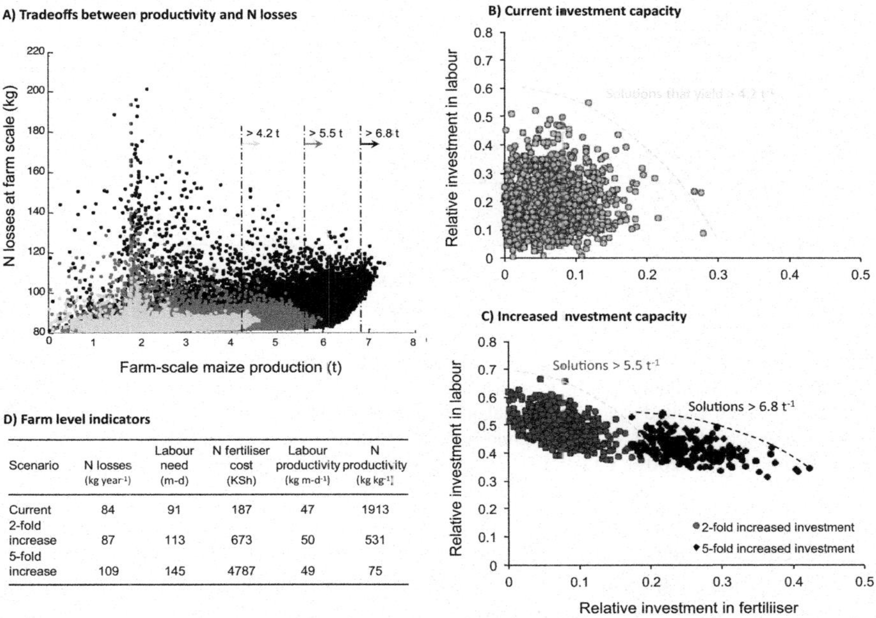

Scenario	N losses (kg year-1)	Labour need (m-d)	N fertiliser cost (KSh)	Labour productivity (kg m-d-1)	N productivity (kg kg-1)
Current	84	91	187	47	1913
2-fold increase	87	113	673	50	531
5-fold increase	109	145	4787	49	75

Fig. 10.9 Farm level simulations using a dynamic crop-soil model linked to a Metropolis type global search algorithm, simulating crop growth, N balances, soil erosion and labour allocation for a case study farm in Kakamega district, Kenya. The farm had nine fields and maize was grown in most of them. The objectives of the simulation were to maximise maize production at farm scale while minimising N losses and soil erosion. A total of 12 management parameters were optimised describing decisions such as land and labour allocation, and investments in labour and fertiliser inputs. Three scenarios of investment capacity were run: current (light grey), two fold increase (dark grey) and five fold increase (black dots) in investment capacity

obtained through empirical data, can be used to explore trade-offs by modifying the value of the independent variable(s) within the range of variability for which the model has been fitted. This category of models includes also econometric models, using linear or log regression models, and the GIS-based land use change models, which are a spatially explicit form of regression models. This type of models is seldom used in farm scale trade-off analysis. The last category (B.3) includes farm scale models in which the various components of the farm are simulated dynamically, with specified feedbacks, and linked among each other to exchange information at predefined intervals that depend on the time steps of each component model (e.g., daily, weekly, monthly). The example in Fig. 10.10 A shows a trade-off between milk production and weight gain of cattle in smallholder farms of Kenya as a result of alternative feeding regimes, exploring different possible scenarios over periods of 10 years (Tittonell et al. 2009). Figure 10.10b reveals different degrees synergy between primary and animal production, based on the same simulation outputs.

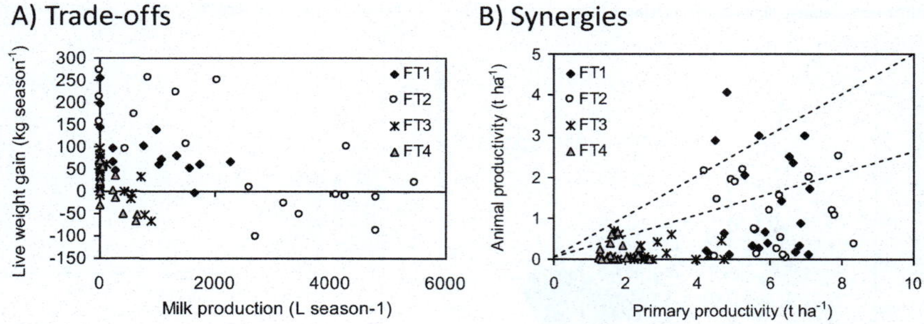

Fig. 10.10 Farm-scale indicators of productivity and resource use efficiency derived from the 10-year baseline simulation for the four case study farms. (**a**) Seasonal changes in live weight of the entire cattle herd plotted against seasonal milk production. (**b**) Primary vs. animal productivity (milk and weight changes), with dotted lines indicating the 1:4 and 1:2 productivity ratios. Modified from Tittonell et al. (2009). FT1 to FT4 represent different farm types

Since these models are not designed to optimize objectives (although they can be linked to global search mechanism as well), the trade-offs can only be analysed by running multiple scenarios and plotting the data against the indicators of interest (each point in Fig. 10.10 is the result of a 10-year simulation). Trade-offs curves or 'clouds' do not emerge as clearly as with the first three types of models described. The main advantage of dynamic farm scale simulation models is that they capture feedbacks at every time step, since the outcome of one component (e.g. animal manure production) can be the input of another component (e.g., manure application to cropping fields). Their main disadvantages are (i) their large requirements of data for the parameterization of the individual components and, (ii) when decisions are not dynamically simulated as in the case of Fig. 10.10, management variables are often 'fixed' for the entire horizon of the simulation, which prevents capturing feedbacks between the biophysical and decisional aspects of the farm system in the long term.

10.3.5 Integrating Methodologies in a Redesign Process

Based on the modelling approaches to support redesign described above, major challenges ahead of analysing trade-offs at farm scale include (i) finding an appropriate balance between detail and data requirements to make tools operational in data-scarce environments; (ii) linking dynamic simulation models with other means of representing evolving decision-making, to avoid running unrealistic long-term simulations using fixed decision parameters; and (ii) incorporating biodiversity mediated processes (such as pest regulation by natural enemies, synergies between crop species in polycultures, effects of soil biota on nutrient availability, etc.) that have received little attention in most current models in spite of their importance in agroecological transitions. This highlights also the importance of linking experimentation, field data gathering (biophysical and socioeconomic) and modelling tools, during the redesign of agroecosystems, in particular when these are increasingly governed by biological processes that cannot be easily controlled in an agroecological transition.

Most of the alternative agricultural systems, namely regenerative agriculture, organic farming or ecologically intensive low input systems, rely largely on biologically mediated processes, and aim to provide ecological services of support and regulation. Experimentation is essential to support redesign in such cases, as deterministic modelling will likely fail to reproduce the specific behaviour of such systems. I showed that field measurements, experiments and modelling can be combined in the diagnosis of a specific problem (cf. Chapter 2). Designing biologically intensive systems such as disease-suppressive cropping patterns or pest-suppressive landscapes is not possible without field data, measurements, and evaluation. On the other hand, scaling up the impact of a certain spatial configuration of biodiversity is not possible without models. We saw that field measurements and experiments may not be enough to validate the results of our bio-economic models of the farm system, and that expert knowledge and farmer perceptions may provide the right type of information we need in such cases. Thus, instead of segregating modelling from field measurements, experiments or expert and lay knowledge, I prefer to propose a framework for their integration, through the five steps in agroecosystem redesign presented as IDEAS in Chap. 8:

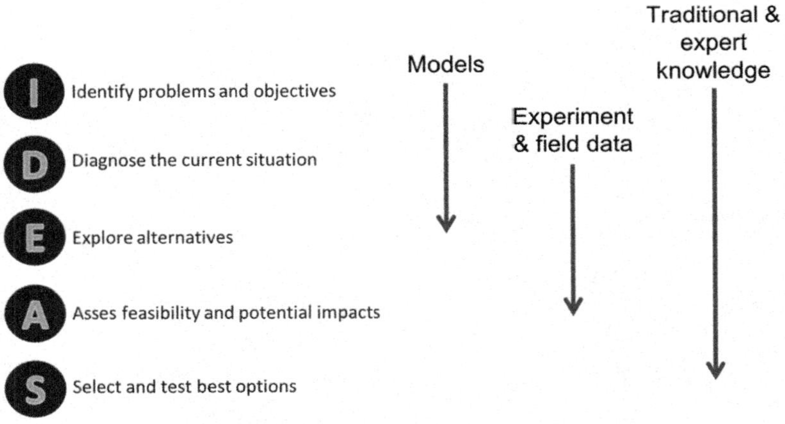

While the steps I, D and E correspond to the phase of analysis, steps E, A and S correspond to what we defined as 'synthesis' (cf. Chap. 2). Both analysis and synthesis are necessary phases in the design of systems that make intensive use of the natural functions of ecosystems. Models, field data, experiments and knowledge must be used in combination throughout this process.

10.4 Vertical Integration: Food Systems and Value Chains

The concepts and examples discussed so far around agroecological transitions and system redesign focused mostly at individual farm scale. Agroecological transitions, as mentioned repeatedly in this book, require transformations at scales broader than the farm. These include the landscape, the rural community, often the region or the entire country, the market, and generally the food system. The food system is defined as the chain of activities connecting food production, processing, distribution, consumption, and waste management. Fig. 10.11 places the agroecosystem, as was represented in Chap. 3 with its ecological and socio-cultural dimensions, at the intersection between these elements. Input markets, policies and institutions are part of the food system but they are also broader than it. Similarly, natural ecosystems are part of the food system but they may also be broader than it. It all depends on where the boundaries chosen to delimit a food system are placed and on the scale at which it is described. Food systems are also scale agnostic: they can be defined at community level, at regional or national level, or even at global level. Analysing agroecological transitions from a food system perspective reveals the degree of interdependence between the elements of the food system (production, processing, distribution, consumption and waste) and their decisive influence on the structure and functioning of the agroecosystem. Value chains, on the other hand, are defined as the various business activities and processes involved in producing a good or delivering a service. When value chains involve food products, they represent a

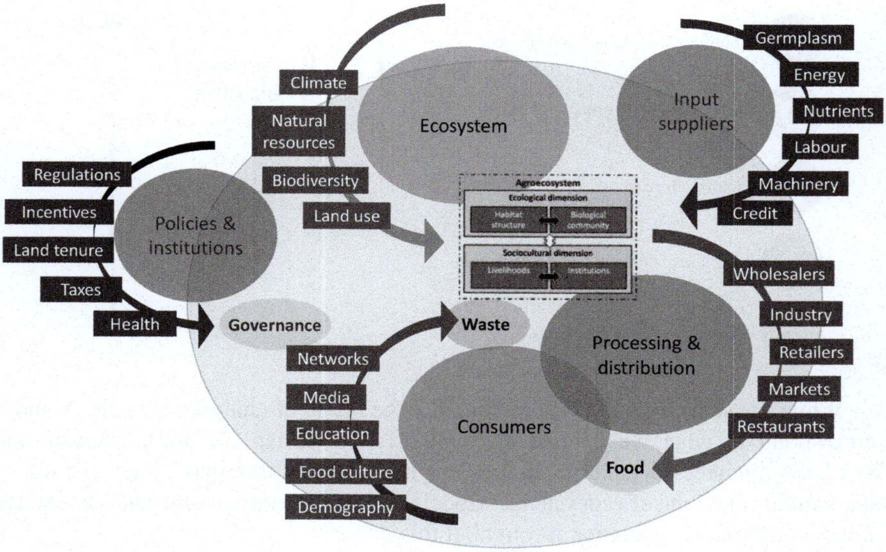

Fig. 10.11 A representation of the food system (grey circle), with the agroecosystem represented at its centre, at the crossroad of the natural and socioeconomic systems

sub-system of the broader food system. In Fig. 10.11, they could be placed approximately around the 'Processing and distribution' domain.

10.4.1 Food System Transitions

Agroecological transitions at food system level are not common, but when they occur, their transformative capacity is enormous. In other terms, they are broad and deep. One of the most conspicuous examples, due to its more than 20 years of existence and its territorial scale, is the case of the *Union de Trabajadores de la Tierra* (UTT: Land Labourers Union) in Argentina. In 2001, the country underwent a profound economic crisis that was the result of neoliberal policies implemented during the 1990s. Millions of people were unemployed and the productive matrix of the country, both rural and industrial, was devastated. The crisis triggered a broad diversity of social innovations, such as non-monetary exchange markets, factories being recovered and managed cooperatively by their workers, new waste recycling businesses, and so on. These were all bottom-up initiatives started up by social organizations, without governmental support. The agricultural counterpart of these organisational innovations was the UTT. As an organisation based in the outskirts of Buenos Aires, that brought together landless farming families from Argentina's hinterlands and neighbouring countries, they started negotiating with municipal authorities the usufruct of land that was abandoned or underutilized. In many cases they had to pay a fee or rental for the land. For a small fee, or for free, landless families were able to access parcels of land where they could produce for self-consumption and the market, after they were trained on agroecological production practices.

The goal of the UTT was, literally: 'to produce healthy food for the poorest'. In other terms, agroecologically produced food at affordable prices, even cheaper than at the supermarket. When the production was enough to start trading, the UTT opened a first 'selling point' in a low-income neighbourhood of Buenos Aires, where urban people could buy food (mostly vegetables at first) directly from farmers, without intermediaries, and at affordable prices. This was an immediate success. Soon, new land was acquired, new farming families joined, and new selling points were opened. Then internet trading came along as well. New production and selling units opened up in different parts of the country. Nowadays, people from all socioeconomic backgrounds buy directly from farmers through these selling points, which are now 412 located in different cities in 20 out of the 24 provinces that the country has. As the food system transformation progressed, agroecological family farmers from different parts of the country joined in, selling their produce through the UTT selling points and internet channels.

Nowadays, 42% of the farmers delivering food through the UTT selling points are smallholder farmers who own land, whereas the other 58% are farming on the collective farmland around urban areas following the original model developed by the organisation. The UTT collective farms initially focused on producing fresh vegetables, and most of them remain specialised in this, due to their proximity to urban markets. Through engaging new members who own land and farm in different parts of the country (different soils and climate), the UTT selling points are now able to deliver other types of agricultural products (meat, subtropical fruits, fish, dairy, legume pulses, etc.), even processed ones (flours, teas, fermented food, dairy, etc.). The entire system is built on trust, on farmer-consumer relations, the only form of certification used is participatory guaranty systems. Farmers who are in 'transition' to agroecological production, meaning even those that are still temporarily and/or exceptionally using pesticides, are also accepted as members, provided they are engaged in a pesticide reduction transition towards pesticide-free production.

The business model is simple: out of the price payed by consumers for the food they acquire in the UTT selling points or through internet, 60% goes to the producer, 20% to cover transport costs, and 20% to finance the organisation and its fixed costs (logistic, selling points, personnel, material, etc.). It is as simple and transparent as that. Why is this agroecological food cheaper than conventionally produced food? Farmers who sell their produce through conventional channels, such as wholesalers or supermarkets, get between 5 to 30% of the price that consumers pay, and they often need to wait until they get paid. When farmers get 60% of the selling price, and through the UTT, then food can be sold at affordable prices to the consumer, certainly much cheaper than by conventional retailers.

The UUT counts nowadays with 22,334 farming families. In the Argentinian context, this can be considered impact at scale, as they represent a large number of farmers when compared with the total smallholder, family farming sector of the country. For example, the last agricultural census of 2018 revealed that there are in Argentina 31,393 family farms smaller than 5 ha. Of course, there are some double-counting's, since many of these smallholder farmers owning land are already members of the UTT. Yet roughly, 22,334 farming families represent 70% of the total number of registered family farms in Argentina. This represent one of the most successful examples of agroecological transitions at scale, worldwide, and it provides an important lesson: the transition at the level of the agroecosystem was fuelled, supported and inspired by a major innovation at the food system level.

The success of the UTT (*Union de Trabajadores de la Tierra*) strongly relies on political activism, engagement, and the ability to navigate through existing administrative and institutional structures. It also emphasizes the importance of continuous learning and training for farmers and technicians in agroecological production practices, as well as postharvest, processing, and trading technologies. More information about the UTT can be found at their website: https://uniondetrabajadoresdelatierra.com.ar. Interestingly, the UTT's expansion has occurred despite changes in political power in Argentina, with both left- and right-wing parties governing the country over the past two decades. This expansion has been possible due to the UTT's independence from governmental aid. This serves as an important lesson for emerging agroecological movements in other parts of the world, where they may be striving for government support as the only way forward. While governmental support can be valuable, it is not the sole determining factor for success. Building strong social movements, fostering community engagement, and developing self-reliance can lead to sustainable growth and resilience, even in the absence of extensive government assistance.

10.4.2 Value Chains and Value Networks

Agroecological transitions require not only technical innovation but also institutional innovation, including market-related institutions (as depicted in Figure 2B – Tittonell 2014a). Institutions, in the broadest sense, encompass markets as well. Agroecological transitions at the farm or landscape level cannot be separated from innovations in markets and value chains. Circular and solidarity economies are integral components of agroecology and are among the ten defining elements of agroecology (Barrios et al. 2020 – see Chap. 1). The conventional approach to value chains, often depicted as linear and unidirectional chains of events (production, transformation or processing, and trade), is not compatible with agroecology. Instead, agroecology promotes circularity, solidarity, recycling, diversification, synergies, and natural regulation. These elements transform value chains into value networks. Figure 10.12 illustrates this concept, albeit in an exaggerated manner, to highlight the notion of value networks.

In a traditional value chain, every step in the chain adds value to the final product, which is ultimately traded on the market. However, in a value network, value is not solely generated through the main chain but also through the utilization and processing of by-products to diversify market outputs, recycling initiatives, the provision of ecosystem services with local and global significance, the creation of local job opportunities, contributions to local food security, and the exploration of new opportunities such as agro-tourism, among others. By embracing the principles of circularity, solidarity, and diversification, agroecology expands the notion of value creation beyond the linear value chain paradigm.

A promising opportunity to promote agroecological transitions, especially in forest frontier landscapes, is the potential integration with bioeconomic approaches. Bioeconomy broadly refers to the utilization of elements from local biodiversity to create value chains. There are different definitions of bioeconomy, ranging from those that focus on sustainable exploitation of products and services from natural ecosystems like forests or wetlands, to broader definitions that encompass resources from land, sea, crops, animals, micro-organisms, etc. The broadest definitions include also biofuels as products of the bioeconomy. However, it is important to note that the bioeconomic approach can be associated with extractive systems, leading to uncontrolled extraction of non-timber forest products or other resources. When development relies solely on a bioeconomic value chain, territories often specialize in a limited number of products and processes, neglecting other production activities, including those related to local food security. This specialization leaves communities vulnerable when prices drop or markets decline, leading to desperate situations.

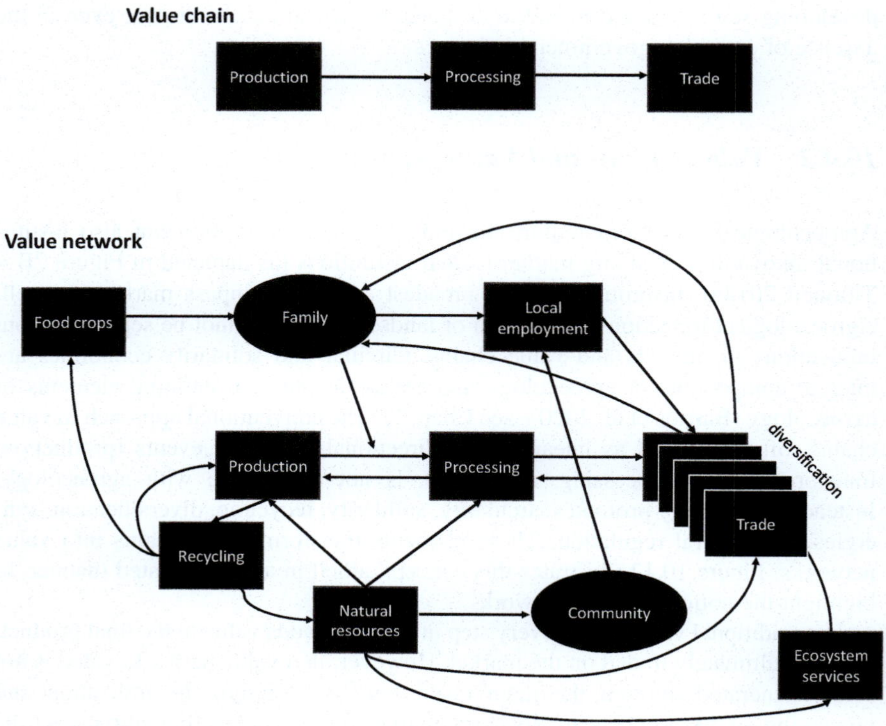

Fig. 10.12 From value chains (linear, unidirectional) to value networks (circular, diversified)

An agroecological approach to the bioeconomy offers a strategy to prevent such vulnerabilities by fostering territorial value networks. By integrating bioeconomic value chains within diverse agroecosystems and landscapes, it is possible to build community resilience and mitigate market volatility. Additionally, agroecological approaches enhance the resilience and efficiency of the bioeconomic value chain through practices like recycling and resource conservation. This integration enables a more holistic and sustainable utilization of local biodiversity while maintaining the long-term viability of both the agroecosystem and the local economy.

Value networks can be designed together with the stakeholders involved. Figure 10.13 illustrates this with an example from the Colombian Amazon region, where farmers produce cacao (*Theobroma cacao*) and a native relative species known locally as Copoazu (*Theobroma grandiflorum*). Copoazu seeds can be used to extract fat that can be turned into a produce that resembles chocolate (sometimes termed chocoazu or cupulate). However, the main interest of copoazu fruits resides in the mucilaginous pulp that surrounds the seeds, which is used to produce juice or ice creams with an exquisite fruity flavour. Copoazu grows well in the Amazon region, from where it originates, and can still be found in the wild. The value network designed by local stakeholders (which is being implemented as these

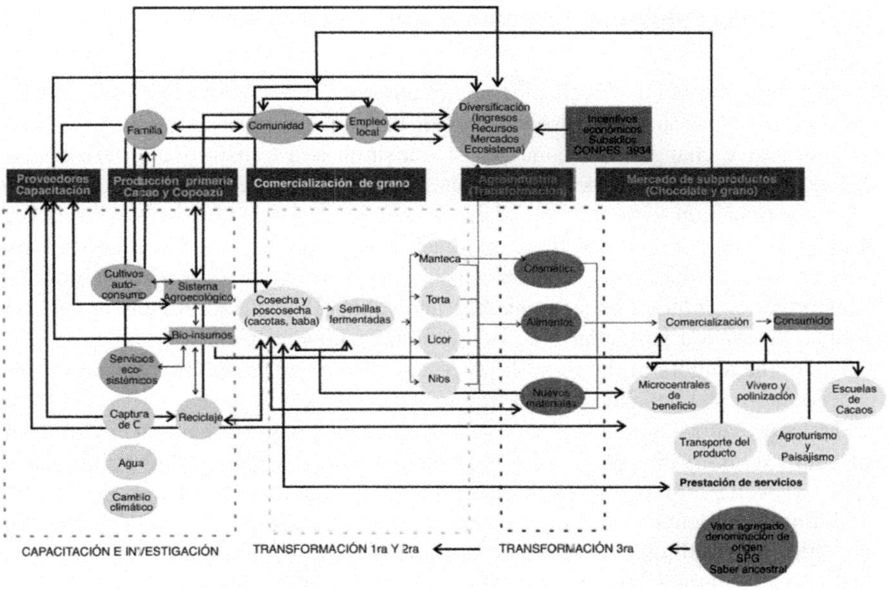

Fig. 10.13 An example of a value network around Cacao (*Theobroma cacao* L.) and Copoazu (*Theobroma grandiflorum, Willd. ex Spreng. – K. Schum.*) designed in a participatory way with farmers and food processors from Caquetá, Colombia. The scheme was developed by Valentina Ramírez, and is part of a progress report of the EU-funded project ABRIGUE (*Agroecología, Bioeconomía, Resiliencia, Innovación y Gobernanza – Unión Europea*): https://www.sinchi.org.co/abrigue

pages are being written – April of 2023) comprises not only the production, processing and commercialisation of copoazu and cacao but also the transformation of their by-products, their link with food crops for self-consumption through bio-inputs, their transformation into cosmetic products, the associated ecosystem services (e.g. carbon sequestration), the development of local infrastructure to support the value network (plant nurseries, cool storage, training facilities, etc.) and the potential for agro-tourism around the bio-cultural diversity of native *Theobromas*. I call this an agroecological approach to the bioeconomy (AEBE).

Agroecology-based Bioeconomy (**AEBE**): By combining the principles of agroecology and bioeconomy, it is possible to create synergies that promote community well-being, enhance environmental sustainability, and ensure the long-term viability of bioeconomic value chains in a way that prevents extractivism, and contributes to local food security and resilience.

10.5 Agroecological Transition and Transformation

Approaching the end of this chapter, and of this book, I would like to discuss one last concept which is strongly associated with the idea of agroecological transitions: transformative change. The concepts of transition and transformation are defined differently across disciplines or schools of thought. In general, it is possible to associate transition with gradual, adaptive change, and transformation with radical, structural change. In politics, these approaches are often described as being 'reformists' versus 'revolutionary'. In some technical or scientific realms, transition and transformation are used interchangeably, and yet in others the term transformation is used to describe a particular type of transition pathway (Geels and Schot 2007 – these are the authors of the multi-level perspective on sustainability transitions, the conceptual model depicted in Fig. 10.3d).

Transformations are defined as fundamental changes in human and environmental interactions in the realms of global environmental change (e.g., Pelling et al. 2015), as well as among the resilience thinking community (e.g., Olsson et al. 2014). Building on the latter, I used the concept of 'transformability' to analyse fundamental shifts in smallholder livelihood systems, following non-linear trajectories of change (cf. Tittonell 2014b). Transitions entail social, technological, institutional and economic change from the current regime to a desirable one, as depicted in Fig. 10.3b (and also earlier in Chap. 2). This definition resonates with the one proposed by Grin et al. (2010) to describe sustainability transitions, and is the one I prefer when referring to agroecological transitions. I do not see transitions and transformations as two separate, mutually exclusive processes. Let me explain.

10.5.1 Transformations as Possible Pathways

In agroecology, Gliessman and Engles (2015) and several follow-up papers refer to five 'levels' in agroecological transitions, which comprise Hill and McRae's Efficiency-Substitution-Redesign model (levels 1 to 3), plus the connection with value chains and consumers (level 4) and the transformation of the global food system (level 5). The Swiss NGO Biovision refers to these levels as being incremental (levels 1 and 2) or transformational (levels 3, 4 and 5), and so do research groups such as Anderson et al. (2019) or Wezel et al. (2018), both from Europe too. But these definitions are all relative. Depending on where one stands, it could be argued that a 'sustainable intensification' of agriculture represents an incremental transition, whereas an 'agroecological transition' is actually not a transition but a transformation that requires more radical, large-scale and long-term changes. On the other hand, as already discussed, the nature and extent of an agroecological transition depends on the starting point of the agroecosystem (or food system). The five levels described above can be easily applied when characterising agroecological transitions that start from conventional, industrial (western) agriculture, but not necessarily so when the transition needs to start from low-input smallholder agriculture on degraded soils and social, political and economically challenged environments.

I had the opportunity to discuss this extensively with Steve Gliessman in the field in Mozambique. In October 2016, the Food and Agriculture Organisation of the United Nations (FAO) invited us to deliver a practical training on agroecology for governmental extension agents in Marracuene, some 50 km north of Maputo. It was of course a pleasure to work together, and for me to spend time in the field with such a referent. This was Steve's first experience in sub-Saharan Africa and there and then, while visiting smallholder farms and communities, I had the opportunity to make my point: that the idea of 'five levels of agroecological transition' as described by him did not make much sense in the African smallholder context, to which he immediately agreed. In areas of extensive soil degradation, for example, soil restauration is a prerequisite to any form of agroecological transition (Félix et al. 2018). Efficiency improvement or input substitution are concepts that make little sense in systems where inputs are not used at all. Similarly, in the context of sub-Saharan Africa, an agroecological transition may start with a redesign, through e.g. agroforestry, and hence be 'transformative' already from level 1. The example of the UTT in Argentina described in the previous section illustrates that the agroecological transition that took place at farm scale was a consequence of an entire food system transformation, so once again a transformational process preceded more gradual changes in the agroecosystem (such as gradual pesticide reduction). All of this reinforces the idea expressed earlier that transformations may be one possible pathway within the more generically defined, comprehensive process of agroecological transition.

10.5.2 Innovation Domains

When considering all the areas that require innovation to support agroecological transitions, including field management practices, value chains, networks, policies, and the E-S-R (Efficiency, Substitution, Redesign) steps in the transition, we can identify four nested innovation domains (as shown in Fig. 10.14). These domains are the Management domain, Technology domain, Agroecosystem Structure domain, and Governance domain. The Management domain encompasses innovations that focus on improving the efficiency of the current agroecosystem without necessarily bringing about radical changes in its structure and functions. These innovations aim to optimize existing practices and processes. The Technology domain includes innovations that not only improve efficiencies but also involve input substitution. Examples include the replacement of pesticides with biological control agents or chemical fertilizers with biofertilizers. Technological innovations primarily aim to enhance specific components or inputs within the agroecosystem without fundamentally altering its structure and functions.

The Agroecosystem Structure domain is where innovations are directed toward a structural redesign of the agroecosystem. These innovations result in new processes,

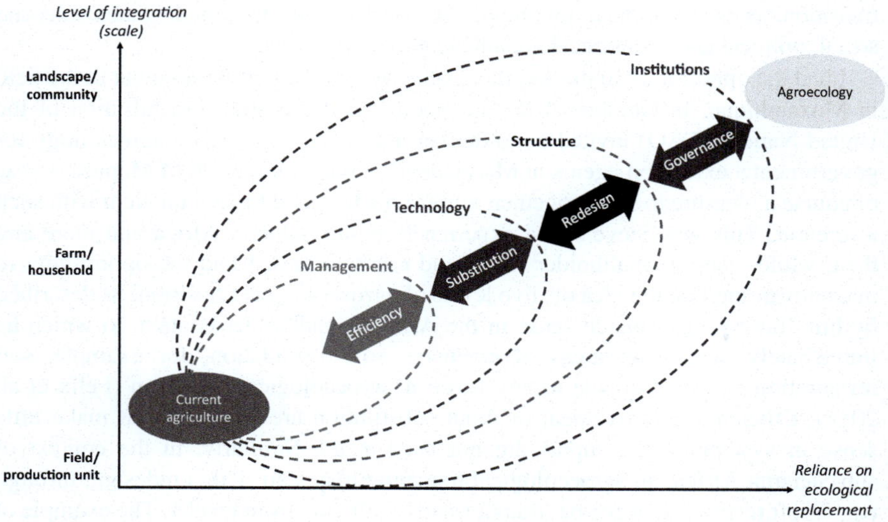

Fig. 10.14 Innovation domains in the transition to agroecology: management, technology, agroecosystem structure, and institutions. These are represented as sets of nested constraints, and the four phases of an agroecological transition as the forces necessary to push such boundaries (i.e., the constraints imposed by each domain). The transition is depicted as a gain in ecological replacement at increasing integration levels (scales). The size of the sets and the distance between the block arrows indicate the relative strength of the constraints to agroecological transitions

functions, and services within the agroecosystem. They go beyond improving efficiencies and input substitution, transforming the fundamental structure and functioning of the agroecosystem itself. Such innovations can also create new frontiers for efficiency gains and provide new opportunities for technological development, thereby enlarging the space for innovations of the two previous domains. Finally, the overarching innovation domain is the Institutional domain. This domain encompasses innovations that not only lead to efficiency gains, input substitution, and agroecosystem redesign but also drive changes in the governance of production factors and natural resources. Institutional innovations involve creating new policies, regulations, and governance mechanisms that support and enable agroecological transitions at a broader societal level (cf. Table 10.1).

The interplay between these innovation domains is crucial for successful agroecological transitions. However, it is important to note that the extent to which institutional innovation is possible greatly impacts the other domains (Fig. 10.14). When the space for institutional innovation is limited, it constrains the possibilities for innovation in other domains, hindering agroecological transitions. Thus, the presence of supportive institutional frameworks is vital for fostering and facilitating innovation across all domains and driving effective agroecological transformations.

Beyond agroecology-based management practices necessary to increase efficiency in crop and animal production, which are developing fast, there are major gaps in the realm of input technologies that constrain further agroecological transitions. For example, in animal production, and although progress is underway, the design of biological solutions to replace traditional antibiotics or vermifuge treatments has still much ground to cover. In crop production, biofertiliser technologies are developing fast but mostly within intensive production sector, such as horticulture, where production areas are relatively small. In large scale production of cereals and pulses, chemical fertilisers are not yet being replaced by biofertilisers; only incipiently in some cases, limited by the large volumes of biofertilisers that need to be handled in large scale operations. But even when bio-based technologies would be available for a complete input substitution, the sustainable multifunctional management of agroecosystems may be hampered by its own structural constraints, which may require a thorough redesign.

Systems that rely on single species of plants and animals, deployed over homogeneous landscapes (e.g. soy monoculture, sugarcane plantations, or dairy cow-perennial ryegrass type of systems, etc.), offer narrow space for manoeuvring in terms of ecological intensification. They require profound redesign of their social and ecological structures. Redesign of cropping systems into strip cropping arrangements, with high diversity of crops and cultivars between strips but also within strips, has shown advantages in terms of reducing pest and diseases incidence in the intensive agroecosystems of The Netherlands (Ditzler et al. 2021). Within the new design, the efficiency of external inputs increases substantially, which shows that management and technology gains are supervenient to agroecosystem structural innovations.

The extent at which innovations can take place in the redesign of current agroecosystems is constrained by their institutional context. Institutions, as said earlier, include markets, regulations, knowledge systems, principles and public or private organisations. For example, the redesign of pastoral systems cannot be regarded in isolation from the policy and legal environments in which pastoralist communities need to operate, especially when land tenure or access to natural resources are at stake (e.g., Kassam et al. 2016). Intercropping in mechanised agriculture will not scale out among farmers if they are penalised in the price they get for their produce when they sell the grain mixture (e.g. wheat and Faba bean) in the local grain cooperative (they are often charged for grain 'separation' costs, even when the grain is later sold in mixtures as livestock feed) (Delmotte et al. 2011). The national laws on agroecology past in countries like Brazil or France create ample opportunities for innovations within the domains of management practices, technologies and structural redesign. All these examples show that management, technology, agroecosystem structure and institutions represent *nested* innovation domains that influence one another, constraining or enlarging their space for manoeuvring of each domain. They are crucial, and incremental, in the development of agroecological transitions and transformations.

10.6 Summary and Concluding Remarks

Agroecological transitions have been described in different ways. In this Chapter I proposed a three-gear engine model, in which the gears represent, from big to small, the enabling conditions, the innovation support system and the agroecosystem. Innovation support systems connect the broader dynamics in terms of enabling conditions (supportive policies, consumer demand, environmental concerns calling for new approaches to food production, etc.) with the dynamics at farm or community level: the agroecological transition. The 'oil' that makes the gears turn and communicate power is represented by three dimensions, knowledge, innovations and policies, which are an alternative way to define agroecology, beyond science, practice and movement.

Agroecological transitions, sooner or later in the process, require agroecosystem redesign. Redesigning together with stakeholders comprises two major phases: prototyping and back-casting. Prototyping, or developing a vision of a desirable agroecosystem, must consider elements of the context, the available resources, the livelihood system and the objectives for the redesign. Once a vision is developed, for example the 'ideal farm' prototype, the pathway to achieve it is delineated through baseline studies, prioritization and piloting exercises, and integration of system components. There are multiple possible transition pathways and sources of inspiration in terms of practices and technologies. The various alternatives to be considered to undertake redesign towards agroecological farming (e.g. agroforestry, polycultures, push-pull systems, etc.) can be categorised according to their reliance on ecological replacement and to the level of integration they comprise, from field to landscape/community.

When making decisions about redesign, there are often multiple trade-offs between objectives. Simulation models allow quantifying and analysing such trade-offs ex-ante, to inform decisions. Multiple goal optimization models, referred to also as bio-economic models, are frequently used to analyse trade-offs. This approach tends to have an economic bias, as objective functions are often expressed in terms of money, and the limitation of being 'static'. These models typically run for a single season, or a single sequence of seasons, and thus they are unable to capture either relevant biophysical feed-backs or temporal variability in the system, particularly in long-term analyses. An alternative is to use inverse modelling techniques by coupling dynamic simulation models with global search algorithms. This combination allows a number of objectives to be optimized by running dynamic models, with a large number of possible combinations of parameters that represent farm management decisions. Models and experiments are not mutually-exclusive. They can be integrated, together with local and expert knowledge, in the various steps of an agroecosystem redesign process (cf. IDEAS).

Agroecological transitions entail changes at different scales and levels of integration. Beyond the global, national or regional 'enabling conditions' represented in the three-gear model, agroecological transitions are greatly motivated by food system transitions or value chain innovations. The food system is the chain of activities connecting food production, processing, distribution, consumption, and waste management, whereas the value chain is the sequence of activities involved in creating a product or a service. Transitioning towards alternative food systems at scale, as in the example of the *Union de Trabajadores de la Tierra* in Argentina, can sparkle sustained, long term transitions to agroecology at farm and landscape levels. One way of creating value while preserving nature, especially in highly diverse environments, is to develop bioeconomic value chains with agroecological bases, so that value chains evolve towards more complex, circular and equitable value networks. Transitions at such higher scales as markets or food systems are often referred to as transformations. In my view, shared by others, transformations are one possible pathway within more comprehensively defined agroecological transitions.

Agroecological transitions require innovation, both technical and institutional, at different levels of integration, involving a gradually increasing reliance on ecological replacement strategies. The ‘space’ for the necessary innovation to undergo the various steps in an agroecological transition (efficiency, substitution, redesign and governance) can be defined as a set of nested domains that entail changes in management, technology, agroecosystem structure and institutions. Such domains influence one another, with each level constraining the space of manoeuvring for the level immediately below, at the same time as exerting ‘pressure’ on the one immediately above. Innovations are crucial at all these levels to foster wide and enduring agroecological transitions.

References

Altieri MA, Nicholls CI (2017) Agroecology: a brief account of its origins and currents of thought in Latin America. Agroecol Sustain Food Syst 41:231–237

Anderson C, Bruil J, Chappell MJ, Kiss C, Pimbert M (2019) From transition to domains of transformation: getting to sustainable and just food systems through agroecology. Sustainability 11:5272. https://doi.org/10.3390/su11195272

Barrios E, Gemmill-Herren B, Bicksler A, Siliprandi E, Brathwaite R, Moller S, Batello C, Tittonell P (2020) The 10 elements of agroecology: enabling transitions towards sustainable agriculture and food systems through visual narratives. Ecosyst People 16(1):230–247. https://doi.org/10.1080/26395916.2020.1808705

Cardoso I, Guijt IM, Franco F, Carvalho AF, Neto PS (2001) Continual learning for agroforestry system design: University, NGO and farmer partnership in Minas Gerais. Agric Syst 69:235–257. https://doi.org/10.1016/S0308-521X(01)00028-2

Delmotte S, Tittonell P, Mouret J-C, Hammond R, Lopez-Ridaura S (2011) On farm assessment of rice yield variability and productivity gaps between organic and conventional cropping systems under Mediterranean climate. Eur J Agron 35:2232–2236

Ditzler L, van Apeldoorn DF, Schulte RPO, Tittonell P, Rossing WAH (2021) Redefining the field to mobilize three-dimensional diversity and ecosystem services on the arable farm. Eur J Agron 122:126197. https://doi.org/10.1016/j.eja.2020.126197

Dogliotti S, García MC, Peluffo S, Dieste JP, Pedemonte AJ, Bacigalupe GF, Scarlato M, Alliaume F, Alvarez J, Chiappe M, Rossing WAH (2014) Coinnovation of family farm systems: a systems approach to sustainable agriculture. Agric Syst 126:76–86. https://doi.org/10.1016/j.agsy.2013.02.009

Duru M, Therond O, Fares M (2015) Designing agroecological transitions, a review. Agron Sustain Dev 35:1237–1257. https://doi.org/10.1007/s13593-015-0318-x

Félix GF, Diedhiou I, Le Garff M, Timmermann C, Clermont-Dauphin C, Cournac L, Groot JCJ, Tittonell P (2018) Use and management of biodiversity by smallholder farmers in semi-arid West Africa. Glob Food Sec 18:76–85. https://doi.org/10.1016/j.gfs.2018.08.005
Geels FW, Schot J (2007) Typology of sociotechnical transition pathways. Res Policy 36(3): 399–417
Giraldo OF, Rosset PM (2018) Agroecology as a territory in dispute: between institutionality and social movements. J Peasant Stud 45(3):545–564. https://doi.org/10.1080/03066150.2017.1353496
Gliessman S (2006) Agroecology: the ecology of sustainable food systems, 2nd edn. CRC Press. https://doi.org/10.1201/b17881
Gliessman SR, Engles E (2015) Agroecology: the ecology of sustainable food systems, 3rd edn. CRC Press Taylor & Francis Group
Grin J, Rotmans J, Schot J (2010) Transitions to sustainable development: new directions in the study of long term transformative change. Routledge. https://doi.org/10.4324/9780203856598
Groot JCJ, Cortez-Arriola J, Rossing WAH, Massiotti RDA, Tittonell P (2016) Capturing agroecosystem vulnerability and resilience. Sustainability (Switzerland) 8(11):1206. https://doi.org/10.3390/su8111206
Hill S, MacRae R (1996) Conceptual framework for the transition from conventional to sustainable agriculture. J Sustain Agric 7:81–87. https://doi.org/10.1300/J064v07n01_07
Kassam K-A, Tourrand JF, Boone R (2016) Building resilience of human-natural systems of pastoralism in the developing world: interdisciplinary perspectives. Springer. https://doi.org/10.1007/978-3-319-30732-9
Lopez-Ridaura S (2005) Multi-scale sustainability evaluation. A framework for the derivation and quantification of indicators for natural resource management systems. PhD Thesis, Wageningen University
Modernel P, Dogliotti S, Alvarez S, Corbeels M, Picasso V, Tittonell P, Rossing WAH (2018) Identification of beef production farms in the Pampas and Campos area that stand out in economic and environmental performance. Ecol Indic 89:755–770
Olsson P, Galaz V, Boonstra WJ (2014) Sustainability transformations: a resilience perspective. Ecol Soc 19(4):1
Pelling M, O'Brien K, Matyas D (2015) Adaptation and transformation. Clim Change 133(1): 113–127. https://doi.org/10.1007/s10584-014-1303-0
Rossing WAH, Albicette MM, Aguerre V, Leoni C, Ruggia A, Dogliotti S (2021) Crafting actionable knowledge on ecological intensification: lessons from co- innovation approaches in Uruguay and Europe. Agric Syst 190:1–16. https://doi.org/10.1016/j.agsy.2021.103103
Sevilla Guzmán E, Woodgate G (2013) Agroecology: foundations in agrarian social thought and sociological theory. Agroecol Sustain Food Syst 37:32–44
Teixeira HM, van den Berg L, Cardoso IM, Vermue AJ, Bianchi FJJA, Pena-Claros M, Tittonell P (2018) Understanding farm diversity to promote agroecological transitions. Sustainability 10(12)., 20 p.:4337
Teixeira HM, Bianchi FJJA, Cardoso IM, Tittonell P, Peña-Claros M (2021) Impact of agroecological management on plant diversity and soil-based ecosystem services in pasture and coffee systems in the Atlantic forest of Brazil, agriculture. Ecosyst Environ 305:107171. https://doi.org/10.1016/j.agee.2020.107171
Timler C, Alvarez S, DeClerck F, Remans R, Raneri J, Estrada Carmona N, Mashingaidze N, Abe Chatterjee S, Chiang TW, Termote C, Yang R-Y, Descheemaeker K, Brouwer ID, Kennedy G, Tittonell PA, Groot JCJ (2020) Exploring solution spaces for nutrition-sensitive agriculture in Kenya and Vietnam. Agric Syst 180:102774. https://doi.org/10.1016/j.agsy.2019.102774
Tittonell P (2013) Towards ecologically intensive smallholder farming systems. Design, scales and tradeoffs evaluation. In: Vanlauwe B et al (eds) Agro-ecological intensification of agricultural systems in the African Highlands. Earthscan\Routledge, Oxon, pp 132–144
Tittonell P (2014a) Ecological intensification – sustainable by nature. Curr Opin Environ Sustain 8:53–61

Tittonell P (2014b) Livelihood strategies, resilience and transformability in African agroecosystems. Agric Syst 126:3–14

Tittonell P (2018) Ecological intensification of agriculture, Ch. 13. In: Serraj R, Pingali P (eds) Agriculture & food systems to 2050. Global trends, challenges and opportunities. World Scientific, pp 437–472. https://doi.org/10.1142/11212

Tittonell P (2019) Las transiciones agroecológicas: múltiples escalas, niveles y desafíos. Rev FCA UNCUYO 51(1):231–246

Tittonell P (2020) Assessing resilience and adaptability in agroecological transitions. Agric Syst 184:102862. https://doi.org/10.1016/j.agsy.2020.102862

Tittonell P, Leffelaar PA, Vanlauwe B, van Wijk MT, Giller KE (2006) Exploring diversity of crop and soil management within smallholder African farms: a dynamic model for simulation of N balances and use efficiencies at field scale. Agric Syst 91:71–101

Tittonell P, van Wijk MT, Rufino MC, Vrugt JA, Giller KE (2007) Analysing trade-offs in resource and labour allocation by smallholder farmers using inverse modelling techniques: a case-study from Kakamega district, western Kenya. Agric Syst 95:76–95

Tittonell P, van Wijk MT, Herrero M, Rufino MC, de Ridder N, Giller KE (2009) Beyond resource constraints – exploring the physical feasibility of options for the intensification of smallholder crop-livestock systems in Vihiga district, Kenya. Agric Syst 101:1–19

Tittonell P, Klerkx L, Baudron F, Félix GF, Ruggia A, van Apeldoorn D, Dogliotti S, Mapfumo P, Rossing WAH (2016) Ecological intensification: local innovation to address global challenges. Sustain Agric Rev 19:1–34

Tittonell P, Piñeiro G, Garibaldi LA, Dogliotti S, Olff H, Jobbagy EG (2020) Agroecology in large scale farming: a research agenda. Front Sustain Food Syst 4:584605. https://doi.org/10.3389/fsufs.2020.584605

Toledo VM, Barrera-Bassols N (2017) Political agroecology in Mexico: a path toward sustainability. Sustainability 9:268. https://doi.org/10.3390/su9020268

Wezel A, Goris M, Bruil J, Félix GF, Peeters A, Bàrberi P, Bellon S, Migliorini P (2018) Challenges and actions points to amplify agroecology in Europe. Sustainability 10:1598. https://doi.org/10.3390/su10051598

The manufacturer's authorised representative in the EU is Springer Nature Customer Service Centre GmbH, Europaplatz 3, 69115 Heidelberg, Germany. If you have any concerns regarding our products, please contact ProductSafety@springernature.com

Printed and bound by CPI Group (UK) Ltd, Croydon, CR0 4YY
01/10/2026
02225853-0011